GEAR

GEAR

Cultures of Audio and Music Technologies

ELIOT BATES AND SAMANTHA BENNETT

The MIT Press
Cambridge, Massachusetts
London, England

The MIT Press
Massachusetts Institute of Technology
77 Massachusetts Avenue, Cambridge, MA 02139
mitpress.mit.edu

The MIT Press would like to thank the anonymous peer reviewers who provided comments on drafts of this book. The generous work of academic experts is essential for establishing the authority and quality of our publications. We acknowledge with gratitude the contributions of these otherwise uncredited readers.

This book was set in Adobe Garamond Pro by Westchester Publishing Services. Printed and bound in the United States of America.

Library of Congress Cataloging-in-Publication Data

Names: Bates, Eliot, author. | Bennett, Sam, 1967- author.
Title: Gear : cultures of audio and music technologies / Eliot Bates and Samantha Bennett.
Description: Cambridge, Massachusetts : The MIT Press, [2025] | Includes bibliographical references and index.
Identifiers: LCCN 2024024738 (print) | LCCN 2024024739 (ebook) | ISBN 9780262551854 | ISBN 9780262382540 (epub) | ISBN 9780262382557 (pdf)
Subjects: LCSH: Sound—Recording and reproducing—Equipment and supplies—History. | Sound studios—Social aspects. | Audio equipment industry—History. | Music fans.
Classification: LCC TK7881.4 .B38 2025 (print) | LCC TK7881.4 (ebook) | DDC 306.4/6—dc23/eng/20241104
LC record available at https://lccn.loc.gov/2024024738
LC ebook record available at https://lccn.loc.gov/2024024739

10 9 8 7 6 5 4 3 2 1

EU product safety and compliance information contact is: mitp-eu-gpsr@mit.edu

Contents

Acknowledgments *vii*

I GEAR CULTURES *1*

1 GEAR *3*

2 FETISHIZATION *27*

II MATERIALIZING GEAR *59*

3 EXTRACTIVISM *61*

4 MANUFACTURING *79*

5 INTERFACE *109*

III STAGING GEAR *135*

6 EVENT *137*

7 PRINT *161*

8 ONLINE *183*

IV ENCOUNTERING GEAR *205*

9 AGENCY *207*

10 TESTING *219*

11 ORGANIZATION *233*

V GEAR ATTITUDES *249*

12 SECRETS *251*

13 HERITAGE *269*

14 OBDURANCE *291*

15 CONCLUSION *313*

Glossary *333*
Notes *337*
References *359*
Index *375*

Acknowledgments

Samantha and Eliot thank Frank Wells for providing permission for us to conduct ethnographic research at AES, and the Australian National University Research School of the Humanities and Arts External Visitors program for making possible our first extensive cowriting session in 2018. We are grateful for the generous sharing of knowledge from the following, with whom we conducted interviews: Kelli Coyne, Larry Crane, Stephen Crane, Tim de Paravicini, Jakob Erland, Ted Fletcher, Dave Harries, Hutch Hutchison, Annie Jamieson, EveAnna Manley, Peter Montessi, Rick Perotta, Robin Porter, Ben Sneesby, Marshall Terry, and Terri Winston. And we extend our gratitude to the peer reviewers, who trusted in the project and provided constructive critique that improved the manuscript.

Earlier versions of this research were presented at the 2020 AES Annual Convention in New York ("When Manley Pride Threatens Manly Pride: Examining Gear Online"), at the 2020 AES Virtual Vienna Convention ("From Audio Technologies to Gear Fetishes"), at the 2022 IASPM-UK/Ireland Branch Conference, Liverpool ("Gassing for Gear: Acquisition, Fetishisation and Technological Exhibitionism in Gear Cultures"), at York University (UK) and Utrecht University ("The Critical Organology of Gear Cultures," 2023), and at the University of Lethbridge Social Distinction in the 21st Recording Studio Symposium ("Interface Aesthetics and Analog Feel: Rematerializing Synthesizer and Recording Studio Gear Cultures," also presented at Wesleyan University (Middletown), 2020). Many thanks to these universities and roundtable organizers for making this possible.

Samantha would like to thank Stephen Crane for his time and insights. Thank you to students in my "Music Recording and Production Techniques" course for your gear thoughts and contributions. Thanks to my ANU School of Music colleagues, particularly Bonnie McConnell and Pat O'Grady, and, of course, eternal gratitude to Matt Barnes: technical manager extraordinaire, "right arm," and fellow gear lover. Thanks to Clem Bennett for validating practically everything in this book just through his day-to-day work. With eternal gratitude to Paul. And for Ronnie: remember it's not about the gear.

Eliot would like to thank Amandine Pras, Ky Brooks, Arseni Troitski, and Antoni Piza for being enthusiastic and critical interlocutors (in the best way possible) during this project's journey, the students of Eliot's "Social Life of Music Technologies" class at the CUNY Graduate Center, and, most of all, Ladi Dell'aira for putting up with the mass of cables and mess of book drafts with unwavering humor while this project was coming to fruition.

The companion website for *Gear* can be found at https://gearcultures.ws.gc.cuny.edu/.

I GEAR CULTURES

"The museum, we call it!" Stephen Crane, Sydney's 301 Studios chief maintenance engineer, proudly gesticulates toward a collection of recording-studio gear stacked floor to ceiling and wrapped around three walls of the studio's interior café. It is late 2019, and Samantha is attending the Australasian Sound Recording Association's annual conference, at which former Abbey Road recording engineer Martin Benge is delivering a keynote speech. 301 Studios—formerly EMI Studios, Sydney—is one of Australia's last remaining large-format recording facilities, and with gold and platinum discs lining the walls, a grand piano in the foyer, and famous musicians dropping in for chats, the studio conjures up an aura of history, mythology, and stature. Stephen enthusiastically leads us on a studio tour. As we enter the main control room, Stephen grins and gestures to the 72-channel Neve 88-R console: "It's the world's most expensive laptop stand!" The console imposes itself on the space in front of a backlit nighttime frieze of the Sydney harbor skyline, with Lexicon 224 and Bricasti reverb consoles perched on its channel strips. We peruse a large cupboard full of microphones, entire rooms dedicated to analog and digital tape machines, and a sprawling, dust-caked loft space shelved out to house old and broken gear. Walking back through the corridors, we pass enormous floor-to-ceiling canvases on the walls: a blown-up photo of Abbey Road Studios' front door, a Neumann U47 condenser microphone in its cradle, extreme close-up shots of knobs and faders on a 1960s EMI TG-series mixing console, and the tape path of an EMI J37 analog recorder.[1]

Stephen finishes up the tour in the "museum." Perched on high stools, tea mugs in hand and eyebrows raised, we scope the surrounding wall-to-wall, floor-to-ceiling display of gear engirding the café. Stephen explains that the studio has accumulated so much gear over the years that it no longer fits in the dedicated recording spaces. A 24-track Studer sits alongside an API console. Racks of signal processors, including an Eventide H3000 Harmonizer and an AMS dm 2–20 tape simulator, sit next to even more racks of Pultec EQP-1A equalizers and Avalon VT-747 preamplifiers. Two Fairchild 670 limiters are stacked next to a Rupert Neve–designed Amek compressor, numerous UREI 1176 compressors, and more dbx, Empirical Labs, Teletronix, and Neve gear than even the most intensively tracked recording session could ever require. Although these racks of vintage gear and more than a dozen tape machines are operational, they serve multiple purposes: this legacy equipment can be brought to sessions on an as-required or as-requested basis; they visually celebrate the studio's technological history and stimulate technological voyeurism; they link to the studio's numerous gold and platinum records (this gear "museum" is adjacent to a wall emblazoned with awards); and, most importantly, as social actors, they function as the main instigator of stories and conversations (e.g., X preamplifier was used on Y session by Z musician and recordist) that inevitably commence when seated in the studio's centralized social space.

Exploiting its visual draw, this display is more than a convenient storage space, more than just a collection by a wealthy investor (the studio is owned by former SAE CEO Tom Misner), and much, *much* more than what the studio needs to run professional recording sessions. Conference attendees, already gathered around these racks and stacks, point and nod enthusiastically over their coffee and cakes as they lean in to take closer looks and pictures: socializing over gear. There are no computers here, no software, and nothing cheap. This vast display is the "best of the best," the "gold standard," as Stephen reinforces. When not surrounded by gear, we are surrounded by gear imagery. But what exactly is this *gear*? Why is there so much of it? If the Neve 88-R only serves to prop up a laptop, why does the studio need it? And why, as it lines the walls of the 301 Studios café, does this gear elicit such intrigue, fascination, and socialization among the studio visitors? 301 Studios is not the only place to have amassed such a hoard of professional audio technology. As it turns out, most of the places with similar gear stashes are not famous, professional recording studios.

1 GEAR

"GEAR R' US!" boomed the all-caps subject line of a registration circular sent by the Audio Engineering Society (AES) in the lead-up to their 2019 New York Convention. In the body of the email, accompanied by an image featuring an API mixing console, Shure SM57 and Neumann U87 Ai microphones, a Focusrite RedNET audio interface and a Genelec studio monitor (loudspeaker):

> **GEAR HERE**
>
> **MORE GEAR * MORE INSPIRATION * MORE REASONS TO ATTEND**

Boasting "Over 300 Exhibiting Pro Audio Brands," this long-standing professional audio industry convention draws thousands of attendees from all over the world—this time, enticing visitors with a "free exhibits-plus badge" upon entering a code at the registration checkout: AES19GEAR.

Since 1948, the AES has been the world's leading professional organization representing all the trades entailed in the manufacturing and use of audio technologies and producing industry-wide technical standards. The annual AES convention hosts a five-day conference featuring numerous predominantly scientific papers, engineering briefs, and posters as well as tutorials, hands-on building sessions, standards committee and regional branch meetings, and plenary talks and panels from audio-industry personnel. However, as the email alludes, these research, professional, or learning opportunities are not the main draw. As professional recordists and as long-term members of the AES, we recognize how, over time, gear has become

Figure 1.1
AES show floor, Javits Center, 2017. Photo by Eliot Bates.

central to the community of professionals in many audio-related trades, from education to manufacturing. For a few days each October, the sprawling exhibition floor inside New York's Javits Center transforms into an audio-technology playground (figure 1.1). Its pulsating, low-frequency rumble vibrates and swells around the cavernous atrium, cacophonously soundtracking the display of hundreds of shiny new audio technologies: gear.

WHAT GEAR IS

Gear: mixing consoles, outboard effects processors, microphones, amplifiers, speakers. These are technological objects for studio recording—the tools of the professional audio-recording industry—yet their (omni)presence in the broader music industries, as well as throughout "prosumer" cultures, transcends the established, professional audio engineer guild. The AES 2019 email shows how gear is taken for granted due to its ubiquity. But what is gear, what does gear do, what does it mean for something to be described as gear (rather than as something else), and who is involved in the gear attribution?

Gear is not the tool itself, but what happens when tools—and ideas *about* tools—articulate discursive formations and inculcate embodied practices/dispositions that structure social and economic relations. A tool does not become *gear* because of the normative uses of the tool, but rather because of the *excesses*—everything above and beyond what is needed for the basic minimum operation of the tool, whether those excesses are practices, embodied dispositions, ideas, or discourses. Accordingly, our study is not of all the technological objects that you find in a recording studio (or on stage, in the field, in a laboratory): not *all* technological objects *become* gear. Rather, this book questions how and why *certain* technological objects become gear. This leads to our first constellation of research questions:

What do we mean when we talk about gear, why do only certain technological objects become gear, and how did *those* objects specifically become regarded as gear?

Gear, as technological objects enmeshed in discursive formations, are irreducible to the words said about them: gear discourse is always relational to the irreducible materiality of gear.[2] In the research process, we wondered whether the book should begin, as most technology studies books do, with invention narratives, continuing to the forces that enabled one invention to succeed where others failed.[3] But within this twenty-eight-year historical period that we are researching, we discovered an increasing amount of consumer awareness about gear provenance, including the source of the metals used in gear and site of product assembly, and the effects of environmental legislation affecting how technologies can and cannot be made today. During the same period, many promotional videos circulated that show how gear and gear components are made. The fetish nature of gear that we theorize in chapter 2, then, is more complex than the customary narratives of alienation and commodity fetishism would suggest. It begins at the site of extractivism, continuing through supply chains and many other sites until ultimately gear is purchased and arrives at the first of many resting places.

Gear materialities and gear discourses, taken together, provide an incomplete picture of gear. Whilst gear can be—and *is*—objectified and acquired for the sake of it, the look of gear and the haptics of gear tactility contribute greatly to its fetish capacity. This leads to our second constellation of research questions:

Why does gear look and feel the way it does, and take the morphological form and material composition that it does? In what ways is gear fetishized? When gear becomes fetishized, especially outside of situations where it is being used for its nominal task, what specifically about it is fetishized, and what is the nature of the fetish relation?

As we will see, recording technologies as gear have accumulated a host of peculiar connotations that go beyond their use value as audile technologies: becoming sexualized objects (where, as technoporn, they stand in for the women that are missing in gear cultures and associated studio settings), associated with weapons and military (arsenals of "secret weapons" stored in studios-as-bunkers), regarded as akin to drugs and narcotics (with all their addictive potential), and ultimately hoarded objects that can be selectively displayed to others.[4] While we ourselves do not wholly share these sentiments about gear—we use technologies to create music that is meaningful to ourselves and others—we recognize that these modes of objectification, acquisition, and hoarding are meaningful to many participants and even provide much of the constitutive force of and market potential for gear cultures.

Technological voyeurism, our term to explain the patterned ways that people gaze on technologies, is intrinsic to both manufacturer (production) and user (reception) discourse. Although voyeurism affects many gear culture practices, notable ones include the techniques manufacturers use to play off heritage tropes to promote technological voyeurism. For example, at the 2020 NAMM (National Association of Music Merchants) convention, brand-new gear was presented as already imbued with vintage aesthetics: Royer microphones displayed in bell jars, UAD audio interfaces presented in museum glass cabinets complete with object labels, and API consoles encircled with elegantly braided, red velvet queue ropes.

That said, none of these extramusical nonessential meanings and practices would have durability if it were not for the continuing power of studio-sited music recordings to captivate listeners, and for this captivating power to be routinely associated with the tools—both recording/mixing tools and instruments-as-technologies.[5] Recording-studio gear *can* help musicians and working engineers make recordings, and in music-focused settings, users will probably primarily think of gear as tools of the trade. Even there, though, we encounter another aspect of fetishism that continues to be arguably the most significant driving force in defining the properties of gear that is made today: a widespread obsession with "vintage" gear and recapturing aspects of "classic" recordings made during a supposed "golden age" of record production. The technostalgia, most often for US-, UK-, and German-made gear originally made in the 1950s–1980s and now considered "vintage," lives on in a used-gear market, in contemporary clones of this gear, in museums, and in skeuomorphically rendered software plugin emulations.[6] All of this contributes to what we term the **obdurance** of gear in the twenty-first century. Hence, we ask:

Which pieces of gear are valorized and why? Why is some gear awarded, iconic, canonical, or used as a museum piece to illustrate an era/concept, while other audio technologies are thrown away and forgotten?

Varying ideologies in manufacturer staging and user discourse suggest gear valorization is beneficial for gear production and reception. Is this a way of connecting to canonized recordings on a deeper level, beyond the Benjaminite aura? Is this simply an attribution of technological agency in the broader realm of (largely popular) music-recording practice? Does owning a piece of this valorized gear replace human agency or at least stand in for the long-term acquisition of skill sets? Perhaps gear is a way of holding on to the past; as the so-called golden age of record production and its concomitant recordists, technologies, and techniques recede into the distance, valorized gear can ground us in that past. Or perhaps UAD's slogan "Analog sound at the speed of light" underscores compelling extramusical associations with space flight and science fiction during an environmentally and politically unstable twenty-first century. This leads to our fourth research question:

What accounts for the explosion in boutique audio manufacturers at a time when digital tools should have rendered obsolete most of this far less convenient, far more expensive, and sometimes operationally temperamental gear?

To examine that, we assess the trajectory of gear, including how it is built, who builds it (and where), how extramusical and technoaesthetic attributes become imbued into it, and how this leads to meanings and practices and embodied dispositions related to gear.[7] We also examine this with regard to the social aspects surrounding gear, by which we do not just mean some social activity that pertains to gear, but instead the formation of durable **gear cultures** around specific technological objects.

WHAT A GEAR CULTURE IS

Cultures all have shared beliefs and practices, mythologies, histories, locally specific argot, and material culture—and this is equally true for gear cultures. But defining a gear culture is complicated. Noting simply that *a gear culture is a culture organized around specific technological objects where the objects serve as active participants* would raise more questions than it answers. It reduces some paradoxes to rationalisms when instead they should be understood as paradoxes.

Here is what gear cultures are *not*. Gear cultures might at first glance seem to be something like subcultures or scenes, but organized around audio engineering practices instead of, say, a musical style or genre. But gear cultures lack the symbolic resistance to capitalism or to hegemonic mainstreams that are characteristic of subcultures, and they lack the plural, fragmentary, and hyperlocal formations that characterize scenes.[8] Additionally, gear cultures can hardly be described as inclusive; in fact, gear cultures are, at best, predominantly (hyper)masculine spaces—where the presence, labor, and impact of women is, at best, concealed—and are, at worst, toxic spaces of intimidation and exclusion.[9]

Gear cultures are also distinct in that they do not just include consumers of commodities, but they also include many of the people and firms involved in the design and manufacturing of gear and some of its components—and those involved in sales, distribution, and demoing. Gear culture participants may live in a cultural city known for its vibrant music scene, or they may instead live on a mountaintop, off the grid, running a studio off solar power that the public never visits and only would know exists after the occasional post to a gear website. As long as the place is sufficiently connected by logistics firms and transportation infrastructure, and the "there" so happens to have accumulated sufficient convertible capital, anywhere could be a node within networked gear cultures.

So what *is* a gear culture? In one sense, it relates to the concept of associated milieu in the philosophy of Gilbert Simondon, especially with regard to his project to understand technologies and technical individuals as cultural actors and "how life can become technological."[10] Simondon's insistence on understanding the co-creation of technical individuals and their associated milieux is instructive. Technological objects do not just enter surreptitiously into a preformed cultural context; rather, gear and gear cultures are co-constructed. This leads us to the fifth constellation of research questions:

What produces, defines, and sustains gear cultures? What holds these communities together, and what kinds of sociability does gear facilitate? Who participates in gear-making in different capacities, and what are their motivations for continuing to make gear?

When we began gear cultures research, we assumed that widespread amateur and professional audio-recording practices held gear cultures together and would define modes of participation; perhaps we would study produsers, or producers as consumers.[11] Although some gear cultures participants are online content/media creators, the technologies we studied are not the same ones that, for example, made podcasting possible or supported the livestreaming boom in the first year of COVID-19. Neither are all producers consumers, nor all consumers producers. Beyond the practical applications, general-purpose technologies (such as computers and phones) lack the distinctive qualities of gear that animate gear cultures by emphasizing the felt but unknowable.[12] This also harkens back to Simondon, who in his essay "Culture and Technique" noted that "Even if techniques had neither utility nor purpose, they would have a sense [meaning]: in the human species they are the most concrete mode of the power to evolve; they express life.[13]

Therefore, one challenge we faced was moving beyond our own assumptions about what these technologies were designed, or scripted, to do.[14] Adding to an inventory of mundane technological uses—for example, noting how loudspeakers can be repurposed as microphones or how the 1176 compressor can act as a microphone preamplifier—is an exercise in chasing ephemera, but it does not inform us that much about gear cultures. To understand gear cultures requires a more radical rethinking of the relations between people and material technologies.

In writing about technoscientific measurement devices ostensibly designed to produce "objective" accounts of the world, such as microscopes and ultrasound machines, Karen Barad states that

> Phenomena are differential patterns of mattering ("diffraction patterns") produced through complex agential intra-actions of multiple material-discursive practices or apparatuses of bodily production, where apparatum are not mere observing instruments but boundary-drawing practices—specific material (re)configurings of the world—which come to matter.[15]

Equalizers, microphones, and mixing consoles might seem to be things with functions other than observing the world, even as we devote a whole chapter to the many ways audio gear is tested. But we argue that gear cultures, too, are best understood by attending to "agential intra-actions" that constitute both "material-discursive" and "boundary-drawing" practices.

We agree with Barad that the poststructuralist legacy of reducing all of life to representations and knowledge practices (at the expense of materiality and ontology) has occluded our understanding of the very things we attempt to study.[16] That said, some phenomena that sustain gear cultures and their discursive frameworks would appear to be largely representational in nature, and neglecting gear-related epistemology would miss some of the constitutive forces in gear culture formation. The capacity of certain technological objects to overflow with extramusical and extra-audio meanings, whether on account of their heritage or their entanglement with the "secrets" of recording studios, leads (partially) to widespread gear attitudes and to the obdurance of gear—which, in turn, leads to the further material reorganization of the world. Gear discourses and gear meanings may take the semiotic form of metaphors and symbolism, or they may entail a "folk" technological determinist kind of conflation of human activities (i.e., audio engineering) where tools are ascribed human-like kinds of agency. This is not surprising: anthropomorphism has long been a means for conceptualizing musical instruments and other technologies.[17] To understand gear semiotics, we explore the sixth constellation of research questions:

How and why does gear become discursive? How do meanings and attitudes get woven into technologies, and why did certain people–gear relations lead to discourses and practices of technostalgia, fetishization, iconicity, and other kinds of gear historicization?

Gear cultures have durability due to their continuing capacity to produce human-social identities for participants. Much as was the case with "hi-fidelity" domestic social formations in the 1940s–1950s, the consumption of recording-studio gear is related to aspirations to buy one's way into a white, masculine, middle-class social identity: indeed, the "home studio" has largely replaced the audiophile "listening room."[18] But the studio-related gear cultures we examine are more similar to the 1930s–1960s ham radio "technical culture" described by Kristen Haring with regard to the "socially distanced" modes of social interaction, the gatekeeping of technical knowledge, and the racialized exclusions that kept ham radio in the US a predominantly white and male pastime. Extending Haring's conclusion that within this "technical culture . . . the technical identities of people and technologies are coproduced," within gear cultures, people's social identities are produced, *and* hegemonic gendered divisions of labor are reproduced.[19]

This gender division plays out in many ways. Historically, women have undertaken assembly labor in audio and music technology manufacturing processes.[20] This continues well into the twenty-first century; we witnessed firsthand the meticulous and skilled work involved in setting delicate ribbons in microphones and hand-wiring transformers, much of which is undertaken by women. Women, however, are rarely found in trade show booths except as the partner of a male equipment designer; neither are women at all welcome in online gear fora. We critique the world's highest traffic online audio and music technology forum, Gearslutz/Gearspace, which not only excludes women via its name and discursive formations but also reinforces hegemonic and toxic masculinities (see chapter 8). In gear cultures, men get to foreground their ideas, designs, companies, and discourses, whilst women—and their work—are concealed or discouraged. In some gear cultures, gear even *replaces* women, standing in for women's missing presence in a variety of social relations (spousal, partner, mother). We interrogate this *manosphere* via recent theorization from masculinity studies.[21] In saying that, the insularity of twenty-first-century consumer-focused, enthusiast gear cultures led to a queering of both masculinities and heterosexualities parallel to similar phenomena in other domains.[22]

METHODOLOGY/DATA

Our research collaboration began in September 2014 in Cork when, following individual presentations at the IASPM-UK/Eire branch conference and a couple of nights in the pub, we identified a shared interest in critically interrogating record production. A marathon brainstorming session in a café ensued, and many of the ideas from that meeting resulted in *Critical Approaches to the Production of Music and Sound.* However, other ideas concerning audio technologies and economics, heritage, materiality, manufacturing, trade shows, and online discourse fell outside the book's remit.

Independently, we had been working for years on projects that all related in some way to gear. Eliot had been an active participant and had followed developments in online gear cultures since 2001. They had worked in studios in the US, in Turkey, and in the UK. Samantha had worked as a recordist in studios in London and had been deeply involved in music technology education in the UK. Most importantly, though, "The Cork Sessions" marked the beginning of a cowriting and coresearch process that has continued ever since. During these sessions, we shared our contrasting experiences with gear—and with different *types* of gear—finding common ground in our individual understandings of gear and shared professional recordist/practitioner/educator/researcher trajectories.

In general, the research for the gear cultures project evolved through an iterative process of method assemblage.[23] Both of us have undertaken participant observation and produced extensive ethnographic fieldnotes and photographs in gear-related milieux ranging from studios to trade shows to museums to DIY hacker/makerspaces to universities to online. But the limitations of "deep hanging out" in local sites for producing more generalizable findings became apparent when research participants provided contradictory accounts of the same phenomenon. We conducted long-form interviews with individuals working in aspects of audio technology ranging from gear designers to manufacturers to technicians and maintenance engineers to marketers to magazine editors to nonprofit organizers—including with important stakeholders who had never previously been interviewed. We conducted many more informal short-form interviews at trade shows and audio-related conferences. Another significant set of data came from fieldnotes, ethnographic photographs, and semi-structured interviews conducted at numerous kinds of gear sites. Individually and together, we have visited recording facilities, gear manufacturers, museums, and other diverse gear sites.[24] The visits enabled us to gather heterogeneous data from sites along the trajectory of gear—the making, staging, normative operation, and heritagization—in ways impossible via straightforward interviewing or archival research.

Since technological voyeurism often involves tactile experiences or haptic ideations (i.e., imagining touching gear), we found that sensuous scholarship and visual anthropology provided touchstones for more relevant methods.[25] We took thousands of photos in an attempt to depict gear cultures from every angle, and our photographic analyses—and sharing of them during focus groups and within informal research settings as a point of discussion—have been especially fruitful. However, neither method was sufficient to fully assess the technological interfaces that are so important to our study (see chapter 5), so we simultaneously honed our theory of the interface whilst rolling our own interface analysis methods. For Eliot, this included ripping open a Focusrite Red microphone preamplifier and reverse engineering its resource supply chains, and trawling through recently declassified army instrumentation studies to uncover the early history of knob and potentiometer standardization. For Sam, this included a detailed, critical, interfacial analysis of studio gear in the recording studio she designed and built at the Australian National University (ANU) School of Music in Canberra. Additionally, we have both been involved in building gear, and we include reference to our own experiences building microphones and breadboarding circuits.

Similarly, we rolled our own multimode analysis of gear manufacturer websites, gear-related web forums, and related online platforms. Our research extends back to 1995 with discussions on the rec.audio.pro newsgroup and subsequent Rec Pit forums (2000–2005), but we focus primarily on gear discussions on Gearslutz. Here, we have different perspectives: Eliot was an active Gearslutz participant from 2003 to 2019, whereas Samantha actively avoided it. GroupDIY (2006–present) serves as a contrasting case study. While we applied codified modes of discourse analysis—such as Foucauldian discourse analysis and grounded theory, conversation analysis to understand turn-based communications, and some natural language processing-style analyses of thread posts—ultimately, we needed a more nuanced framing for the *materialities* and digital materialities that were the center of online gear discourse. Hence, for us, multimode analysis means subjecting the same corpuses to sequences of codified *and* experimental methodologies. Beyond discursive moves, we have also tried to identify flashpoints—for example, moments when a gear culture, or a subset thereof, was brought together or fell apart in an unexpected way, moments when the social rules and discursive norms permanently changed, or, as have become our favorite flashpoints, moments when gear discourse just heads in the weirdest and wildest directions neither of us ever thought possible.

This book is based on previously unpublished research. Our source material is part of an extensive archive of data that has not been used in scholarship on technology, audio, or music—with the exception of recent articles of ours that are part of our broader gear cultures project.[26] We do not provide an exhaustive account of any one of these kinds of materials, but we do argue that the kinds of "studying up" we did were essential to appropriately frame the gear and gear cultures phenomena.[27] The integrative methodology we devised is not intended to become a prescriptive template; rather, it grounds our approach toward studying up, affirms our commitment to multimode study, and represents our attempts to ethnographically represent sociotechnical formations.

SCOPE OF BOOK

Gear examines recording-studio technologies from 1995 to 2022. We begin in 1995 since that was the first year when the nascent rec.audio.pro newsgroup had expanded to include professional gear designers and gear enthusiasts—inaugurating new means of socializing about gear that substantively transformed professional audio technologies and who engaged with them. We ended our in-person fieldwork in January 2020, with both of us enduring the sensory onslaught of the NAMM show in Anaheim—the last major trade show preceding the first shutdowns due to the COVID-19 pandemic. That said, many contemporary events arose due to historical ones, and our supplemental archival research examines select gear-cultural issues as early as the mid-nineteenth century and builds upon major historical works that cover technologies from the inception of sound recording until the turn of the millennium.[28]

When considering *which* technologies to examine, we faced a considerable problem. Obviously, if our goal was to explain the technologies that contributed toward an artist making a hit single in a studio, then we would need to account for the full signal chain and everything used for the editing and mixing stages—which today would weigh heavily toward a digital audio workstation (DAW) running plugins and an audio interface that facilitated analog electrical signals coming and going from the computer. But, even in 2020, a visit to a trade show, a perusal of a gear magazine, or a peek at an online gear forum shows a very different picture: a world that is dominated by other kinds of hardware technologies, most of which should have been rendered obsolete by the DAW, plugins, and interface. We realized that the "gear" descriptor does not consistently neatly map onto the most important, ubiquitous, or current technologies, but onto what, from the outside, might seem like an arbitrary list.

So, what constitutes gear, and what does not? Let us think back to the AES 2019 email. Certain well-known microphones and mixing consoles are clearly gear exemplars; others less so. Outboard hardware objects such as equalizers, compressors, microphone preamplifiers, and a few early digital reverb units continue to be sought after and copied, but equally—if not *more*—important patchbays, computer audio interfaces, and monitoring controllers fail to attract equivalent attention despite being the same size and shape. Technically, the latter technologies should be gear. But they are not *gear* because they have not yet absorbed the same kinds of extramusical and extra-audible meanings, practices, and contexts that the former technologies do.

Higher-end studio monitoring speakers and amplifiers typically are gear but also overlap with audiophile (home stereo) technologies—another gear culture, but not the one we are primarily studying in this book. We do not devote much space to software tools including effects plugins, except for the hundreds of virtual or modeled versions of coveted hardware objects, most of which additionally sport skeuomorphically represented interfaces. This is not because plugins are "worse" than their hardware equivalents, or less relevant to recording workflows, but because they are not perceived of as gear in the same ways: in the twenty-first century, software is ubiquitous and available, exists as codes and licenses, occupies less space, and is not nearly as economically prohibitive as hardware counterparts.

As professional recordists, we disagree with this opinion vis-à-vis the purported quality or usability of software tools—we both own and use software regularly, even when we have access to high-quality hardware outboard gear. Despite the apparent lack of taxonomic or morphological logic, this book sticks with a "ground-up" conceptualization of gear, since assessing this assemblage, contradictions and all, provides the most insights into how gear becomes socially constitutive within its associated milieu—gear cultures.

Our study is not about "American" or "British" audio engineering, but rather about an international phenomenon dominated by an Anglophone hegemonic cultural formation.[29] The central nodes of the geographically propagated networks we analyze include trade shows, conferences, recording studios, museums, manufacturing sites, and university music technology programs located in the US, the UK, Australia, Aotearoa, Canada, Turkey, Greece, Sweden, Switzerland, Ireland, and Germany. We also analyzed Anglophone online and print milieux that include residents in those locations and those from many more countries, too; for many of the participants in these milieux, English is a second or third language.

Here, we focus on the sites, people, ideas, ideologies, materialities, images, sounds, and discourses that determine *why* gear looks, feels, and sounds the way it does—and why many people feel the way they do about gear *outside* of normative use scenarios. Our emphasis, then, is not on music or music-making, although everything we discuss has substantive implications on recorded music-making worldwide. Non-normative gear uses determine which tools are made, what form they take and material they use, what they cost, who is interested in them and why, where they are made and by whom. We also deliberately avoid perpetuating "invention narratives," since their teleological reductions fail to help us produce generalizable social science findings about how and why technologies become meaningful, or how meanings and practices and dispositions change. One peculiarity about gear cultures: they are, by definition, organized *around technological objects*, unlike local or regional cultures where material culture may be perceived as being more or less important than other aspects in cohering that culture. The ways these objects are fetishized, in every sense of the term (see chapter 2), means that there is widespread confusion about what these objects are—particularly that they are fully functioning social actors like people. We find the fetishization of gear fascinating yet weird. Since we wish to understand gear "from the inside," considerable passages of the book may seem to be devoid of people as we try to unpack the material, discursive, and embodied perspectives toward technological objects.

WHY WE NEED A BOOK ON GEAR CULTURES

We have gear on the brain. If you are reading this book, you likely have gear on the brain too. Gear is a niche yet fascinating world. Having been immersed in gear worlds ourselves for years, we recognized the need for a book that explores gear as a phenomenon beyond its presence in the professional audio industry or its default application as components of recording and production signal chains.

On our various gear travels, we noticed how audio technologies are routinely taken for granted. Indeed, as young recordists in studios in London and San Francisco and in Turkey, we understood gear workings and operational modes, and applied gear without much—if *any*—critical thought in our day-to-day routine recording sessions. We arrived at the studio. We switched on, prepared, operated, and packed down gear. In these instances, we focused almost entirely on music and spoken word: performances, instruments, and musicians as well as the takes, mistakes, and breaks. In our experiences, particularly during Samantha's tenure as a Director of the UK's Music Producer's Guild, we understood this tacit knowledge as shared throughout the industry, which has amply been documented by recordists in specific works and in extensive interview collections.[30] Similarly, where scholarly works situate singular technological objects as the study, they do so in relation to their normative music-related purposes.[31] And even when technological objects are considered historically, they are positioned in relation to the recordists and/or to studio workplaces that housed them,[32] albeit with a reverence that has, in recent years, helped reinforce a technological canon. Curiously, when technologies are centered in films (*Sound City*, *808*), the object becomes a conduit for the stories of musical agents and agency, as opposed to recordists, studios, or even the gear itself.[33] In *Sound City*, for example, interviewer Dave Grohl asks designer Rupert Neve questions about the 8028 mixing console. As Neve details the inner workings of the desk, Grohl is filmed smirking and gurning as a sarcastically subtitled text narrates Grohl's "switching off" from the interview, alluding to how Neve's description of the technology is boring, confusing, and beyond his comprehension. Here, Grohl knows the console's importance to musical and recording historiography but has little interest in its technical or operational aspects, despite its foregrounding in the film. Regardless of including Rupert Neve's interview, the film foregrounds the technostalgic potential of the Neve 8028 as opposed to its operational capacity.

All kinds of weird and wonderful ideas and ideologies are embedded in technological objects. Consider compressors. The Fairchild 660 Vari-mu limiter, for example, is often regarded as the "holy grail" of compressors—perhaps not all that surprising when considering its associations with canonized workplaces (Abbey Road), musicians (The Beatles), recordists, and recordings, as well as other factors including age, operational, and build quality. Yet other ideas—for example, that using the Shadow Hills mastering compressor is akin to becoming a WWII code breaker—is a much greater leap into a fantasy world where the gear is removed from its use context and becomes a fetish object. Indeed, gear embodies and constructs historical, cultural, geographical, musical, economical, gendered, ludic, weaponized, sexualized, astronomical, and fantastical ideas: some authentic, some bizarre, some utterly *delusional.* All extend far beyond the basic conditions of what the technology is *for* and how the technology was designed and scripted to accomplish a narrowly defined task (e.g., a compressor processes and controls the dynamic range of an audio signal).

In the twenty-first century, these sorts of extra-audible and extramusical ideas are not exceptional. Rather, they are rife, to the extent that such marvelous ideological constructs of what gear is and what it means dominates gear discourse. To that end, we see the potential in unravelling how gear came to be gear in this way and how such ideas and ideologies materialize in newly crafted technological objects. How did so many non-contextual attitudes form around its existence, acquisition, and application? And why do these attitudes exist and proliferate in the form and vehemence that they do?

Despite being insiders as practitioners and educators, we find ourselves on the periphery of many manifestations of gear cultures, especially those surrounding the gendering, sexualization, and weaponization of gear, and the incessant "gassing" for more gear. That said, we wish to understand how this excess of thought, discourse, representation, and activity around gear is productive—of sociocultural *capital*, of individual and collective *identity*, and of a sociocultural *milieu* that is organized around gear. These complex gear worlds connect communities, discourses, and people—the designers, builders, users, and gear lovers. Throughout gear cultures, many gear users do not engage in practice-based, contextual gear talk but rather in fetishistic thinking; hence, gear-as-fetish is a central theme throughout. When considering these gear worlds together, we have a framework for considering gear trajectories: the making/staging/encountering/attitudes arc that traces the life of gear. Rather than reiterating linear chronologies, we recognize the temporal disjunctures that happen in the process of gear historiography. In consolidating these areas, we hope to present a more cohesive understanding of gear cultures. Our primary goal is to explicate a portable framework through recourse to specific examples, even though we have amassed considerably more material and data than can be adequately presented in this book. Hence, *Gear* is designed to be a springboard for future research on gear cultures. We want to illuminate how new ways of analyzing gear and its cultures expand our field and move beyond philosophy of technology, determinism, or hermeneutical analyses that tend to obscure rather than illustrate what actually goes on with gear.

Here's what we are *not* attempting to achieve with this book. We cannot possibly represent all technologies of all eras in sound-recording history, neither can we deal with all manners of relations with audio technologies in all the places in the world that are engaged in sound recording; such a book would be impossible, especially in light of the work required just to accomplish city-specific recording-studio ethnographies or genre-specific tech-processual analyses.[34] We do not champion the retention of analog technologies in a largely digital technology landscape, or hardware production processes versus in-the-box, computer-based workflows, and we try to avoid most "versus" discourse, whether that be analog/digital, hardware/software, or technology/process. In saying that, we acknowledge that most of the gear we discuss is gear we have used ourselves.

Many of these caveats and questions are revealed and resolved over the course of the book. One of the most common questions we hear from gear-interested people is "Why bother with hardware gear when anyone can record on a laptop?" Embedded in this book are some answers. Because, as we have discovered, gear is not simply about recording. In fact, gear is rarely primarily about recording at all.

WHY WE WROTE THE BOOK THE WAY WE DID

Following publication of our coedited book, *Critical Approaches to the Production of Music and Sound*, we decided we still had plenty to say about recording and production, but our discussions kept returning to gear.[35] Coauthoring this book was a no-brainer; we simply carried on from *Critical Approaches* and realized that our two minds together, with our different musical, engineering, studio, geographical, experiential, teaching, research, and gendered perspectives, would result in far more interesting and useful work than what we could achieve on our own. Additionally, we have mixed experiences in the academy, and although fortunate to be employed, and to be part of wonderfully supportive and creative networks, we both feel uncomfortable with the invasive competitiveness and individualism that discourages shared, collaborative work. We both found that writing sole-authored monographs was rewarding but isolating. Blabbing on about gear while circumnavigating the AES convention or while on dozens of four-hour Skype and Zoom sessions was much more fun, and out-weirding each other with hyperlinks to the more extreme corners of gear cultures has become sport.[36] That said, we chose not to alternate between our contrasting writing styles and areas of focus. To ensure a cohesive, singular voice throughout, we edited each other's writing, or cowrote in real time via Google Docs.

Coauthoring is surprisingly uncommon in our academic fields. We hope our findings will encourage scholars to share and pool their ideas for the greater benefit of the humanistic social sciences. Writing about gear and gear cultures has been fun, stimulating, and productive, but always fraught with problems. Indeed, much of our research has elicited a self-reflexive response, whereby we review and critique our own and each other's responses to our findings. To represent the *mangle* of this,[37] our consolidated writing style experiments with voice and register, blending analysis and arguments with more colloquial, observational, and allegorical approaches to critique. As in gear cultures, deliberate ambiguities feature in our prose. Gear cultures are exciting, funny, and weird; we try to reflect this in our experimental writing style.

Speaking of diversity and reinforcing literature, we have found that recent studies interrogating sound, technology, and materiality from feminist, Indigenous, and critical race theory, and from class and queer positionalities, ask the most salient questions applicable to how *we* aim to understand gear and gear cultures—and we foreground this work in *Gear Cultures*. If readers find some of "the usual suspects" missing from this book, it is unlikely because we are unaware of the work or author; theoretically and methodologically, we found it most productive to engage with contemporary, innovative, and refreshing ways of thinking, and to recuperate historically neglected texts that are equally pertinent today. Furthermore, much of the work of white, male, heteronormative scholars, which dominates music technology studies, is useful within their assumed epistemic purposes (even if those are only implicit), but the episteme we analyze here is substantively different. Our book is more concerned with introducing a body of new primary source material and deriving findings about technology and culture from that than deconstructing longstanding arguments within technology studies and popular music studies. Men are cited more often than women across all academic disciplines, and male-authored papers are far more likely to be cited than female-authored papers, regardless of the quality of the scholarship.[38] Such scholars are often in privileged positions as tenured faculty, and they often have female partners willing to undertake unpaid domestic work, particularly childcare, in order for them to advance prolific academic careers.[39] Our interest in amplifying neglected voices and heterogeneous voices in scholarship mirrors our interest in analyzing neglected labor within gear cultures. Considering these in tandem, we make recommendations in the conclusion on how to make gear cultures, and the fields in which we study, more equitable.[40]

A WALK DOWN THE CIRCUIT OF GEAR

Electrons do not flow, orderly, in one direction through a circuit—neither does our journey through gear cultures. To try to make some sense of the mangle, we have coalesced our analysis around thirteen problems that loosely fit into five domains: gear cultures, materializing gear, staging gear, encountering gear, and gear attitudes.

Gear has time travel potential akin to a DMC DeLorean. It might transport us back to World War II or launch us into space. Regardless of whether any recording of musicians is going on, gear is wanted, needed, and acquired in extraordinary quantities to gaze upon and fondle. That is because gear is *sexy*. Some gear culture participants hoard gear, fetishize it, describe it in militaristic, ludic, astronomical, narcotic, and sexual languages, and, in doing so, develop such severe cases of GAS (gear acquisition syndrome) that they need professional therapy. Works that examine both the fetish nature of commodities (fetishization within political economies) and the materiality of sexual fetishes (deviant sexualities) are uncommon. Chapter 2 shows how, in gear cultures, the two are intricately linked and in practice are the same fetish. In analyzing all these bizarre embodiments in gear, we set up our theory of **gear fetishes**. Here, we also connect the resource dependencies and materiality, morphology, and design of technological objects (part II) with the meanings, practices, and embodied dispositions that are accumulated through gear stagings in different media (part III). In turn, this introduces our methodical exploration of the processual, ideological, mythological, objectophillic, and valorized aspects of gear cultures that we investigate in parts IV and V.

What connects Yo'oko, the jaguar of the Santa Rita Mountains, with a Canadian mining company, environmental activism, and the continuing market for boutique microphone cables purported to have pseudo-magical properties? How do we put our finger on gear cultures, especially in light of the much-sought "nimble fingers" of female assembly labor and the normative male fingers in the US Army that tested the first standardized control knobs? Our project begins to take form, in part II, at extractivist sites and extends through the use of materials to make components, the design of circuits, and the use of components to make technologies, to the design and features and haptics of technological interfaces.

Much of the planet is mined and refined to produce the raw materials for studio gear and the energy to power it. In addition to exploring audio technologies and their operation as contributors to the Anthropocene, in chapter 3, we also examine the cultural effects of this extractivism and consider the ways everyone along the global mineral supply chains contributes, in some form, to gear cultures. Chapter 4 continues the story at the sites of component and technology assembly, where we reveal some surprising stories about electronic components (capacitors, op amps, transformers, vacuum tubes, mic capsules) that would otherwise typically be taken for granted. As we will show, in the twenty-first century, increased user and consumer interest in materials, components, and manufacturing (including many amateur video mini-documentaries) means that extractivist and assembly gear cultures are not wholly distinct from user-centric ones. But why does gear look and feel the way it does? How do human–machine relationships manifest in technological interfaces? What is and what is not being interfaced? In chapter 5, we develop a nuanced theory of interface that encompasses the variety of different modes of interfaciality present in twenty-first-century gear. We analyze the sensory modalities with which gear users approach gear, and we find out how brand-new interfaces are engraved from templates stored on an Apple] [and why Neve modules color coordinate with the Royal Airforce.

After their making, technological objects still do not possess their fullest set of meanings, practices, and embodied dispositions; they have not yet become *gear*. Part III compares the staging of technological objects in different media. At trade show events (chapter 6), we are deafened by the cacophony of accumulated gear noise, while also encountering mannequins wearing dresses made of cables and photo opportunities with Foxy Stardust, who wants to sell us an audio interface and DAW platform via a mock space launch. Gear-focused magazines vary in their relationships to gear cultures. In chapter 7, we see how *Tape Op* has evolved to become the foremost print media on recording-studio practices, practitioners, and tools. In contrast, *Sound on Sound* routinely demarcates "professional" and "amateur" gear users, while technostalgically harkening back to the "golden age" of canonized gear and "legendary" studios. The weird and wonderful gear worlds of trade shows and print, however, pale in comparison to *online* gear cultures, where gear thrives in all its extramusical and extra-audible glory. In chapter 8, we demonstrate how the online staging of technological objects has the capacity to transform them into gear. In this transformation, we witness the broadest range of fetishistic and extramusical associations that characterize gear today. In addition to ubiquitous sex and military tropes, we find relentless technobabble: microphone dissections salivated over and debated in forensic-level detail, hundreds of users gathering to view the "unboxing" of newly acquired gear, and users challenging each other to "preamplifier-duels"—my gear versus yours—occasionally to the point of making threats of physical violence. While all this obviously affects attitudes toward and practices around technologies, it also is socially constitutive, and we examine what kinds of sociability are made possible within each media space.

By now, it might appear that gear cultures are out of control. So what would it mean to be *in* control of gear? What agencies—attributed, displaced, or otherwise—are projected on gear encounters? Part IV considers aspects of agency and control that frame all gear encounters and two contrasting ways in which twenty-first-century gear culture participants encounter gear: testing scenarios and organizing and displaying gear collections. Within the gear cultures we examine, no singular ideal relation exists between technologies that are more or less controllable and those where the user feels more or less sense of agency when using them. In chapter 9, we interrogate what we term *displaced agency* and *personified technologies*, which relate to products that claim to provide the consumer with the hit-making sound of famous recordists—marketing technologies in a way that blurs the lines between technology and expert user.

What happens when a Neve 1073 and a Pultec EQP-1A walk into a bar? Gear users have concocted all kinds of elaborate gear "shootouts" and amateur ABX tests to out-sound, out-do, and out-*gear* each other. Such gear testing and comparisons reside at the nexus of fetishized gear acquisition and gear discourse, and, as such, chapter 10 flows on from our chapter 2 gear fetishization theory. Even though (surprisingly) the emphasis is *not* primarily on sound or music, the test becomes another means by which male participants display their understandings of discursive norms and appropriate vocabularies and perpetuate the competitive nature of gear cultures; testing in those milieux becomes, ultimately, about performative masculinity.

Gear is organized in specific, almost ritualized ways, sometimes in situ with lava lamps, fairy lights, Persian rugs, and "studio cats." In chapter 11, we recognize the "racking and stacking" ideology around gear positioning on a macro level. The desire to own and exhibit "racks," or massive quantities, of gear contributes toward *gear acquisition syndrome*. "Show us your rack" imagery, sexual connotations and all, pervades trade shows, print media, online spheres, learning institutions, and project and professional studios. Most studios—whether conceived as workplaces, laboratories, "wombs," "nests," "bunkers," or "man caves"—are primarily built to house gear and are organized and configured to coerce certain kinds of human-gear encounters.

Part V considers gear attitudes—not just the attitudes that people have towards gear, but rather attitudes that people impose on or attribute to gear itself. Chapter 12 interrogates secrets and concomitant revelatory discourse. The secret to recording success. Secret weapons. Dirty little secrets. Phrases so ubiquitous in gear cultures that they hinge—almost *rely*—upon them.

While part of a continuum of secrecy around practice and tacit knowledge, secrets amplify technological fetishization. One of the most puzzling problems in gear cultures today is how new gear is routinely shackled to the old. In chapter 13, "Heritage," we consider gear as historiographical tools. We find brand-new interfaces shipped with 1960s black-and-white images of men in white coats. Plugins emulating analog tape machines also emulate the wow and flutter, tape hiss, and saturation that digital audio standards committees worked for decades to erase. We see plenty of gear displayed in museums—trophy-like, mummified, and inanimate in a mode we call **technological taxidermy**. Here, we show how the twenty-first-century normalization of gear as inanimate fetish objects reflects its position in the museum. Why bother with such resource-intensive equipment in the twenty-first century? As we discover in chapter 14, gear is going nowhere. It is not only present but thrives as fetish objects, symbolic of both the labor of love imbued in it by designers and manufacturers as well as the aspirations projected onto it by its users. Gear is resilient and *obdurate*. These attitudes—secrets, heritage, and obdurance—revolve around a central problem: how old, anachronistic, or even obsolete hardware maintains its value today in light of the much more usable, affordable, configurable, and lightweight software options within a digital audio-centric landscape.

If you are still reading, you might have arrived at the same questions we have. Clearly, gear cultures have a lot of problems. We are keen to offer thoughts on how some of those problems might be fixed. In our conclusion, we argue why gear cultures need to be fundamentally reframed, repositioned, and restructured—and why gear needs to change too. Fetishized gear, when taken to its bizarre and illogical extremes, becomes utterly unproductive, akin to the piles of waste of expired media.[41] To advance the important initiatives that are underway to increase diversity, inclusion, and equity in all trades related to the creation of recorded music,[42] we offer approaches that we hope address some of the problematic issues discussed in this book, including the exclusive patriarchal domains and boundary work that happens as gear cultures are self-policed. This will create new opportunities for design. Gear was designed to do work: *it can be redesigned for the future.*

2 FETISHIZATION

GEAR AND WAR

"Victory is our business," states the landing page of Shadow Hills's website.[1] The page flickers under a Super 8 mm black-and-white grain backdrop. A morse code loop taps out over an industrial rumble. We appear to be watching a silent movie when the following notifications appear:

SHADOW HILLS INDUSTRIES

VERIFICATION OF CREDENTIALS REQUIRED

Access limited to the office of strategic services.

:REQUEST ACCESS:

Once clicked, a message appears.

LOOSE LIPS SINK SHIPS

YOUR IP ADDRESS IS

*** *** *** **

Our presence is recorded. We are being surveilled. And Shadow Hills wants us to know this.

All the while, a low-end hollow rumble, as if the sound of a bunker plant room, plays in the background. When we enter the website, we see a curious black-and-white military office location through a shadowed vignette, an insignia depicting two eagles facing to the left and right, a map of the world, and another framed image featuring two older white men in suits and ties: "OUR FOUNDER" and "OUR PRESIDENT" (Franklin D. Roosevelt, president of the US during World War II). The desk features an encryption machine, complete with typewriter, antenna, screen (with white noise), and cylinder. An old-style telephone sits to its right, and a bright desk lamp, complete with animated fly, illuminates a large old green book to the left of the machine. "SHADOW HILLS ENIGMA CYPHER" adorns its cover in gold script. None of the equipment is labeled. Navigating this "room" is a little like navigating a military computer puzzle game: where to look first?

USE YOUR KEYBOARD TO ENTER ENIGMA CODES

It soon transpires that anything typed in is the wrong answer (red hieroglyphs appear over the typewriter); however, clicking on the book zooms in to reveal its title: *SHADOW HILLS ENIGMA CYPHER.* Clicking on the book turns the page to reveal "SHADOW HILLS INDUSTRIES" and "AN INDUSTRIAL MILITARY CONTRACTOR." The next page is even more curious, titled "VICTORY IS OUR BUSINESS" and "AN INTRODUCTION BY WAY OF FILM." Beneath, we see a graphic, positioned like a photograph inserted into a scrapbook, titled "OFFICIAL WAR FILM" and "WAR DEPARTMENT" with a militaristic icon in its center, followed by "This short film explains the mission of Shadow Hills Industries and their plan to utilize Victory! to become both an industrial leader and unmatched ally in the fight towards quality and reliability." A red stamp beneath the text, as if dipped in wax, reads "ENIGMA CODE—LEVEL 3 CLEARANCE GRANTED" with a code in the center.

Further into the book, we see the first reference to technology: "TECHNOLOGY: OUR ALLY—BLAZING NEW PATHS TOWARDS VICTORY!"—with seven more red stamps featuring the same labeling. Each stamp has a title: "MASTERING COMPRESSOR," "EQUINOX," "QUAD GOLDEN AGE MIC AMP," "MONO GAMA," "OPTOGRAPH," "MONO OPTOGRAPH," "DUAL VANDERGRAPH." Only now, after a complicated web navigation, do we learn that Shadow Hills relates to *audio* technology. After we enter an "enigma code" into the "encryption machine," the WAR FILM appears as the music retreats to a Herrmann-esque, spiraling quarter-note motif over which an older US male narrates:

> Behind these locked doors, the office of strategic services guards thousands of secrets. Military secrets that the enemy would give anything to know.
>
> The OSS has the duty of informing operatives deployed undercover of the specifics of their mission, as well as devising methods of sending intelligence back to headquarters without our foes discovering the slightest detail. When it came time for the OSS to devise a system for securely sending top secret data over great distances, the office of strategic services partnered with the industrial military contractor Shadow Hills Industries. Together, they created the most ardent, battle-proven technology for this purpose: the Shadow Hills Mastering Compressor. This is the story of Shadow Hills Wartime Production to make victory our business!

Military montage footage plays through the narrated vignette: a guarded office door, sonar radars, a close-up Hitchcockian "eye," broadcasts including depictions of microphones and headphone-adorned personnel, morse code ticker tape, radio antenna, phone switchboards, and men in white coats drawing on a blackboard and congratulating themselves. Suddenly, "O Fortuna" from Carl Orff's *Carmina Burana* erupts from another military imagery montage, alluding to the mise-en-scéne in Sergei Eisenstein's silent epic, *The Battleship Potemkin* (1925):

VICTORY IS OUR BUSINESS

A FILM STORY OF PRODUCTION FOR VICTORY

A map of the US is overlain with people walking. Crowds of people funnel through a gate, reminiscent of Eisenstein's famous "Odessa Steps" sequence (1925). Flashes of industrial mechanics—cogs and machinery—evoke the industrial revolution. A close-up submarine sonar radar fades into warship barometers and dials. A hand and screwdriver repair broadcasting equipment. For a split second, a large swastika fades in and out of the footage, only just visible. Furnaces, welders, and iron pokers morph into sine waves bouncing along the screen of an oscilloscope. Women sort and pack military equipment. Large crowds of running people morph back into footage of women building equipment and cutting electrical wires. Additional images of male military staff organizing and fixing equipment in position are overlain with more sine wave oscilloscope footage. The film concludes with images of soldiers cut over images of tanks until, finally, the camera rests on a close-up image of a company logo between two VU meters: the Shadow Hills mastering compressor.

FROM TWENTY-FIRST-CENTURY GEAR TO A WORLD WAR II BUNKER: HOW GEAR GETS FETISHIZED

This elaborate context overwhelms Shadow Hills as a company, their technological objects, and their real-world use. We present this description in detail since it exemplifies fetishization in gear cultures, specifically military fetishization. Semiotically, we can "read" the website as epitomizing gear's extrication from its recording and production workflows, rendered inanimate and situated in altogether different eras and environments. Here, Shadow Hills gear is an anachronism: twenty-first-century technology situated in the context of World War II. On one level, the site evokes allied forces; the emphasis on "code crackers" and women military personnel footage conjures up the Bletchley Park code breakers responsible for decrypting Nazi Germany's communications. The Churchillian focus on "Victory!" situates the gear manufacturer among British allies. Yet the centrality of the encryption machine, which visually resembles the Enigma Machine, the sudden eruption of Orff's *Carmina Burana*—a piece epitomizing the music of Nazi Germany—and, for a moment, the appearance of a swastika blurs this paratext into a more ambiguous, generic depiction of World War II. Military aesthetics are reduced to pieces of gear, and the imagery, sound, and visitor interactivity reinforce Shadow Hills's assertion that they hold military "secrets"—and that this web experience fetishizes war. But what does this site not feature? Where are the musicians, recordists, studios, or related recording processes? We never hear the mastering compressor or encounter its specifications. Basic technological matters are delivered through military metaphor. In fact, the names of the gear and manufacturer are barely mentioned at all. The gear is inanimate; each of the shorter videos depicts a

360-degree surround view of the gear without a power cable in sight. Why is such an elaborate context constructed for this gear? And why might these gear framings appeal to users?

We frame gear fetishes through a combination of economic and critical theories that help elucidate why and how gear is imbued with extramusical meanings, and how through commodification it becomes symbolic, fetishized, idealized, and consumed.[2] As was the case with Shadow Hills, a prevalent trend in gear cultures is to strip technologies of their use value. Gear's extramusical details are, therefore, imperative to gear's consumption, evident not just in marketing prose (as above) but also in print media, trade shows, and online interaction.

Gear also becomes a substitute for (inter)human relations: it can be gendered, objectified, sexualized, or queered. In the hegemonic masculinities of gear cultures, which reflect and respond to masculinity in crisis, extensive maintenance work is needed to maintain the hegemony. Discourses around "sexy" gear entail some of the most extreme conceptualizations of what gear can mean. But connoisseurship and value are intrinsic to how gear is fetishized. Only specific technologies become gear, and for those, the economic and sexual fetish is one and the same. Gear users love gear very much, sometimes till it hurts. Using William Cheng's work as a lens to read extreme gear love manifesting in technophilia, objectophilia, and hoarding, we consider how the amassing of gear—quantity and quality—can end in tears. Twenty-first-century gear is meant for display, not just in recording workplaces but also in domestic spaces, museums, and online. Gear displays relate to three interrelated phenomena—gear ownership/acquisition, technological voyeurism, and technological exhibition—and gear culture interactions revolve around gazing at and longing for other people's gear. The ownership/acquisition, voyeurism, and exhibition triumvirate fuel GAS (gear acquisition syndrome), a condition marked by excessive and unending gear collection.

With its extraordinary militaristic metaphorical staging, the Shadow Hills example, dramatic though it may be, could mislead. The technological object's correlation with military and Teutonic fetishes indicates the range of symbols and metaphors outboard gear can index. But the technology has become so naturalized within gear cultures that any story, no matter how implausible, could be told about it. The public that Shadow Hills is hoping to assemble around this peculiar object is primed to believe that studio gear has magical superpowers: *this gear kills.*

COMMODITY FETISHISM, SITUATING VALUE

According to William Pietz, the modern usage of the term *fetish* is derived from a pidgin word *fetisso*, which referred to material/technical objects encountered along coastal West Africa, from modern-day Ghana to Nigeria, starting in the mid 1400s. We know about *fetissos* (also known as *fatish-oaths* and *fetische*) initially due to copious voyage writings by Christian traders. Pietz concludes that the fetisso "originated in a mercantile intercultural space" and is inextricable from Protestant miscomprehension of local cultural meanings, practices, and material culture within West Africa. Although related to the earlier Latin term *facticium*, especially in the sense that it described something akin to the idolatry of manufactured objects, the fetish object of colonial encounters is distinctive in five ways.[3]

First, fetishes are manufactured, irreducibly material, and extend personal or social power rather than being religious icons or embodiments of a divinity.[4] Second, fetishes were regarded as acting in an "impersonal" and "transcultural" way, capable of their effects regardless of belief in or cultural interpretations of the object.[5] Pietz notes a number of Portuguese traders who implemented fatish-oaths on their boats. Third, through having sufficient economic value, fetishes could mediate social relationships. While many earlier fetishes would today be described as jewelry or as amulets, subsequently, objects such as guns and composites of locally sourced and colonizer materialities commonly became fetishes. An object wasn't necessarily initially crafted to be a fetish; subject to the appropriate socioeconomic relations, it became so. Fourth, the objects were personified or anthropomorphized, or they could be spoken about as pseudo-humans. This is obvious in "judicial fetishes" such as the Congolese *ankishi*, which obviously resemble human form.[6] Fifth, in the absence of other interpretations, European traders assumed (incorrectly) that fetish objects were the primary force that structured social order within African societies, although as noted earlier, Europeans experimented with fetish crafting and social ordering, too.[7]

This predates European theorization of fetishization by de Brosses, Kant, and Marx, and therefore predates associations with terms such as alienation, reification, or sexual desire. *Fetishization* became a theoretical keyword of sorts, used by Marx to explain that

> There is a physical relation between physical things. But it is different with commodities. There, the existence of the things quâ commodities, and the value relation between the products of labour which stamps them as commodities, have absolutely no connection with their physical properties and with the material relations arising therefrom. There it is a definite social relation between men, that assumes, in their eyes, the fantastic form of a relation between things.[8]

That said, *fetishes themselves* were not theorized by these writers; Marx is clear elsewhere in *Capital* that he was uninterested in the material properties and qualities of fetishes (technological or otherwise). But as we now know (again), the fetishes of colonial encounters were technologies intended to extend personal or small-scale collective power. While they could be traded or sometimes sold, they did not gain their fetish status solely due to their position within a capitalist economic system. And not all fetish objects within these encounters were reified or commodified in the same way or to the same extent.

When audile technologies become gear, they too are attributed magical powers and take on a fantastic form—like the Shadow Hills example. A microphone preamplifier can be implemented with a single small integrated circuit chip, or it can entail 10 kg of esoteric and expensive materials with terroir-like origin associations and a user interface that far exceeds its instrumental needs; the former is "just" a technological object, whereas the latter has the potential to be regarded as gear. Therefore, while aspects of Marx's theories of alienation and commodity fetishism do apply to an extent when considering general-purpose money and certain commodities, the irreducible materiality of crafted objects is not a marginal property and is not adequately explainable through an economic account alone. Like West African fetishes, gear may be crafted with a normative-use case in mind (i.e., recording sound) *and* crafted to produce magical effects: these effects shift the site of agency for an individual user and structure the social formations around gear. We build upon the earlier understanding of fetishes since the same five principles apply to gear, too.

Like many technologies, fetishes are designed to stand in for a human within some existing social relation, meaning that they are not primarily prostheses of the human body or augmentations of human power (like wands, rings, or other talismans). However, to the extent they continue to exert such effects, rather than simply performing once-human modes of labor, they do so in a world that has been subject to a widespread process of rational disenchantment. Modern-era magical fetishes such as the voodoo doll or *nazar boncuk* (Eastern Mediterranean evil eye) appeared to gain their power *because* of, or in *spite* of, Enlightenment-era discourses and practices of disenchantment.[9] In the twenty-eight-year period we studied, practical knowledge about electronics and engineering, including the most esoteric and complex aspects of its design, circulated more than ever before: instead of rational disenchantment, the same period featured the largest-scale public experiments in fetishized pro audio gear ever seen.

Gear, as magical fetishes, share many traits, including being materializations of colonialism and modernity. Although until the 1950s most technologies used for electrical audio recording were built with components, cases, and screws that were sourced in the same country that assembled the final technology, the origins of the raw materials used to make those components or hardware were more geographically diverse—largely overlain on the map of colonial trade routes. Assessing "the global political economy of material flows," the continual effort to render resource flows and supply chains as unknowable, complex, and abstract phenomena is a paradigmatic feature of contemporary capitalism.[10] The result of this Alf Hornborg theorizes as *time-space appropriation*: a correlation between the "unequal exchange of embodied labour" and "the unequal exchange of embodied land."[11] Hornborg, like Marx, is not interested in the nuanced meanings, uses, or material properties of specific technologies, but instead on how their fetish nature is productive of a global political economy that has always been dependent upon slavery and the destructive appropriation of the new world's land and resources. Our contribution to this point is to demonstrate how the promises of gear for the creation of "music"—alongside the assumptions about the ability of art to produce meaningful emotional-aesthetic experiences—serve as a smokescreen for capitalism's destruction of nature and perpetuation of income inequality. Additionally, the perceived inability of any specific technological object to satiate a professional need means that excess quantities of gear, and excesses along the entire supply chain from mining pit to retail, lead to far more gear being made than is ever used.

As Peter Pels notes, elucidating the "magic" of commodities requires revealing the "social relationships and the history that turned them into positive representations of a certain product capacity, need, or subjectivity."[12] Marxian perspectives fall short, therefore, when those nuanced meanings, uses, and materialities are more important than use value or exchange value. Put another way, simply studying the "function" of a technology results in a sense of "misplaced concreteness"[13] and obfuscates the social relations *always* inherent in technological objects. In our case, *as gear's importance for music-making declined, its importance for mediating homosocial relations increased.*[14] Thorstein Veblen's insights into how conspicuous consumption serves as a marker of conspicuous leisure and disavowal of labor are relevant here but with an unusual twist, since the very objects that connote a "canon of reputability" are the same that connote modes of labor historically described in working-class, blue-collar terms (engineer, operator). Leisure, as the "non-productive consumption of time," is nonetheless dependent upon the kinds of appropriation of space and time critiqued by Marx and his interlocutors, including Hornborg.[15]

To recap, we situate gear within a long history of colonial encounters where crafted objects are widely attributed agency, and where through fetishism, social relations are at once mediated *and* alienated. Gear ceases to primarily be a labor-saving device and becomes the site of gendering practices, sexual desire, symbolic visual representations, nostalgia, heritage, and compulsive spending—which cumulatively define a social space centered on gear. As we will show, gear is not well equipped to handle the burden of these additional expectations and falls short in many ways, but these many modes of fetishization persist nonetheless.

MASCULINITIES AND GEAR CULTURES

Who participates in gear cultures, in which fora, and where? Who is included or excluded, and why? Attending nearly any professional and trade association event related to audio, or stepping into most multiroom studios, the considerable majority of people you encounter are men. During our research period, however, with increasing socialization around gear made possible by online fora, magazine reader communities, trade shows, and professional events, we observed something new: publicly staged fights over people's gear opinions and tech-speak proficiencies, and relentless message forum threads urging users to "show us your racks" and to post "gear porn" pics. Since none of this is necessary or useful for engaging in any professional audio trade, why did gear cultures manifest this way? The **hegemonic masculinity** concept best articulates what we are analyzing, but since it has not been widely used for understanding technologies such as these, or music cultures for that matter, a brief introduction is in order.

What are hegemonic masculinities? In Raewyn Connell and James Messerschmidt's updated version of the concept, "masculinities are configurations of practice that are accomplished in social action, and, therefore, can differ according to the gender relations in a particular social setting."[16] Scholarship in the late 1990s recognized how heterosexuality privileges masculinity, and, as such, the notion of a "normal" man is signified by power over—and control over—women. "Successful" masculinity, defined through male (hetero)sexuality, therefore does not "acknowledge and engage with female sexuality and their (women's) own emotions."[17] Rather, the competitiveness inherent to defining one's masculinity leaves no room for the acknowledgment of female sexualities independent from male desires. In saying all this, hegemonic masculinities may not represent the majority view in gear cultures. Therefore, by being hegemonic, in the Gramscian sense, such hegemonic masculinities accomplish the "demobilization of whole [social] classes" and precipitate an "active struggle for dominance."[18] Specifically, hegemonic masculinities subordinate less- and non-competitive masculinities as well as femininities, and many subordinate groups—for example, women, LGBTQIA+ participants, and men who identify with masculinities outside of that hegemony (for example, non-first world, culture-specific masculinities)—must consent to the hegemony by proxy if they wish to participate in the space.[19] Hegemonic masculinity research builds on the "multiple masculinities" concept to show how each masculine formation draws upon characteristics of a particular social space. In gear cultures, distinct hegemonic masculinities are found in each online forum, each trade show, each print media, and each regional recording-studio culture; by representing distinct spaces and being entangled with different gear, these masculinities come to subtly differ.

How are hegemonic masculinities maintained? One primary purpose of hegemonic masculinity is to maintain boundaries: they constitute gatekeeping and exclusionary labor. Despite their rhetorical force, they in fact are fragile and require continuous maintenance work lest they fall apart on account of being a minority belief and extreme disposition. If extreme ideologies are enacted too frequently, the milieu becomes too toxic for even its own die-hard participants. But participants in the milieu do not need to be reminded every moment; the cumulative effect of intermittently injecting extreme/fringe ideas into the milieu provides an ambient sense that the extreme ideology is ever present. When those most active participants in a milieu perceive any threat to the hegemonic masculine formation, they ramp up the level of the extremity—for example, using greater profanity, threats of harm, or actual harm—to "test" the members: are people in the milieu *really* committed to the local manosphere? But extreme practices are not only reactive. Hegemonic masculine formations cannot help but be influenced by a world around them where increasingly visible queer masculinities inflect popular culture. Perhaps since the spaces lack overt egalitarian participation by most of the world (e.g., women, LGBTQIA+ individuals, and cisgendered men outside the first world), gear cultures participants often "role-play" through imitations of their perceptions of womanhood or queer or subordinate masculinities. This never implies any intent of inclusion or acceptance, however, and the mocking conversational tone and conspicuous display of sexual imagery—even on products or in advertising—is straightforward bigotry done for boundary-maintenance purposes.

What do hegemonic masculinities do? First and foremost, hegemonic masculinities define who gets to participate in gear cultures, the terms and conditions of that participation, and the values and beliefs that participants are allowed to publicly discuss and express among each other. Commodity fetishization and valuation lie at the social core of gear cultures, meaning the performance of connoisseurship to exemplify knowledge of valuable technological objects and to assert one's status is key to the hegemony. As recognized in scholarship on fine-art connoisseurship, the display of such knowledges is a social practice involving accurate identification of specific historical and stylistic attributes of objects.[20] However, rather than create a knowledge base for professional recording technologies, such discourse instead serves as a "game."[21] In the game of gear sports, object valuation is a "move" for performing conspicuous leisure and conspicuous consumption.

However, hegemonic masculinities also have significant ramifications: they (re)inform the design, materiality, use, attitudes, and consumption patterns of technologies. The presence of sexualized discourse always instantiates practices of connoisseurship and valuation in gear cultures. In Brian Sweeney's extensive examination of masculinity and status among collegiate men, the "sorting" of women according to masculine ideas of "worth" and "impact," as well as competitive approaches to objectification, is commonplace.[22] In gear cultures, such preferences are often calibrated to group ideals that orbit around various "manhood acts" and "sex talk" to signal certain forms of heterosexual masculinity.[23] These prevailing and competitive calibrations of "sorting" gear, which parallel practices of "sorting women," lead many gear users to project all manner of sexual fantasies onto technological objects. The process of figuring out and reaching a consensus on which particular gear is *sexiest* and *the best* does double duty as an attempt to stake the terms of engagement within the manosphere.

Masculinity performances do, however, run a contradictory gambit between expressing gender-defined solidarity—"we're all men in this together"—and individualism—"I'm looking out for myself." In gear cultures, we find similar dynamics to those observed by Dennis Mumby in that "an ostensibly shared culture of solidarity is undermined by the contradictory privileging of individual self-interest."[24] For example, in the online sphere, group solidarity forms around "injured" forum members who have been pulled into line, yet simultaneously, members give individualized and autonomous expressions of gear display.

The Shadow Hills "War Film" mashed up militaristic imagery, including multiple shots of women sorting and testing equipment. This imagery, fleeting though it was, illuminates one key role that women have always played in gear cultures. As discussed by Carolyn Marvin in electricity culture in the late nineteenth century and by Jefferson Cowie regarding RCA's post-1930s industrial labor relations, the hegemonic masculinities we analyzed are only possible because of the behind-the-scenes labor undertaken, predominantly by women, in extractivism, component manufacturing, and gear assembly.[25] As the Shadow Hills film alludes to, women work on gear assembly lines and factory floors. Women routinely beta-test technologies, and behind nearly all the "one-man-band" boutique audio technology businesses is a female partner running the engine of the gear business. Women routinely manage the operations of recording studios and run the receptions and the marketing and promotional activities of gear businesses. Women moderate online fora, teach and administer audio technology pedagogy programs, and manage professional organization logistics. Women work in audio technology journalism, in advertising, and in trade show organization.

However, women are typically excluded from participating in—or in shaping the nature of—discourses and practices in hegemonic masculine formations. The formation of the AES's Diversity and Inclusion Committee has addressed and continues to address inclusivity in the audio industry. Their introduction of a code of conduct and accessibility and inclusivity guidelines for conferences and events, as well as a broadening of panel memberships to promote visibility of marginalized groups, has been effective in bringing positive change to the organization and promoting a more fair and inclusive industry. Both authors have actively participated in inclusivity initiatives; related to that, we also strive to foreground women's labor throughout the book. Indeed, to be accepted into gear cultures as an equal participant, women must be exceptional at their craft and outperform the average male participant by a considerable margin. As such, only a few women are accepted as equal participants. EveAnna Manley, Emily Lazar, and Sylvia Massy are three examples of accepted female gear culture participants; in each case, their work and contribution to the industry far exceeds that of most of their male counterparts. Hegemonic masculinities work hard to downplay—at best—and to ignore altogether—at worst—the *actual* contributions made by women to gear cultures. By keeping this evidence concealed from the gear culture hegemon, the primary function of women may be upheld as an invisible yet insidious focus of objectification and fetishization.

Hegemonic masculinities have so far been studied through discourses, beliefs, and practices. Our main contribution is to extend the concept to consider materiality, technological objects, images, and the online circulation of and socialization around the above. We also expand the range of sites where hegemonic masculinities are found, from online fora and blue-collar workplaces to trade shows, professional associations, print media, and museums. If hegemonic masculinities are essential for perpetuating the fetishization of technologies, then the fetishization of technologies is the key activity for maintaining hegemonic masculinities.

Gear cultures are brimming with hegemonic masculinities, and we have seen and participated in many of them as part of this research. For example, Samantha was for a brief time one of the only female directors of the UK's Music Producer's Guild, a highly competitive and, until recently, exclusionary organization. A variety of hegemonic masculinities are found at the AES: these are reflected in the majority-male representation in the journal and convention, in the conference technical program, and among the exhibitors and attendees on the trade show floor. Hegemonic masculinities are particularly foregrounded in the music technology press via sexualized and value-oriented discourses and propped up by advertising that encourages and upholds both valuation and sexual aspects of technological fetishism.[26] The vestiges of hegemonic masculinity in the outward projection of gear manufacturing contrast with the "backstages" of workplaces, where women are responsible for so much behind-the-scenes labor. All this said, each gear culture has its own hegemonic masculine formation with its own styles and traits and terms and conditions of participation.

Hegemonic masculinities in gear cultures are deeply problematic, and rather than passively observing and commenting on them, we regard our role to be one of exposure and dismantling; we include several ideas in the conclusion as to how this dismantling may be expedited. Gear cultures reflect masculinity in crisis. Compounded by the "feminist revolution" and the ever-increasing visibility of LGBTQIA+ politicians, businesspeople, and celebrities, twenty-first-century masculinity has found itself embattled with a society giving increasing space and platform to nonmale and nonheterosexual participants.[27] While feminism and LGBTQIA+ cultures are obvious perceived threats to hegemonic masculinities, other threats to established masculinities come from within. For example, the rise (no pun intended) of Viagra foregrounds techno-scientific advancement of the "normal" male body and, in doing so, upholds the "patriarchal machine."[28] More broadly, masculinity in crisis has been reflected in character portrayals in film, on television, and in music videos since the mid 1990s, including the foregrounding of Korean "soft masculinities" through popular music and film.[29] In the same period, the Western world has increasingly drawn attention to its marginalized groups and has platformed majority-ethnic people, and colonial status quos have been disrupted by Indigenous truth telling.[30] Alongside circulated quick-fire and widely adopted terminologies that critique hegemonic masculinity—"mansplaining" and "manel" being just two—language policing on social media has also increased.[31] The widespread fetishization in gear cultures—often accompanied by and informed by toxic masculinities—can therefore be perceived as retaliation against the perceived threats to hegemonic masculinities.

Gear cultures coalesce around one fetish theme more than any other, and, to the surprise of probably no one, that theme is sex. At least since the 1980s, gear manufacturers have routinely used sex to sell audio technologies. However, one distinctive vocabulary that began in 2002 on the nascent Gearslutz forum entails a specific style of sexualizing gear: describing gear as "sexy," gear users self-describing as "gear sluts," "slut-shaming" other milieu participants who don't buy enough gear, and replicating adolescent locker-room speak within the context of men looking at technological objects in studios. The bodies of gear culture participants can be understood, in one sense, through Jean Baudrillard's concept of the body as "cultural fact," where "the mode of organization of the relation to the body reflects the mode of organization of the relation to things and of social relations."[32] Unsurprisingly, gear culture participants create homologies between fetishized objects and fetishized bodies, and technological objects end up standing in for the missing female bodies. As Paul Théberge noted, "Consumption will be made even more attractive through the creation of new desires," and these new desires play out in the contexts of online fora, trade shows, and the music technology press.[33]

What, exactly, makes some gear "sexy"? Gear sexiness can refer to an intrinsic Boolean property (the object is sexy / not sexy), to a binary comparator (A is sexier than B), to a superlative (A is the sexiest of them all—in other words, sexual supremacy), or to a continuously variable quantity (degrees and qualities of sexiness). When marked as an intrinsic Boolean property, it is typically connected to the social performance of connoisseurship. Sexiness used as a binary comparator riffs on the male game of "who's hotter," conflating notions like "which is the sexiest mic pre for electric guitar" with the social practice of "sorting" women on apps like Tinder. When sexiness instead is marked as a property with variable degrees and qualities, often the technology has been anthropomorphized to such an extent that it has come to replace people (women always and everyone whose labor power is expended while working along the many supply chains necessary for gear). Unlike a blow-up sex doll which at least morphologically resembles a human, you can't really have sex with recording-studio gear—except, perhaps, for phallus-shaped microphones, but the implication of being penetrated undermines the norms of a hegemonic masculinity that mocks and disavows queer and gay culture and so-called beta masculinity.

FOR THE LOVE OF TECHNOLOGY (TILL IT HURTS)

"We love music so much that we might talk about it as an animate, sentient being," explains William Cheng in his seminal *Loving Music Till It Hurts*.[34] The same can be said for gear and its lovers. The eternal needing, getting, and yearning for gear. The metaphorical, analogical, and allegorical lengths to which gear-lovers go to justify their wanton gear lust, to reaffirm their gear culture status, and to amplify their gear accomplishments. As Cheng suggests, the anthropomorphizing of music often ends in tears—bands break up, rock stars die, trends come and go—and the anthropomorphizing of gear produces similar effects on its many lovers. How and why do fixations form around gear, what kinds of obsessions pervade gear cultures, and what are the economic, relational, and social consequences therein? Loving gear often hurts due to **displaced agency**, that being the excessive, false, and, in some cases, delusional attribution of technologies as the primary agents that produce "successful" music recordings. These ideas of success tend to orbit around specific types of musical output—namely, commercially recognized, Anglophone popular and rock music, particularly records made between 1955 and 2000 by (mostly white and often dead) male musicians and/or recordists at canonized studios such as Abbey Road (UK), Muscle Shoals (US), and Ton Hansa (Germany).

This chronology, combined with these types of studios and recordists, constitutes a tiny fraction—a microcosm, even—of the world's available music recordings. Yet from this tiny pool, enormous ideas, understandings, and mythologies have formed and spiraled, not simply around the music and how it sounds, but around the means by which these recordings were created. In gear cultures, gear is staged front and center and is attributed excessive responsibility for how recordings are made and the success by which they can be measured. Forget about Lennon and McCartney's combined songwriting prowess, Martin's all-seeing production eye, Emerick's engineering, Abbey Road studio's distinctive acoustics, Apple Records—or even the individual performances of John, Paul, George, and Ringo. In the gear culture narrative, *Revolver* is *Revolver* because of the Fairchild 660 limiter. To us—and to you, reader—this may seem far-fetched. Yet when we unpack how gear obsessions manifest in single technological objects, we can quickly trace them back to canonized musicians, recordists, production workplaces, or combinations thereof. In other words, detached from recording processes or use value, a shedding of the contextual skin(s) serves to lay the technology bare, following a "reverse engineering" of sorts:

> Fan loves music—band/artist no longer exists—fan cannot see artist perform—fan owns records and obsesses over recordings—fan wonders how recording was made—fan seeks information on "behind-the-scenes" aspects of recording finding production location / agents—fan locates technological means and fetishizes technologies as embodiment of recorded musical success—fan sheds context, focuses on gear.

This is, of course, closely related to an insatiable appetite for *more*: more details, more knowledge, more stories about favorite artists and albums—a kind of hoarding in itself (we will return to that shortly). This is, however, only one of many ways that gear culture participants arrive at fetishizing technological objects. Wherever we look in gear cultures—at trade shows, in the advertising campaigns of gear manufacturers, in online forums, in the gear spiels of retail stores—we find this "Russian dolling": the methodical removing and discarding of gear's contextual layers until the focus narrows on its ever-shrinking parts. But why stop at the technological object, the "black box" itself? Many gear lovers don't. Chapters 3 and 4 remove the lid and find the circuit board being "technobabbled" among gear lovers. Disassembling the circuit board, we find very rare and prized electrical components associated with the highest build quality of boutique manufacturers. Opening the components, we find even scarcer raw materials of specific terroir. Never content with just the gear, gear lovers dismantle and disembowel their gear, fussing over its circuitry and electronic detail. Fetishization, at the very heart of gear cultures, lies deep inside the guts of gear circuitry.

Gear is loved not only because it functions as important cogs in the wheel of music production. In gear cultures, the role of gear in music production is simply a boundary that the love of gear transcends. Love conquers all, and the love of gear knows no bounds. We break gear love down into three recognizable phenomena—technophilia, objectophilia, and hoarding—and find out what happens when gear lovers love their gear, till it hurts.

Technophilia

As people who have dedicated large portions of our lives to working with and studying audio and music technologies, we are often considered technophiles (Samantha approves, Eliot regards themselves as a luddite). Technophilia, in its most basic sense, means an "attraction to technology."[35] Technophiles positively receive and embrace new technologies, possessing an enthusiasm that may extend into and combine with futurism and utopianism. They engage directly and affirmatively with technologies and look for the benefits. And there is no doubt that technophiles enjoy looking at gear: they even routinely engage in technological voyeurism, looking at others looking at gear. Maria-Elena Osiceanu links technophilia with egocentrism; the role technologies play in one's self-fulfillment is intrinsic to this attraction. In an early article, Richards identified a "technophilic state of mind" in car advertisements in that they promise "invulnerability and transcendence."[36] He understood car technophilia as "narcissistic phantasy," where ideals of power, magic, and perfect bodies may be projected onto an object. In recording culture, the technophile, as a distinct 1980s–1990s recordist type, has been linked to tech-utopianism and the insatiable appetite some professional recordists have for new, cutting-edge technologies.[37] During this era, technophiles focused on music production through a technological (as opposed to a processual or personnel) frame, and during a contemporaneous technological arms race in commercial music recording, objects were specifically marketed to technophiles.

In the mid 2020s, technophiles are found throughout gear cultures. Attractions to gear surface among professional recordists and amateurs, gear designers and sellers, and online gassers. In fact, gear cultures are built on the default understanding that users must be attracted to gear: technophilia's logical other, "technophobia," is rarely found. Interestingly, gear culture technophiles are not the same professional recordist technophiles Bennett identified in the late twentieth century. Gear culture technophiles are attracted to gear of the past, whether that is original models, clones, or new variations on established gear themes. The fetishized notions of past music and audio technophilia—power, speed, labor and cost saving, downsizing—have given way to fetishization around expensive technologies affiliated with canonized recordings. The "best" technologies, therefore, are not necessarily those in widespread use today. Computers, plugins, and control surfaces can be high-performing and expensive too, but these are not gear: nobody made a canonized record using computer-based digital technology prior to 1980.

Objectophilia and "Sexy" Gear

Gear cultures are full of objectified technologies, including ones deemed "sexy." All manners of sexualized discourse circulate around gear. However, this is not necessarily evidence of objectophilia, which is distinctively individual, relational, and sexual. *Objectùm-sexuality*, a term coined by

Eija-Riitta Eklöf, refers to an individual's erotic desire for inanimate objects. This extends farther than technophilia, where attraction to technology could be economic, social, or status-oriented; for the objectophile, the attraction to technology is additionally sexual. Although rare in gear cultures, we do find objectophilic tendencies among those participants who move beyond canon-informed attraction and toward sexualization and haptic ideation. Our research has yielded similar findings to those in queer scholarship. While objectophilia is understood as a "non-normative" form of sexuality, it is "genuine though rare."[38]

Evidence of objectophilia in gear cultures is mainly found online in audio technology forum threads. Voyeuristic vernacular including "looking" and "seeing" extends into haptic ideation: "touching" and "feeling." Gear is often adorned with knobs, switches, buttons, and flashing lights and finished off with sleek, smooth, metallic surfaces. Such interfaces invite haptic encounters: no surprise, for some gear lovers, these haptic encounters can be desired more than gear looks. Objectophilia can seem like a misplaced sexual attraction—for example, in the 638 posts on Gearslutz where a user described having a "boner" for gear. But such ideations move into dangerous territory when forum users describe sexual interactions with gear without consequence: in one instance, a user concluded an objectophilic post with the comment that "gear never says no." Here, loving technology till it hurts manifests as toxic forms of hegemonic masculinities—including perpetuating rape culture.

Objectophilic tendencies in gear cultures are part of a broader queering of heterosexual realms. Heiko Mortschenbacher noted that "various other sexual normativities may be effective and practices that have a non-normative status from the perspective of the social macro-level may be considered to have the status of a locally ritualised norm."[39] In other words, although heterosexuality represents the dominant sexuality, as the presence of lesbian, gay, and bisexual individuals is normalized and accepted, so too are uncommon forms of heterosexuality. Additionally, "strays"—a conflation of "straight" and "gays"—are straight men interested in other men and men's lifestyle, image, and activities.[40] Strays became especially visible in the twenty-first century alongside an abundance of men's magazines, blogs, and television shows, including *Queer Eye*. The presence of "professionalised queerness" in popular culture promotes gay culture to both gay and straight men, thus queering dominant heterosocial norms.[41] Objectophile tendencies result in several consequences for gear cultures. Participants often posit real or perceived barriers to obtaining and interacting with gear (that must be "overcome" in the future), and, most often, a female partner is defined as that barrier, whether it is "the girlfriend," "the wife," or "the missus" who restricts access to gear. While these barriers are not exclusive to participants with objectophilic tendencies, they show how loving gear till it hurts impacts real-life relationships in both perceived and real ways.

Hoarding #notallhoarders

While the fetish nature of gear as we have theorized it (via Pietz, Pels, and Hornborg) applies broadly toward understanding individual objects and their types, gear stagings in recording studios, audio events, and elsewhere often emphasize racks and stacks of gear—to great excess. A few complex recording scenarios do demand a significant amount of gear to be performed well, but in most scenarios, most of the gear lies around, effectively unused.

Personal hoards of gear—thousands or even millions of dollars of unused or underused technological objects—are compelling to many consumers, to corporate and media entities who profit off promoting their desirability, and to academic and trade organizations. They relate in some ways to often-noted collections of things such as comic books, vinyl records, baseball cards, or postage stamps—and to museums, where only a fraction of the material culture that has been assembled in one place is publicly on display. Most importantly, gear hoards have a key role in inculcating the specific kinds of sociability around gear that define gear cultures.

Hoarding, formerly listed as one symptom within obsessive-compulsive disorder and recognized as a "profound public health burden,"[42] gained its own diagnosis in 2013 in DSM-5. However, beyond a pervasive "emotional attachment to possessions,"[43] we know little about the diagnosis and pathology of audio technology–specific hoarding, since the objects in question are specialized and outside the material lists of common diagnostic tools, and most gear hoarders have not considered themselves to have a disorder that necessitated psychiatric intervention. But not everything about hoarding and hoarded objects *should* be pathologized as a "disorder."[44] Hoarded objects can, in some situations, be significant containers for memory and kinship—especially after the object is subjected to a "metonymic accretion of everyday associations with persons and experiences,"[45] or stated more simply, when the object was used by people and became transformed in the process.

Gear objects are always part of a larger collection of material objects related to audio technology. Within an assemblage, gear typically comprises the most prominently displayed objects; these objects *are* intended for private display and make up the primary visual impression of any studio. Out of view, however, will be many other objects: broken gear that has yet to be repaired, audio technologies that in another milieu might be regarded as gear but here are perceived as less desirable, boxes of working and broken cables, electrical components such as vacuum tubes or capacitors intended for repairing gear, equipment and supplies for testing and repair, spare channels for large format analog consoles, and hoards of gear accessories such as mic clips and mounting adaptors, power supplies, USB dongles, CD installers for legacy software products, backup hard drives, original boxes in which gear was sold, and broken headphones and speakers. Not all this material culture will be treated the same; some cables and supplies might be carefully organized in well-labeled boxes or conveniently hung on cable harnesses, while other bins contain seemingly random collections of heterogenous things. This all suggests that gear-related hoarding is more akin to the materiality of museums than it is to the individuals with attics and storage sheds full of personally meaningful waste as have been featured in "hoarding disorder" literature. Moreover, gear's place within broader assemblages of stuff that nearly always overrun rooms and closets in homes, studios, universities, and elsewhere is what motivates us to use the term "hoarding" rather than the more typical "collecting."[46]

Recent hoarding literature has limited its discussion of hoarding's social aspects to the way that each material object comes to stand in for a person who had either given or formerly owned the object. This is less relevant here where we are analyzing objects that may have been bought new or used and, with those few notable exceptions of "legendary" gear owned by famous producers, are not specifically valued for the fact that other people once possessed these specific objects. But gear hoarding *is* social, especially when curated collections of these objects are photographed and shared, when gear hoarders transport their objects to a trade show to showcase unusual historical gear, and when/if musicians are invited to studios to record.

The scarcity of working examples of certain pieces of gear leads to a competitive kind of hoarding that is characteristic of luxury-item collecting. Some of the most popular threads on Gearslutz/Gearspace pertain to gear that users managed to acquire for far less money than the current price—buying a sought-after microphone for $200 from a studio that was getting rid of their old mics, purchasing a thrift-store find, or coercing the family of a deceased recording engineer to sell items from the attic for "pennies on the dollar"—indicating a joy in personal gain that comes from the inexperience and loss of others.

To convey the sentiment of this, we found an evocative example from graphic novels. In Chris Ware's *Acme Novelty Library*, especially volumes 10–11, we meet the character of Rusty Brown, a collector of G.I. Jim action figures. He is a chronic liar, he is very competitive, and he covers up his private life of whimpering sadness with his abuse of and profiteering off his collecting archnemesis, Chalky White. In one episode, Rusty Brown makes a bet that if he can "score" with a voluptuous woman, he wins an action figure from Chalky; if he cannot, he gives one up. Through trickery, he wins the bet, but thought bubbles show the reader what Chalky will never know: that Rusty is a virgin. The correlation between adolescent male sexuality frozen in a state of arrested development and fetish-collecting is noteworthy, as is the structuring of the milieu around G.I. Jim action figures—war toys for boys. Audio gear's frequent description through weapon/military/war metaphors helps inculcate a similar kind of sexually charged competition among gear collectors. And while Karl Marx wrote of hoarding primarily in relation to money, his pathologization of hoarding is equally relevant here: "This antagonism between the quantitative limits of money and its qualitative boundlessness, continually acts as a spur to the hoarder in his Sisyphus-like labor of accumulating. . . . The hoarder, therefore, makes a sacrifice of the lusts of the flesh to his gold fetish."[47]

TECHNOLOGICAL EXHIBITIONISM

These ways of loving gear—technophilia, objectophilia, hoarding, and gassing—are central to gear cultures. The deepest unconditional gear love is expressed through *performance*. This goes far beyond the act of sharing a few photographs, pointing a forum user in the direction of a web page, or even visiting a studio and admiring the racks. Distinct social practices orbit around the staging and voyeurism of quantities of gear that has a particular or special significance. Such social practices are not confined to one gear culture either; we have witnessed insatiable appetites for ostentatious displays of gear in myriad ways, where significant labor has gone into both presentation and performance.

Take, for example, the API booth (figure 2.1) at the 2019 AES Convention. Red museum ropes cordon off a vintage console, which sits beneath a divider covered in black-and-white historical snapshots of historical API imagery. "50 Years of Timeless Sound" is the somewhat oxymoronic banner.

Figure 2.1
API booth at the 2019 AES. Photo by Eliot Bates.

Men in yellow shirts demo numerous large-format consoles, positioned to be accessible to visitors, who stand—hands on hips or arms folded—to listen to gear stories as they stare at illuminated VU meters. Atop meter bridges sit monitors, and atop monitors sit TEC awards: industry prizes awarded to API gear over the years. For no apparent *functional* reason, an Ampex 8-track tape machine overlooks these proceedings from the corner of the booth; Ampex was not a part of API, nor is 8-track tape a current or widespread recording medium, yet this is part of the display, the performance, and the (re)construction of historical context. All this elaborate staging amounts to what we term ***technological exhibitionism***: the excessive staging of gear conflating technology, history, display, and public access that when combined invites and facilitates primarily male socialization. There is often, but not always, some effort to canonize technologies at play, too.

Technological exhibitionism is, therefore, an *outward* projection of technological–social relations and a key outcome of the technophilia, objectophilia, hoarding, and gassing. It goes beyond simply sharing an image or creating an "unboxing" YouTube video, and, as such, we distinguish it from technological voyeurism. Technological exhibitionism transposes the male gaze onto gear and projects masculinist fantasies on both the look and/or feel of technological objects. It might be individual, or it might be social. It is most often found constructed in professional domains for aspirational consumption by amateur and semiprofessional agents. When done at large-scale gear events, it requires significant labor to plan, organize, display, and maintain the exhibit so that it is conducive to inculcating the right kinds of social relations. Technological exhibitionism is intentional and measured, gratuitous and ostentatious, and designed to perpetuate mythologies and to gatekeep the so-called golden age of sound recording. Gear becomes the focal point for the telling of anecdotes and stories.

GEAR ACQUISITION SYNDROME

one more piece of gear
will help me cope with the world
in my solitude (user t3h, "GAS haiku thread")[48]

Gear fetishization produces social outcomes in that technological exhibitionism serves as a nexus for practices of hegemonic masculinity, and it produces economic outcomes that revolve around GAS—gear acquisition syndrome. Exhibitionism and GAS work together to define an aesthetics of conspicuous consumption *and* to circumscribe the kinds of individual subjectivity that gear culture participants are allowed to possess. But what is GAS, and how does it differ from people just buying technological objects that they want or need for work and/or play?

Gear acquisition syndrome, a term attributed to Steely Dan guitarist Walter Becker, has become so widely known it's become a noun—GAS—and a verb—*gassing*.[49] When recording-studio musicians, especially guitarists and synthesists, experience GAS, it is usually related to their never-ending pursuit of achieving a distinctive tone—chasing the unicorn, for example, of the "perfect" distorted guitar tone.[50] To an extent, this makes sense: a Fender Stratocaster guitar into an Ibanez Tube Screamer pedal into a Fender Princeton Reverb amp will sound quite different than a Gibson SG guitar into a Boss DS-1 pedal into a Marshall JCM900 amplifier. Whereas owning the means for timbral flexibility *can* be pragmatic, amateur guitarists who "suffer" from GAS may own dozens of guitars and amps and hundreds of pedals—most of which are never used. As hundreds of testimonials show, most of GAS sufferers' energy goes not into creating music but into "lusting" over the next gear purchase. Not just academic writers have noticed this: even trade publications such as *Music Radar* published guides to recognizing and dealing with GAS (their seven-step process begins with "dissatisfaction" and ends with "relapse").[51]

To show how gassing articulates fetishization in all the aspects discussed above, and how it relates to the construction of self in a social milieu, we will examine GAS at two different scales. The first is a microscale, common sequence of transactional events where a gear-interested person goes from not knowing about an object to acquiring it and ultimately exhibiting it in some social milieu. The second is a macroscale look at the varying "states" that comprise GAS processes, where modes of voyeurism, display, and exhibitionism are key. Here, we see how gassing relates to the construction of self in a social milieu and how fetishization is central to that construct.[52]

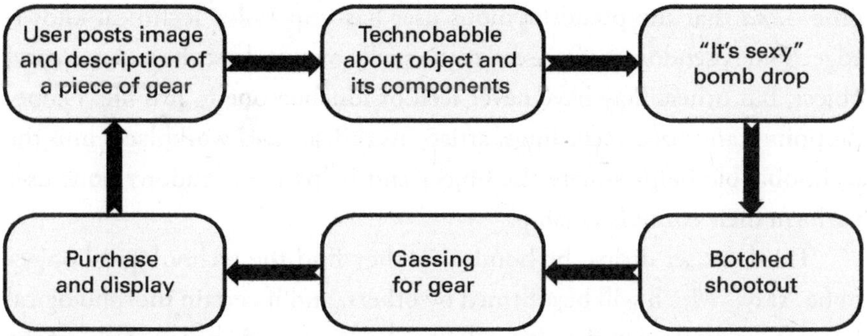

Figure 2.2
Turn-based flow of GAS.

Let's take the example of a typical audio-recording workflow. Nearly all recordists have experienced a process where the results sound, to them, far short of what they imagined. Might purchasing one or more technologies fix the problem? With several thousand microphones and at least five hundred different makes each of mic preamplifiers, EQs, and compressors available on the new market (with just as many vintage ones on the used market), what factors are at play when making important decisions about signal chains and flow? A six-step process is commonly at play (figure 2.2).

First, GAS triggering begins with a picture of gear surrounded by contextual advertising information. Perhaps on the product website, in a gear magazine, at a trade show, or on a gear forum new product listing, a gear picture will be juxtaposed with marketing prose that veers between the empirical (e.g., premium new old stock components, no capacitors in the signal path), the technical (dozens of technical specifications), the terroir (some bit of it is "genuinely" made in the US/UK), and the subjective (it will achieve the "authentic" sound of some historical gear, or will be your secret weapon in your arsenal).

Second, pseudonymous users on the internet opine about how amazing the unit is, blurring the distinction between subjective opinion and empirical fact through copious technobabble (i.e., fake technical terminology culled from audiophile nonprofessional equipment magazines such as *Stereophile* or real terminology that is misused in this context) that simultaneously makes it seem that there's something special engineering-wise going on inside the unit—*and* that the pseudonymous user has impressive technical knowledge. Some pseudonymous users may actually own and use the technological object, but others may have never seen or touched one in real life. Name-dropping canonized recordings, artists, recordists, and workplaces into the technobabble helps situate the object and helps the pseudonymous user perform their connoisseurship.

Third, a user drops the bomb that they find the technological object to be "sexy," which will be affirmed by others, and if certain morphological features are in place in the object, it may lead to a brief "ramping up" of the sexualized discourse. Fourth, inspired by this, one pseudonymous user runs a nonscientific, "botched" shootout that demonstrates that said technology goes "head-to-head" with other technologies "ten times the price." Fifth, having proved that the technological object a) looks sexy and b) can be favorably described using milieu-specific vocabularies, and c) that some consensus of users with different albeit pseudonymous names concluded in a test that the object "blows away" the competition, other users chime in that they are gassing for this technology.

Sixth, one or more users claim to purchase the technology, and others demand to see "pics or it didn't happen." With that personal act of technological exhibitionism, that user completes the cycle of GAS—until their next recording that falls short of their imagined outcome, wherein the cycle begins again.

By analyzing these social games around materiality and these turn-based discursive moves that keep the thread active, we see again and again the indelible links between the sexual fetishization and economic fetishization of gear. Each produces the other, and in both, the fetish is the same: the technological object abstracted from a musical context. These threads can generate hundreds of posts before there is any meaningful consideration of music, musicians, or audio engineering techniques.

The preceding considered gassing as a transactional journey. But how does gear fetishization—both economic and sexual—and technological exhibitionism manifest in the gassing process? We break down the gassing process into several states, valuations, and practices (figure 2.3). This illuminates how gassing practices relate to the construction of self in a social milieu

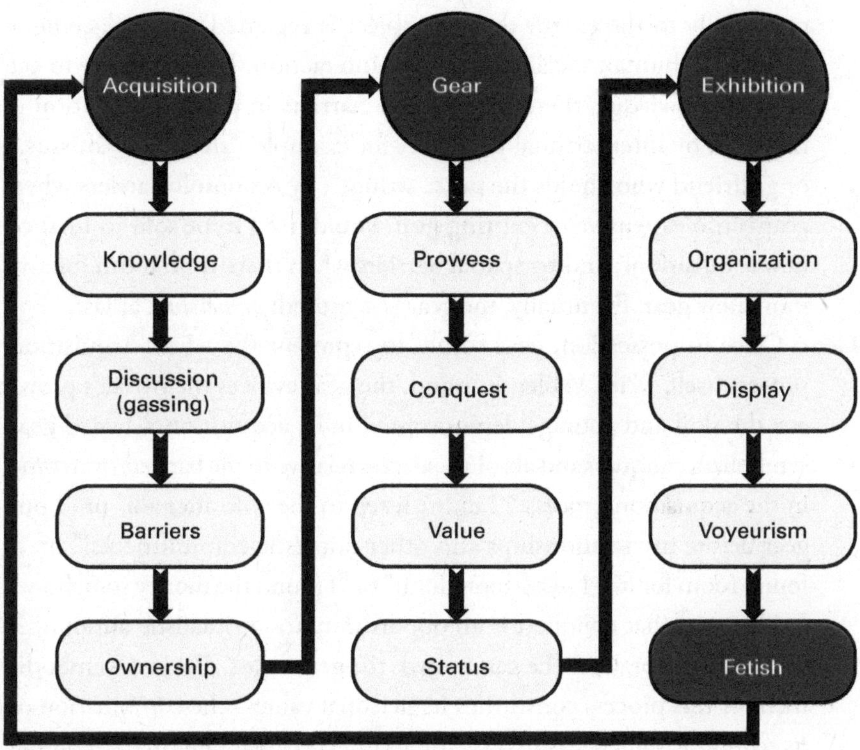

Figure 2.3
Gear acquisition states.

and, most importantly, how fetishization and exhibitionism are central to that construct.

Acquisition: The wanting, needing, and getting of gear entails a multistage process. Early on, there must be prior knowledge of some kind: that the gear exists, that it is valuable and meaningful, that it is connected to other important things (workplaces, recordings, artists), that it purports some benefit, and that talking about it with others is beneficial and rewarding. Gear knowledge leads to community discussion: discourses, gassing, and social interactions around a particular piece of gear. When this veers toward components and interfacial elements, we find "technobabble" transactions. When gear "unboxing" becomes the nexus, conspicuous consumption and human–material relations matter instead, potentially to the extent that the object is regarded as a replacement for normal human social contact and interaction. But to be able to act on this knowledge, there are so many barriers in place. Acquisition is hindered by interpersonal barriers—for example, "the wife," "missus," or girlfriend who "holds the purse strings"; by economic barriers when gear is too expensive or existing gear would need to be sold to finance new acquisitions; and by spatial barriers when there is no room for any more new gear. Eventually, the gear is acquired: *ownership*, at last.

Gear: Once in possession, gear comes to represent the whole acquisition process itself. With Veblen in mind, the gear evinces the owner's prowess: the skill and courage demonstrated in its acquisition. Owned gear symbolizes conquest and absolute success relative to the barriers overcome in the acquisition process: "I didn't listen to the wife after all!" or "I put gear before my relationships and other domestic commitments!" or "I found room for it—I *made* room for it" or "I found the money somehow." Evidence of that conquest is an opportunity for a ritualistic announcement and celebration: the gasser owns the gear at last. The gear's embodiment of this process constitutes its personal value—the combination of its economic value measured as cost, its interpersonal value in overcoming adversity, and its social value in the spectacle for those who bore witness to the conquest. "Winning" and the gaining of the prize, beyond value per se, will hopefully mark the owner as possessing a higher social status. Gassers are willing to do what it takes to take possession of the object, no matter how difficult. Such status, symbolic of bravery, quite often is discussed through metaphors of battle and war.[53]

Exhibition: Gear, now owned and imbued with values and status, is *organized*. This later stage of the GAS process usually includes small-scope technological exhibitionism, including imagery of unboxing, of situating the gear object next to others and in proximity to a DAW computer. The gear does not have to be turned on, but sometimes it is wired up, powered on, with meters and LEDs clearly glowing. Photographed gear provides evidence to be exhibited to other users of the proof of ownership, values, and status. In turn, socially shared imagery invites voyeurism, the gaze of other gear users: look at this gear, I own it, look at what I went through to get it, I am courageous and brave, I triumphed over adversity and now I hereby present to you the ultimate prize—I *won*. Fetishization is deeply entangled in all stages of the process, but the performance around winning the battle strongly reinforces the fetish. Exhibited gear embodies lustful gazing to the extent that it inspires others to want to acquire the fetish object for themselves.

CONCLUSION

We have never met a person that, at least in small ways, does not behave irrationally around certain kinds of technological objects. Whether it is talking to a kettle that has yet to boil, patting a misbehaving computer on its "head" to prevent it from crashing again, or kicking a noisy washing machine, technological objects continue to be the recipients of misplaced human social attention. The capacity for this is differential, though: not all objects are socialized to human norms to the same extent, in all places. The creation of whole social formations around this appears to be a relatively new phenomenon though.

By analyzing the turn-based discursive moves that keep a gear discussion thread active and these social games around materiality, we see again and again the indelible links between the sexual fetishization and economic fetishization of gear: each produces the other, and in both, the fetish is the same: the technological object abstracted from a musical context. Forum threads can generate hundreds of posts before there is any meaningful consideration of music, musicians, or audio engineering techniques that are in the service of meaningful music. Hence, widespread fetishistic practices that abstract technologies from their primary-use cases are necessary conditions for having a gear culture. For individuals, however, the small amount of social capital they hope to gain by converting their sexual fetish into an economic purchase of new gear to be showcased comes at a significant cost. In this regard, gear is just like any collectible, and gassing for gear—loving tech till it hurts—is no different than other modes of technophilia and socially shared hoarding, whether the fetishized object might be postage stamps or garden gnomes.

II MATERIALIZING GEAR

As we have explored, gear consists of those technologies that, due to being fetishized, gain a whole host of practices, meanings, and significations that extend well beyond the most obvious kinds of use or exchange value, and that do so to such an extent that people create cultural formations around them. Unsurprisingly, the specific materialities of the gear—not just the chosen electrical components but also the provenance of various metals, where and by whom the gear was assembled, and how materialities are presented to consumers through aestheticized equipment interfaces—are essential for gear's fetishistic properties to work.

Our analysis of gear cultures begins with mines and material processing sites. Gear could not be made without the labor and environmental destruction being conducted in these many sites. The price of material commodities would be orders of magnitude higher than they are today without extractivist firms' deep entanglements with international organizations, especially the IMF and World Bank, national governments, foreign trade treaties, and the global supply chains that connect ore to buyers. That said, extractivist and materials processing regimes impart divergent kinds of political and socioeconomic consequences depending on the region. The story of any piece of gear, then, involves reciprocal landscapes between a recording studio, typically in the global north, and the more than twenty different countries that were mined or drilled to produce the substances of gear.

In gear cultures, fetishized objects are not black-boxed to the extent that other consumer electronics are; many consumers open up their technologies and even "mod" them by replacing selective components with expensive and esoteric alternatives ("modding" refers to practices of altering the electrical components of premade technologies; within gear cultures, people who do this are known as "modders"). Others keep them intact but enjoy discussing certain technical and component-level aspects with gear designers. Some manufacturers are influenced to use certain costly components and manufacturing processes—for example, hand-soldered through-hole or point-to-point wiring—due to perceived consumer demand. Gear design and manufacturing are certainly not sequestered from the consumer side of gear cultures; many professionally active designers and/or manufacturers regularly participate in broader gear cultures, both online and at trade shows. That said, few gear culture participants understand how gear is *actually* made and by whom, and a subtheme of our analysis of manufacturing and design is the considerable invisible labor without which gear would not exist.

The power of gear to captivate users and prospective consumers is attributable only in part to its sound or use in making quality music recordings. Its interface—the surface of contact between user and object—is even *more* compelling to gear culture participants. But what is being interfaced, and how is the act of interfacing experienced? We consider the interfacing sensorium and how five modalities—material, visual, haptic, audible, ludic—work together to inspire the lusting and gassing for gear and, ultimately, its fetishization.

3 EXTRACTIVISM

JAGUARS AND COPPER

In February 2016, a 41-second video of El Jefe, the only known jaguar living in the wild in the US at the time, "went viral . . . briefly making the Rosemont Mine a global issue."[1] It was recorded by Chris Bugbee, an employee for the Center for Biological Diversity and volunteer for the University of Arizona's $771,000 jaguar research project, who traversed the mountains in southeast Arizona's Sky Islands with his trained dog Mayke in search of jaguar and ocelot scat. The video showed El Jefe "roaming the oak woodlands of the Santa Rita Mountains, walking downstream and strolling through the forest by night and by day."[2] The release of the video was controversial: although Bugbee intended for the material to bring additional awareness to the plight of endangered mammals on the US–Mexico border, conservationists feared that it might provide too many clues as to El Jefe's precise location. Indeed, two years later, another jaguar named Yo'oko was killed by a poacher.[3] Prior to being driven to near extinction by colonial settlers, the range of jaguars extended through southern Arizona, New Mexico, and Texas. The last female jaguar was documented in the US in 1963, but male jaguars—such as El Jefe, Yo'oko, and, before them, Macho B (who was euthanized in 2009)—continue to be occasionally seen north of the US–Mexico border.[4]

The Rosemont-Helvetia Mining District of the Santa Rita Mountains has been the site of colonial mining operations since 1874 and Indigenous artisanal mining prior to that, though active commercial mining ceased in

1951.[5] Hudbay, the Canadian mining company that owns the patents on Rosemont, estimates that Rosemont's hydraulic and electric shovels would extract enough rock for the planned nearby copper-molybdenum concentrator to produce 2.15 million tons of copper and 71,000 tons of molybdenum over a twenty-year period.[6] While over 30 percent of global copper demand is met by recycled materials and 95 percent of the copper in waste can be recycled,[7] demand for "virgin" copper continues—especially in audio technology manufacturing. One of the most likely customers for Rosemont's copper, if it does begin to be extracted, would be high-end audio—especially for the high-purity copper demanded for gear's internal wiring and the cables that connect studio gear.

Jaguars are unlikely protagonists in the story of audio gear, its resource dependencies, and the extractivism that makes technology possible. Due to the complexities of US land use laws and the customarily wide-ranging right to mine enshrined in the General Mining Act of 1872, endangered species protections often stand as the sole legal bulwark against new mining and development.[8] Jaguars attract considerably more public sympathy than other endangered species in the area such as the Gila topminnow or Chiricahua leopard frog. We know considerably more about the potential harms this copper mine might cause—to wildlife, to aquifers, to environments, or to the tangible and intangible cultural heritage of Indigenous groups—due to the location of the proposed mine near an affluent city (Tucson) and major research university and from Tony Davis's investigative reporting for the *Arizona Daily Star* on the twenty-five-year-long battle between government agencies, NGOs, Indigenous nations, and mining companies.[9]

Although the mine has yet to produce any copper for high-end gear wiring (though Hudbay mines just across the mountaintop are running full steam), the Rosemont saga unearthed arguments about the biological "significance" of jaguars in the US—the three jaguars spotted in the region allegedly traveled from Mexico.[10] This is pitted against Hudbay's escalating claims about just how many "high-paying jobs" the Canadian-owned mine will produce (first they claimed three hundred, then six hundred). The mine would destroy seventy Tohono O'Odham burial sites and sacred springs and a centuries-old retreat site for Yaqui medicine men, but some claim that the mine's output is essential for America's transition to renewable energy (a savvier "sell" than audio cabling). Would ecotourism to natural habitats for migratory birds produce more net economic impact in the region than a short bout of open-pit mining plus the pipe dream of post-mine mining tourism? How much would property values decline in Green Valley if Hudbay started mining in view of the communities? Does Arizona, a perpetually drought-ridden, land-locked state, even have enough water to support a mine whose freshwater consumption would be at least 3 percent of that of the entire Tucson metropolitan area? Are the dozens of washes into which mining tails would be dumped really "ephemeral" waterways, and thus not subject to Clean Water Act provisions?[11]

These are difficult questions that underscore impossible trade-offs. How can one compare ethical goods across seemingly unrelated domains: landscape aesthetics, Indigenous sovereignty, economics, employment, conservation, hydrology, and mineral resources? These trade-offs are not specific to Rosemont or to Arizona. We tend only to learn about mining problems when something goes horribly wrong, whether it is a mine collapse leading to considerable loss of life, an attempted cover-up of massive groundwater pollution, or, in the case of tin, tantalum, and gold mining in the Congo, a war that took the lives of 5.5 million people that was largely financed by legal and illegal mining.[12] Compared to such disasters, Hudbay's plans for high-paying jobs, sophisticated land rehabilitation post-mine, and water conservation could appear to be exemplary. If Rosemont does not end up happening, Hudbay will find another location to pursue a copper mine of similar scope—most likely in a country with looser environmental and human rights laws.

EXTRACTIVISM IN THE ANTHROPOCENE

Copper is the main material of gear's wires and circuit board traces, and it is used in power transformers and in the bronze and brass alloys of microphone casings. Copper is one of the few metals to be singled out in marketing prose. "Two-ounce copper PCBs" and "99.9% oxygen-free copper wires" are purported as markers of quality that justify a higher expense; the values of these are frequently espoused by gear designers and consumers alike. Rosemont copper is a specific and local case study about a particular mineral that is essential for audio technology and the material culture of music—a mineral that impacts the environment and contributes to ecocide.

That said, our choice to begin with Rosemont stems from our refusal to reduce gear to abstracts of indeterminate stuff through "global supply chains" that traverse arbitrary space.[13] For us, extractivism constitutes the fundamental reorganization of the physical world; we do not see materiality and the Anthropocene as primarily metaphors, discourses, or ideas. Previously, Eliot researched the specific supply chains that made Bangka (Indonesia) tin and Kivu (Congo) tantalum part of the electronics we use. It took a year and a half for Eliot to determine where just two of the dozens of metals used in contemporary electronics actually come from and to uncover a few of the ethnographic stories that accompanied the circuitous journey from mine to manufacturing.[14] Kyle Devine undertook a similarly long and convoluted journey when trying to uncover the origins of recorded music media.[15] Just as Devine hit a dead end when attempting, unsuccessfully, to pinpoint the origins of the barrels of crude oil that were to be converted into vinyl record "pellets," we were unable to determine where many of the *specific* chemicals, solvents, and epoxies used throughout component manufacturing were mined or produced. Unsurprisingly, none of the electrical/mechanical engineers we interviewed knew where the raw materials for the components that they used in their designs came from. Many did not even know where the components they used were made, only where the component manufacturers were headquartered. Concomitantly, as the anthropology of mining literature has shown, few mine workers know in which technologies and in which countries the ores of their labor will end up—and therefore what the value of their labor might be.[16] The true cost of gear is concealed for everyone along the supply chain.

Gear is only possible due to mass-scale extractivist economies around dozens of distinct substances, due to energy-intensive and toxic-waste-producing mineral smelters, due to components manufacturers headquartered in North America, Europe, and East Asia who subcontract their manufacturing to numerous developing countries, and due to global logistics and transportation companies that consume fossil fuels to keep supply chains running. The ease with which we can ignore the resource ecologies and political economy that make this possible—in other words, the present and future cost and consequences to human, animal, and plant life—exemplifies classic Marxian alienation and reification. Gear's story is truly planetary in scale, but not in the sense of "global," that, say, people in all countries now listen to hip-hop or drink caffeinated caramel-colored carbonated colas. The flow is one way: the first world appropriates the natural resources, multispecies ecologies, cultural heritage, and labor power of the peripheries in order to produce gizmos that may or may not be used to make music. The realization of just how much of the planet is being mined, and just how much of the world's population is working in some capacity to make technological production possible, is at once extraordinary and frightening.

So how much mining is needed to make one average piece of outboard gear? It is only ten kilograms worth of metal. A car weighs one hundred times more, right? But the material impact of every recording studio is far greater than that of one piece of gear. A ground-up build of a professional single-room recording studio—a space designed to accommodate a single project at a time that features an acoustically treated "live" tracking room and a front-to-back design control room for critical mixing—with a typical amount of gear entails fifty to seventy *tons* of construction material—including concrete, gypsum, wood, galvanized steel, fiberglass insulation, green glue, and fire-resistant polyester fabric covering for acoustic treatments—and one ton of permanently installed cabling, including copper, nickel, PVC, and irradiated polyethelenes. Electrical devices would consist of a half ton of studio monitoring such as speakers and amplification, headphones and hearback systems, a 32+ channel mixing console weighing one-third ton, and more than one ton of additional audio plus computer gear—necessitating steel, aluminum, tin-based solders, the many minerals and metals and rare earths used for circuit components, FR4 fiberglass PCBs, and all the epoxies, resins, ceramics, and petroleum-derived insulators and component packages that typically go unremarked but are in all electronics, too. It adds up quickly and begs the question, What would that studio need to do to justify that many tons of extractivist legacy?

Can extractivism and mineral smelting ever work for good? In the rare cases that a mine is located where it is not dispossessing people of their land or their way of life; not causing pollution to the surrounding land, water, and air; not destroying essential habitat for threatened animal species; and producing positive benefits such as employment, wages, and collective belonging for mine workers, then the mine *could* be perceived as an unqualified good. However, that is almost never the case in the twenty-first century: the questions are *how* bad is any particular extractivist project, in which ways, and whether it is ultimately worth it. Moreover, the harms caused through extractivism and subsequent smelting and component manufacturing stages should be weighed against the benefits incurred by maintaining that level of industrial activity. In practice, this never happens: the costs and benefits of one set of extractivist regimes (e.g., petroleum) are weighed against those of alternative extractivist regimes (e.g., the lithium + silicon + cadmium telluride + gallium arsenide + aluminum + copper that comprise solar power's material dependencies), but the scale of industrial activity itself is not called into question.

The impact of mining, at and around the mining pit, can be measured in the air, in groundwater, and on land.[17] In mining vocabulary, to create a mining pit first requires removing **overburden**: the layer above the desired ore, usually topsoil. Technically, this overburden could be replaced once the mining operation has concluded, but in many cases—especially in **ASM mining** (artisanal and small-scale mining, also termed "informal mining")—this does not happen, leaving the site incapable of growing anything and reducing the hectares of available topsoil-rich land in a region. Once overburden has been removed and miners can access rock containing the desired ore, the mining process continuously extracts far more rock and gravel than actual ore; this gravel and rock, termed **mine tailings**, must be dumped somewhere. For example, in the case of tin mining in Bangka, Indonesia, 99.985% of the pit extractions did not contain sufficient tin and became mine tailings.[18] We highlight Bangka since it was one of the primary sources of electronics solder during the period we are analyzing—and their tin is used in several composites to make component leads and other essential parts of electronics. Also, in proximity to gold-mining sites, ore-processing facilities wash ores over a **sluice** and **amalgamate** the washed gold ore with mercury.[19]

The leftover waste, known as **slurry**, may be itself further processed to extract gold but often leaks or is dumped from the processing site, creating a large-scale toxic mercury and chemical dump. If mercury-rich slurry contaminates a working mine area, ASM miners run a high risk of mercury poisoning due to their lack of protective gear. Further kinds of environmental degradation can happen, ranging from deforestation, seabed destruction from suction/dredge mining, toxic particulate matter in the air, groundwater pollution, and open-pit mine collapses, the latter a leading cause of death amongst Bangka's tin miners.[20]

In the early 2020s, public attention has focused primarily on the human-caused climate crisis and fossil fuel extractivist regimes, also known as the Anthropocene, and secondarily on waste pollution—including discarded plastics forming garbage patches in the oceans and e-waste being dumped in Nigeria, Zheijiang, and elsewhere.[21] Only very recently have such questions begun to be asked about the materiality and energy consumption of recorded music or the making of electronic instruments.[22] Gear consumes energy in its making and use and is rarely designed to be energy efficient; gear certainly contributes to the climate crisis. Even though much vintage and used gear continues to circulate on the used market, not all technology used in studios is designed to resist obsolescence. Low-cost audio equipment like studio computer DAWs, which are routinely discarded, all constitute pro audio-specific e-waste.

GLOBAL ACCOUNTING OF RECIPROCAL LANDSCAPES

James Smith and Jeffrey Mantz theorized *global accounting* as a means for rethinking the "productive foundation of the digital age, neoliberalism, and postmodernism," while also accounting for the widest range of understandings of the same irreducible material resources that make technology possible.[23] Coltan became well known due to its importance as the main source of tantalum, which is essential for semiconductor manufacture and for the tantalum capacitors used in some audio gear, and due to its role in financing the Second Congo War, which claimed 5.4 million lives. Tantalum, along with tin and tungsten, gained the moniker of "conflict minerals" and became more stringently regulated than others due to international attention. They continue to be subject to "bag and tag" schemes managed through the iTRI scheme. Untagged tantalum became branded by foreign media and NGOs as "blood minerals" or as "dirty," which contrasted with Congolese understandings of the same as sources of familial livelihood.[24] This leads to a question, then: How might gear-related extractivism be good for local economies in least developed countries (LDCs)? Certain actors in the DR Congo undeniably benefited from coltan mining. But in most cases, the gross domestic product of LDCs is improved by increasing manufacturing and service industry jobs, not by ramping up extractivism.[25]

Global accounting, then, brings the anthropology of mining—ethnographic work centered on giving voice to local mining communities—to bear on questions around development, technological innovation, manufacturing, and the constitution of the global north itself. This approach bolsters theories within resource ecologies and the anthropology of money concerning how technological development in "the center" is dependent upon extractivism in "the periphery."[26] However, think of the Rosemont copper mine: not all "peripheries" are in countries that are paying off IMF structural adjustment loans. Some are just a few kilometers away from major urban areas in the US, Canada, and Australia, on land facing contested management policies.

Here, we find Jane Hutton's concept of *reciprocal landscapes* to be a useful heuristic for articulating specific linkages between sites that otherwise would be unconnected. Hutton analyzes five nonlocal materials that define the unique aesthetics of the New York City "landscape" and aggregates stories and ethnographic analysis of what happened along that pathway of reciprocity. She does this around five "materials-in-motion": bat guano harvested in the Chincha Islands of Peru that nourished the first iteration of Central Park; granite from Vinalhaven Island in Maine that became the sidewalks of Broadway in the late 1800s; steel from Alleghany County in Pennsylvania that buttressed Riverside Drive; London plane trees that ended up on 7th Avenue in Harlem by way of Riker's Island (and prison labor); and ipe logs from Brazil that ended up creating the distinctive look of the High Line park in 2009.[27]

Our first reciprocal landscape of gear entails aluminum, where electrolytic capacitors and the reflective gleam of shiny aluminum fascia represent materials that are fetishized in gear cultures. Since 2000, "aircraft-grade aluminum" has been highlighted in the marketing materials for microphones by Vanguard Audio Labs, Korby, innerTUBE Audio, Ashman Acoustics, and Ehrlund; for the cases of Coil Audio's mic preamps and the more expensive outboard gear product lines of dbx; and for mic accessories by Grace Design. The Focusrite Red series of outboard gear, introduced in the 1990s, helped aestheticize anodized and brushed aluminum faceplates and extruded aluminum cases as indicators of premium gear—where they constitute most of the cost of the units in question.[28] Aluminum faceplates are prominent in gear by Manley Labs Avalon, Elysia, Audient, Universal Audio/Urei, A-Designs, and Daking. Besides aluminum and the cheaper "cold-rolled steel" options, we have encountered very little experimentation with alternative materials—magnesium, nickel, and brass could work just as well—with the exception of Madrid-based Tierra Audio, who since 2020 has explicitly framed their "eco" products around their use of "cherry wood from sustainable forests" as well as Paulownia and Kiri woods and recycled paper for packaging.[29] Why is aluminum desired for the cases and knobs of rackmount audio gear, seeing as it lacks any particular properties that make it especially desirable for this application? Let's explore aluminum's reciprocal landscapes.

An Avalon VT-737sp preamplifier, like much rackmount gear, is adorned with numerous shiny aluminum knobs. Reflecting the glow of LED indicators, they are a tactile point of contact and control, conveying the starlit promise of musical possibility. In her book *Aluminum Dreams*, Mimi Sheller contrasts the way that aluminum producers, especially Alcoa, aggressively marketed the "hopes and idealistic projections of modernization, prosperity, and leisure-filled future utopias" with the concomitant "false hopes that led to failed technologies, pollution, social dislocation, and environmental devastation."[30] Post–World War I military aircraft, which replaced wooden frames with aluminum, helped instantiate aluminum's associations with discourses and practices of speed and "mobile modernity." However, this came at a cost: speed, mobility, and lightness in the first world "pivoted on the co-production of other regions of the world as backward, slow, and relatively immobile".[31] For products being made since the 1980s, Suriname (the major site for Alcoa's bauxite mines and aluminum smelters) and Guinea (managed by Reynolds Metal Company and Alcoa) are the primary extractivist halves of the reciprocal landscapes. Taking the case of Guinea, at least two billion dollars in mining contracts were attached to the military junta, and much of the increase in the alumina industry was justified in relation to the austerity measures imposed by IMF structural adjustment programs. In other words, "development" was dependent upon privatization-imposed reductions to the state sector and the erosion of democratic institutions.[32] We will not detail here the ethnographic accounts about changes to ways of life in Suriname and Guinea, although Sheller's book references many such accounts. However, we do wish to point out the paradox between the realities of open-pit bauxite mining, which necessitate disproportionate amounts of electricity and lead to wide-ranging groundwater pollution and land destruction, and the more recent marketing of aluminum as a "green" material that is essential to addressing the world's global warming problems.

WHERE AUDIO GEAR COMES FROM

During research, we came to realize just how alienated electronics manufacturers, consumers, and researchers had become from technologies and their constitutive components, and how profoundly hamstrung we collectively had become in overcoming their apparent complexity and unknowability—despite their use of ubiquitous electronic components. We had to make a choice, one that many researchers make when they confront a spectacular mess of poorly documented problems: do we throw up our hands and give into magical thinking (a problem so fantastic it cannot be understood) or persist and share our findings, even if incomplete and messy? We chose the latter and focused on one passive component that we *were* able to document reasonably well: the electrolytic capacitor.[33] Electrolytic capacitors are essential for equipment power supplies, for power filtering, for timed voltage changes such as those that comprise the attack/release of compressors, and for other applications; the Japanese corporations Nichicon and Panasonic, in their Taiwan plants, manufacture nearly all the "audio-grade" electrolytic capacitors used today. You can recognize them as the soda can–shaped components that range from small (the size of a pencil eraser) to large (the size of a pill bottle).

Our story of the electrolytic capacitor and its constitutive substances has three intended outcomes. First, we share facts to help demystify a ubiquitous thing that is empirical and knowable. Second, we share our process so that others can start to demystify technologies and their origins through empirical research. Third and most important, at a time when all life on Earth faces numerous crises related to Anthropocenic environmental and climatic change, what is needed more than ever are problem-solving approaches that traverse the borders between siloed fields of knowledge and tech-scientific specialties. Neither academic researchers nor tech companies routinely engage in anything approaching the global accounting paradigm as advocated by Smith and Mantz, let alone the reciprocal landscapes approach of Hutton. How can we know if any technological solution to a man-made problem will have a net gain if we cannot even fully account for the consequences of that solution?

Even basic electronic components that feature in most gear—capacitors, resistors, transistors, operational amplifiers, vacuum tubes, potentiometers—contain a large variety of substances. No one trade possesses the necessary knowledge to understand—or skill to perform—the entirety of their manufacture, as it draws upon many kinds of expertise undertaken by discrete specialists: materials scientists (including metallurgists), chemical engineers, electrical engineers, and mechanical engineers. A purchasing agent at a component manufacturer would know from whom they buy materials, but materials bought from mineral wholesalers or regional chemical suppliers would have been purchased from geographically disparate—unknown—sources. While the mining operations and supply chains of certain minerals are somewhat well established and documented in academic and journalist accounts, it is very difficult to track the origins and supply chains of chemicals, epoxies, resins, and acids, some of which are produced at petroleum, fertilizer, or natural-gas facilities. Unlike ores and metals, these are more likely to be regionally produced, although they are dependent upon raw materials like crude oil, natural gas, and boron, whose supply chains are better documented.

Undeterred, we persisted on the goose chase of the electrolytic capacitor and, as of 2022, its many reciprocal landscapes. We first had to figure out what is in them. Initial data sources for this included the product guides and technical notes published by Nichicon and Panasonic, the two primary manufacturers of audio-grade EL caps,[34] and numerous component manufacture texts that shed light on the aspects that the manufacturers only generically describe.

As Nichicon notes, "An aluminum electrolytic capacitor consists of cathode aluminum foil, capacitor paper (electrolytic paper), electrolyte, and an aluminum oxide film, which acts as the dielectric, formed on the anode foil surface" (figure 3.1).[35] The anode's aluminum foil is etched to rough up the surface (accomplished with a chloride solution) and then subject to electrolysis, which requires boric acid, to create the oxide layer. Metal leads,

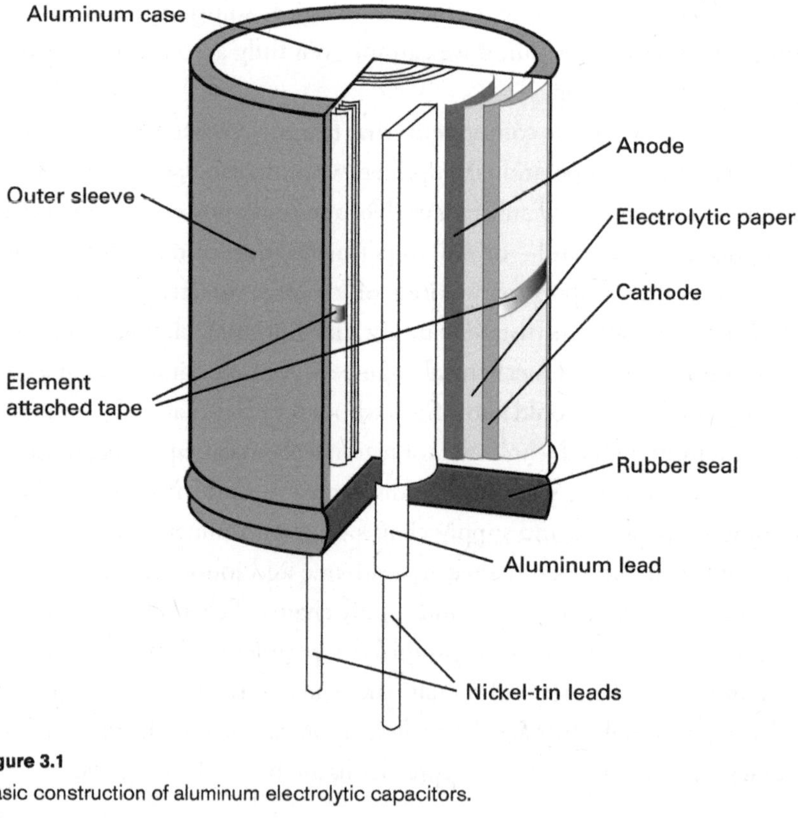

Figure 3.1
Basic construction of aluminum electrolytic capacitors.

typically consisting of a nickel-tin amalgam, are welded to the aluminum foil layers. Panasonic is more specific about the construction process: The two etched aluminum layers, anode and cathode, are sandwiched between layers of kraft paper, which is then rolled up. Then, the paper is impregnated with the liquid electrolyte—the substances of which can vary considerably. The wound-up unit is inserted in a resin-coated metal case that is sealed with rubber or a rubber-bakelite mix to prevent the liquid electrolyte from leaking out. A printed vinyl sleeve around the capacitor indicates the brand, model number, and electrical characteristics of the component. We are leaving out additional steps that do not introduce new substances into the equation, but the above gives a sense of the process and materials. In the solid tantalum electrolytic capacitor variant, sintered tantalum and manganese oxide are used instead of the sandwiched aluminum-paper layers.

Armed with a list of substances, we further investigated each one. Raw minerals are easier to trace since the US Geological Survey publishes their biannual *Mineral Commodities Summaries*, which detail the top ten producers of that mineral per year and, in some cases, detail supply chain issues that might affect US-based manufacturers (most of those issues would face a Taiwan-based manufacturer, too). But most of the substances in electronics are not raw minerals. For example, bauxite from the ground is not directly usable in electronics; it must be purified and oxidized before it takes the form of aluminum oxide. Thus, we investigated material science texts to determine the most common processes used today and what other substances (e.g., acids, nonorganic compounds) were necessary to arrive at aluminum oxide or other key substances—and how *those* were made.

Most electrolytic capacitors found in gear today are made in Taiwan, with a lesser number made in Japan or mainland China. To support the necessary economies of scale—manufacturers annually produce *tens of billions* of capacitors—manufacturers need a mix of regionally sourced organic and inorganic compounds and a variety of specialty metals that come from smelters farther afield. Each capacitor is dependent upon products and byproducts from facilities whose primary output is soil fertilizers—most notably, nitroso and nitro group compounds. They use resins, epoxies, and ethylene glycol typically produced by regional petrochemical facilities that process crude oil.

One common source of ammonium is natural-gas processing plants; this ammonium is directly employed, and it is a key base ingredient for those soil fertilizer plants, too. Other chemicals, particularly chlorine and the range of acids used in intermediate processes, come from regional general chemical facilities.

From the perspective of the capacitor manufacturer, most of the metals nominally come "from China" since China-based businesses manage most metal-specific global supply chains today, but only a fraction of that metal actually came *from* China. Therefore, to account for the metals and chemicals in our capacitors, we needed to expand our production map to include eighteen other countries: two in North America (the US and Canada), two in South America (Peru and Brazil), seven in Asia (Russia, Turkey, Kazakhstan, Philippines, Thailand, Vietnam, Indonesia), six in sub-Saharan Africa (DR Congo, Guinea, Gabon, South Africa, Nigeria, Rwanda), and one in Oceania (Australia)—see table 3.1. So, what is done in these reciprocal landscapes bound by material relation to the global north studio with its thousands of electrolytic capacitors? Australia is a significant source of the bauxite used in aluminum production, and of manganese dioxide. Other countries also produce bauxite and manganese dioxide, though: Guinea is the second-largest bauxite producer, and South Africa and Gabon also have significant manganese mines. Indonesia is one of the top three producers of nickel, tin, and rubber. India and Russia, in addition to China, are the primary aluminum-smelting facilities today. The US and Turkey host most of the boron mining that feeds the boric acid producers of China. The Kinshasa region of the DR Congo, Rwanda, and Nigeria are the main sources for coltan mining that produces tantalum, although Brazil is quickly emerging as a primary producer, in part due to the continuing implication of coltan mining in Congolese conflicts stemming from the Second Congo War and subsequent tracking/certification scheme for "conflict minerals." Coltan ore must be processed at very high heat to result in usable tantalum, though, which happens largely in China and Kazakhstan, although Rwanda now hosts some processing facilities. To complete the list, the Philippines and Russia are, after Indonesia, the primary nickel producers; Peru continues to be a significant tin producer; Thailand and Vietnam contain large rubber

Table 3.1

Material origins and processing locations of 16 materials essential to electrolytic capacitor manufacture in Taiwan

Application	Material	Where it is extracted	Where it is smelted/processed
dielectric, anode, casing	aluminum	Australia, China, Guinea	China, India, Russia
dielectric oxidation	boric acid (from boron)	US, Turkey	China
dielectric oxidation	ammonium	various locations	fertilizer plant near manufacturing facility (East/Southeast Asia)
aluminum foil	chloride	various locations	local to manufacturing facility (Taiwan/Japan)
dielectric	sintered tantalum	Kinshasa region of the DR Congo, Brazil, Rwanda, Nigeria	China, Kazakhstan, Rwanda
separator	kraft paper	forests in China, Russia, Northern Europe, Canada	Paper mills in various locations
liquid electrolyte	nitroso	various locations	fertilizer plant near manufacturing facility (East/Southeast Asia)
liquid electrolyte	water	local to manufacturing facility (Taiwan/Japan)	n/a
liquid electrolyte	ammonium	likely natural-gas reserves in China and Russia	natural-gas facility near manufacturing facility (East/Southeast Asia)
liquid electrolyte	acid (various)	various locations	local to manufacturing facility (Taiwan/Japan)
liquid electrolyte	ethylene glycol	OPEC crude oil supply chain	petrochemical facility near manufacturing facility (East/Southeast Asia)
tantalum cap electrolyte	manganese dioxide	South Africa, Gabon, Australia	unknown
metal leads	nickel	Indonesia, Philippines, Russia	local to extraction site
metal leads	tin	China, Indonesia, Peru	local to extraction site
casing	resin	OPEC crude oil supply chain	petrochemical facility near manufacturing facility (East/Southeast Asia)
bottom seal	rubber	Thailand, Indonesia, Vietnam	local to extraction site

plantations; and forests and paper mills in China, Russia, Northern Europe, and Canada all produce kraft paper. There are likely substances involved in EL-caps that are not listed above, and other countries who play a lesser or occasional role that are not reflected above, but this is a start, at least.

Let's assess the economic impacts of participating in a gear-related economy to one of the many reciprocal sites. Despite having one-quarter of the world's available bauxite deposits and being one of the top three bauxite exporters, the Republic of Guinea has among the lowest GDPs of any African country ($2,014 per capita in 2017). This is largely explained by the level of government debt and the extensive involvement of the IMF and World Bank in the national economy: in 2005, government debt was 98 percent of the national GDP and inflation was officially 31.4 percent; since then, debt has decreased and GDP has increased, but increases in GDP have been lower than the inflation rate, meaning that purchasing power for Guineans has been worsening (2015 was the first year where GDP increases outpaced inflation). As Johannes Knierzinger notes, ignoring the environmental consequences of the extractive economy, the key impediments to improving Guineans' economic situation include the opacity of the global aluminum industry and its largest companies (including its use of tax havens and opposition to anti-trust legislation) and the lack of an integrated aluminum industry in Guinea (processed aluminum is much more profitable to export than bauxite ore).[36] Various governments since 2003 have experimented with "transparency initiatives" and stronger negotiating tactics with aluminum multinationals, but the looming national debt greatly weakened any potential negotiating power.

CONCLUSION

In our first stop on the pathway to explicating the fetish nature of gear, we attempted to "blow up" the argument, to provide a sense of the many material flows that are essential to contemporary audio electronics, and to show how much of the planet is working to make gear possible to exist. The resource supply chains of all materials in gear are conveniently concealed from gear cultures. As we will see later, rarely do gear culture participants acknowledge the extractivist legacy their gear. However, even if gear cultures gained awareness of Indigenous displacement, environmental and conservation disasters, and the economic and labor ramifications of mining, it would unlikely shift perspectives enough to demand lower-impact alternatives. Much of what is fetishized about gear is the raw materials used to make it. All that virgin copper, shiny aluminum, and brushed steel must come from somewhere, and, as we know with the true costs of more ubiquitous technologies such as the iPhone, it takes more than a humanitarian or environmental scandal for these precious gear metals to lose their shine. In North America and the EU, any greengrocer is required to show the country of origin of each fruit and vegetable. But the gear consumer lacks any awareness of material supply chains beyond the location of the headquarters of some component manufacturers; we are far away from being able to perform global accounting on gear.

The wanton desire for aircraft-grade aluminum, its perennial shininess, and its mirror-like aesthetics reflects the technological utopianism at the heart of gear cultures. Operative here are aluminum's affiliation to aircraft and military history and to space flight, and the desire of many consumers to create their own "spaceship." Gear is built from mined raw materials like aluminum, copper, and tin, not because it cannot be built from anything else, but because of the political economy of resource ecologies and the meanings and symbolic associations that industry consortia use to market particular materials to manufacturers and the public. When the above are naturalized, we forget that it could be otherwise; it seems obvious why such materials fit so well within broader gear culture histories, norms, and values, and why such materials are necessary to feed the insatiable appetites of gear culture participants, hungry to get their hands on shiny new gear.

4 MANUFACTURING

BUILDING GEAR

On a chilly October Saturday morning at the 2019 AES show, we gathered at the southeast corner of New York's Javits Center to participate in a workshop cosponsored by the Audio Builders Workshop (the outreach project of AES's Boston chapter) and the popular DIY kit company Microphone Parts. The prospect of this workshop had us excited for months; both of us, along with ANU School of Music technical manager Matt Barnes and Black Mountain Studios engineer Clem Bennett, had been anticipating the workshop as a highlight of the convention. We were assembled there to assemble mics—namely, a medium-diaphragm electret condenser with a Schoeps-style transformerless circuit board. At least, that is how our object-to-be was described to us. Arriving at the impromptu DIY stations, we were greeted with a soldering iron, and Microphone Parts proprietor Matt McGlynn dropped off by each soldering station a playing card–sized color booklet with instructions and a box containing all the stuff to build the microphone: film capacitors, resistors, a JFET, a mic capsule, a small printed circuit board (PCB), some wire, and all the metal that comprises the exoderm of the mic. We have three hours to build it.

Situated on the concourse between the conference happening downstairs and the trade show floor upstairs, the wafting odors of rosin core solder fumes and the visual impression of twenty of us, along with a large group of high school students, sitting there assembling electronics attracted attention (figure 4.1). David Josephson, famed mic designer and Vice-Chair of the

Figure 4.1
Eliot and Samantha at the mic build. AES 2019. Photo by Matt Barnes.

AES Microphone Committee, wandered by looking bemused but stopped to chat and seemed to be genuinely pleased that people were interested in learning about making microphones. Samantha asked if her assembly would make the Josephson cut and was delighted when David replied that it would. We were on track: $200 in cost and three hours later, despite the hand jitters from having downed our second or third morning cup of caffeine, most of us had working microphones, good enough to use on vocals or instruments for professional recording sessions.

Most of those three hours went into hand-soldering **through-hole** components to a PCB. This type of assembly is difficult to automate; therefore, most of the labor is done by hand, just as we were doing—albeit done with better assembly tools and a greater degree of precision and care. Through-hole assembly has a seventy-year history in audio but continues to be a preferred way of assembling much gear. Point-to-point wiring is even more time-consuming and labor-intensive: in the absence of a circuit board, components are either soldered to each other directly or to terminal strips. This method continues to be used in some of the most expensive boutique audio gear, too—D. W. Fearn is as known for this as for the bright red faceplates of his tube gear.

Both assembly modes are *technically* obsolete, having been replaced by surface-mount technology (SMT), where components float within tape reels that feed a pick-and-place machine programmed to place them in precise spots on a circuit board, later to be reflow- or wave- soldered, also by machine. Contrary to popular conceptions, surface-mount assembly is not fully automated. Programming the pick-and-place and reflow soldering machines are learned skills, touch-up manual soldering of joints that did not flow quite right may be necessary, boards need to be cleaned properly with flux remover, and everything beyond the surface-mounted parts on the PCB requires manual component placing and soldering. Nevertheless, much gear continues to be made in "traditional" ways despite labor costs and the increasing obsolescence of components made in through-hole packages.

Seeing twenty people and a school group studiously assembling microphones in New York's best-known convention center might seem unusual, especially if we assumed that the popular circulating narratives about technological change applied across the board. But gear is not just made in factories in urban industrial or warehouse districts; some gear was *professionally* made in homes, including in designers' garages, sheds, kitchens, and spare bedrooms, and in one case, outside a camping tent on a mountainside in Colorado with a solar-powered soldering iron. Tracing the "material movements" of supply chains prior to assembly reveals an equally ad hoc network of component origins.[1] We have heard stories about television repair shops in Ukraine with stocks of now-obsolete vacuum tubes. We know of a corner store in Kansas with someone's grandfather's hoard of vintage hi fi and ham radio "junk." In a hotel in Shenzhen, China, foreign parts buyers stay for a few days, leaving their room only to seek out the side streets with rows of panel trucks, where they pick up a tray of samples to test back in the hotel room, returning to haggle prices for that truck full of unbranded components. Rather than evoking smooth flows through a well-oiled supply chain supplying informatics-driven just-in-time manufacturing,[2] the recurring emails we get about shipments stuck in containers in a port in Hong Kong evoke supply chain friction, instability, and the many human, material, and environmental dependencies at every stage.

KEY ISSUES IN GEAR MANUFACTURING

Any single gear object exists simultaneously as an abstract concept and as a material instantiation. The abstract concept consists of one or more circuits represented as *schematics* and a set of CAD files or instruction sets, typically known as *gerbers*, provided to the people and/or machines that produce the metalwork and printed circuit board. The material instantiation refers to the specific electrical components and other materials that are used to render those abstract concepts into a physical object. Whereas a "Neve 1073" might to a recordist signify a box with a particular look and feel and sound, to an electrical engineer it is a collection of circuit diagrams with an accompanying bill of materials (BOMs). In this regard, a circuit is more like a musical score than like a piece of music: it is designed to be *interpreted* in different ways. The sound of gear, then, results from a process of materializing an abstract circuit through the choice and layout of components and other materials.

The rapidly shifting market for electrical components means that components may become obsolete overnight or disappear in particular package sizes. Parts sourcing is, therefore, a major issue all manufacturers face, whether they are trying to bring a new product to market with the intention of manufacturing it for a decade, trying to keep old product lines active, or reissuing classic gear. Even gear companies that appear successful from the outside are much smaller than manufacturers in automotive, medical, computer, or telecommunications industries, they are more vulnerable to component shortages—and there is fierce competition for parts. A few kinds of parts used in professional audio are industry-specific—for example, microphone capsules, custom audio-optimized transformers, and a few discrete operational amplifiers or integrated circuits. However, audio is generally at the mercy of transnational component manufacturers whose primary clients are in other much larger industries. Some designers respond to these problems by redesigning circuits around available components, while others go on wild goose chases to find stashes of new old stock (NOS) components collecting dust in small-town TV repair shops or on eBay.

In this chapter, we place gear's circuits and components front and center, primarily from the perspective of contemporary designers and manufacturers in the US, UK, Europe, and Australia, but correlated with the complexities and peculiarities of a widespread consumer gaze. Although we interviewed renowned inventors, we are less concerned with invention stories than with iterative redesigns and the remaking of the past in the present. The NAMM Foundation, too, has recognized the stagnation of the professional audio technology market, although they put a rosy spin on it: "As technologies mature, the natural tendency of manufacturers is to shift emphasis from innovation to efficiency."[3]

Each component we chose to analyze purportedly affects musical sound; has interesting material, technological, and historical ramifications; results from a problematic production process (e.g., toxic chemicals, concealed and/or gendered labor); and, for different reasons, has become a social problem. **Capacitors** are a core building block of all electronics but differ from other passive components because they are consistently implicated in environmental and extractivist problems. For example, the ground and water pollution from polychlorinated biphenyl compounds precipitated some of the largest Superfund waste cleanup sites in the US, and conflict mineral mining for tantalum in the Democratic Republic of Congo was a significant revenue stream for armed militias during the Second Congo War. Capacitors became social problems when many consumers chose to "recap" electronics in order to achieve a different sonic signature, albeit with varying results. **Transformers** have long been used to solve technical problems—for example, impedance mismatches and balancing unbalanced circuits—and in these applications, they impart a distinctive sonic signature to audio. They are also of social interest since they provide a window into the gendered division of labor in audio manufacturing and became a site for arguments about nationalism, "purity" and "loss." **Operational amplifiers** (opamps) are transistorized differential amplifiers that were initially important for miniaturizing electronics and reducing the complexities of circuit design and manufacturing. However, the history of their design and manufacture demonstrates the discrepancies between the idiosyncratic lives of their designers and the "nimble fingers" of the women who enabled them to be made;[4] today they are the most fetishized and misunderstood components amongst electronics amateurs. **Microphone capsules**, arguably the most important contributor to the sound and operability of condenser microphones, unlike the above, are audio-specific designs; today, despite decades of design experimentation involving new capsules and different materials, a handful of designs originating in the 1940s–1960s by German and Austrian companies account for nearly all capsules used in new mic designs today. They open up social questions about heritage and nostalgia at the intersection of design and consumption, and also whose knowledge about these "vintage" designs should be considered valid today. **Vacuum tubes,** or thermionic valves, are glass bulbs featured in numerous technologies of the pre-transistor era—including radios, televisions, and instrument amplifiers—and in studio gear—including condenser microphones, preamplifiers, and equalizers. Like opamps, vacuum tubes are highly fetishized, with some NOS tubes selling for thousands of dollars on the secondhand market.

When considering questions of "invention" or "innovation" in audio technology, two terms more common than merited, we must keep in mind that, for many gear businesses, there are either none or only limited intellectual property protections on circuits. There is therefore little incentive to invent a new kind of object with the purpose of licensing the design to other firms. Public presentations of patent applications require such extensive information that competing firms might be able to copy the innovation and potentially beat the innovator to market with a variant product. Firms that we have talked with instead use strategies such as protecting their trademarks, including their product names and distinctive interfacial elements like knobs, to dissuade competitors from confusing consumers about which object is the real one. Others maintain secret assembly processes that cannot be readily seen by competitors, and/or they attempt to maintain an exclusive relationship with a component supplier for a commissioned component: this is especially the case for transformers and mic capsules. For example, US-based API has consistently attempted to protect their trademark on the API name, on their module color schemes, and on their iconic custom knobs.

When Classic Audio Products Inc. (CAPI) released a line of DIY kits that brought the sound and usability of 1970s-era API circuits into the 500-series modular format, API asserted their knob-specific trademark, which led to CAPI's owner, Jeff Steiger, needing to design a new custom knob that did not infringe on the trademark.

Truly new circuit designs trickle into the market, but the most ubiquitous new objects are clones of circuits that were originally introduced in the 1930s–1970s. For example, we found 119 different hardware products currently on the market by 31 different companies that all claimed to have some relation to a Neve 1073 mic preamplifier or equalizer, which is surprising considering that the Neve 1073 was not even a standalone product when Neve made them in the 1970s for their 8000-series consoles.[5] Occasionally, we see a new circuit design and/or a new way of making microphone preamplifiers and equalizers. Designed by Grant Carpenter, the Gordon preamp springs to mind as an exquisite product where no corners were cut and no expenses were spared regarding component selection. Few designs have, however, had wide adoption or commanded the kind of mass appeal as the "vintage" designs of Neve and API. The same tendency toward cloning is true for many types of gear—microphones, compressors, tape machines, and mixing consoles. That said, not *all* studio-specific audio technologies entail cloning, as technological advances continue to be made with audio interfaces for DAW computers and digital-analog converters, although much of the innovation there is attributable to the semiconductor companies that produce new integrated circuits. However, the rapid obsolescence of audio interfaces means they are less fetishized or perceived *as gear* than longer-standing categories of audio technology.

The cultures of audio technology manufacturing work become visible manifestations of gear cultures through three vantage points, each of which presents a contrasting picture. First and most familiar are the frontstage depictions of boutique gear manufacturing routinely on display at trade shows, in gear magazines, and in brand promotional material (e.g., the genre of heavily curated factory walkthrough videos). Second and least familiar are the backstages of workplaces—the real social interactions and labor that the public rarely sees necessary for gear production. Third is the consumer reception,

Figure 4.2
Manley Labs factory floor, Chino, CA. Photo by Eliot Bates.

impression, or imagination about what audio manufacturing is like, which at times relates to manufacturers' frontstage depictions, but often does not. Most gear culture consumers do not understand why gear costs what it does, and they extrapolate from the pictures of gear innards (observed online or at trade shows) their own fantasies about device quality, labor types, and national cultures of work.

What we found when we visited key manufacturers presents a very different face of audio today, including substantive ethnic and gender diversity that is otherwise invisible at trade shows and in the online and magazine reception of gear (figure 4.2). This is not surprising considering the long history of women doing the bulk of electronics assembly, but it contrasts with the conventional narrative of increasing offshoring to cheaper labor markets or of robotic and human-less manufacturing.

We are especially interested in understanding the continuation of certain labor formations such as educated and/or trained women in specialized assembly roles. We are also intrigued by the emergence of lesser-studied labor formations, including the relocation of a normative sexual division of labor to *expensive* labor markets, and the transformation of tedious and repetitive assembly tasks to a cross-trained heterogeneous job description.

COMPONENTS AS PROTAGONISTS

Open up any piece of gear and you will find many electronics components that you would see in non-audio gear, too. In fact, much of contemporary audio is inconceivable without general-purpose electronics components that were not originally, or at least exclusively, designed for audio applications; pro audio gear design always entails a lot of creative work with repurposing generic parts. Unlike other audio-related applications—for example, mass-produced synthesizers—quite a lot of gear relies upon larger and older parts rather than on modernized ones, even though many new parts are considerably more precise and accurate than old ones. Designers cited a range of reasons for this. Some noted a nostalgia for classic circuits and entrenched habits of making and assembling gear. Others pointed to perceived consumer biases—in some cases, an aesthetic preference for parts that might measure as less accurate in some aspect but are perceived as being more euphonic. And we occasionally found what seemed to be the aftermath of the designer being "redpilled" or having fallen victim to a snake-oil salesperson.

We use the term *components* to refer to those electrical elements that make up circuits (what a manufacturer would group under the BOM), not in the home stereo equipment sense where the term refers to receivers, CD players, and so on. While we are primarily interested in manufacturer perspectives, many consumers today possess selective knowledge about the specific components in the gear they buy and own, although some gear consumers possess incomplete and misinformed knowledge. Secondhand knowledge about components, typically linked to a "forensic analysis" of historical pieces of gear that are prized for their sound, is not only a topic of interest to enthusiasts, but in many cases drives the design and marketing of new gear, or at least new clones of old gear.

Which aspects move beyond designer-specific engineering concerns and become sociocultural problems? To the experienced electrical engineer, some of these aspects may seem puzzling since they are not in all cases the most important considerations regarding the selection of the component in question. From a media archeology point of view, the components we are *ignoring*—resistors, diodes, relays—might seem odd as these are numerous and significant parts, but, for whatever reasons, they have not (yet) become sociomaterial problems in the way other components are. This is not the story of all components, then, but rather stories where five kinds of components become the protagonists and where the social and cultural formations around them reward investigating.

Capacitors

Capacitors are one of the elementary passive components in all electronics, and, in action, they store and release electrical charge. Many different materials have been used to make capacitors over the years, from water in a jar, to silver mica, to sintered tantalum, to aluminum foils coiled up into a cylinder. Electrical engineers, when considering what kind of capacitor to use, would most likely assess four interrelated factors: equivalent series resistance (ESR), dissipation factor, loss tangent, and quality factor (Q). They would care how many hours a capacitor is rated to operate (mean time before failures, MTBF) when running at the maximum rated temperature, which would indicate durability/longevity. They also consider what happens when the dielectrics of capacitors experience too strong an electrical load: film capacitors "self-heal" (good), while some kinds of tantalum electrolytic capacitors explode when they are overloaded (very bad). Some but not all kinds of ceramic capacitors become microphonic, literally buzzing like the piezoelectric disc you find inside a holiday greeting card or a cell phone on vibrate mode.

Knowledge about capacitors outside of the engineering domain tends to be different. In the marketing of some expensive boutique gear, we find phrases like "no coupling/electrolytic capacitors in the audio path," for example, in the John Hardy servo microphone preamplifiers or the Knif V804 tube preamplifier. These are both highly regarded designs. The design decision is not for question, but we wondered why *this* technical detail was selected for marketing material rather than others; there are equally highly regarded designs that *do* have coupling caps, and other significant design details in the Hardy and Knif preamplifiers went unmentioned. Enough other designers have mentioned this that the phrase, decoupled from its original usage, has entered the folksonomies of consumers; this becomes absorbed into the technobabble lexicon, which is then mistakenly applied to the wrong products—meaning ones *with* capacitors in the audio path.

Capacitors are also a component that some consumers care about enough that they are willing to pay a premium for ones believed to be better quality. From designers and engineers, we have heard that since around 2013, there has been a viciously competitive market for NOS American and German capacitors made before 1978 (the year when toxic PCB-laden oil was banned from use in paper-and-oil capacitors). These pre-1978 capacitors are coveted not just for pro audio applications but also for home stereos and guitar amplifiers. In 2020, Microphone Parts, hosts of our microphone build, released a special version of one of their microphones where much of the marketing prose described a no-compromise approach, including the use of a coveted capacitor that cost $115 (a polyester film cap would have cost no more than $0.10).[6] The Gordon microphone preamplifier, acclaimed for classical-music recording and precision-test environments is, to our knowledge, the object featuring the highest percentage of US-manufactured components. The preamplifier is so expensive partly due to it featuring more than sixty Utah-made polystyrene and Teflon capacitors costing $10 each—one hundred times the cost of generic capacitors used for similar applications.[7] In our discussions with Grant Carpenter, it was clear that his choice of capacitors was driven by extensive listening and testing, and was integral to his design; no mass-market components would work equally well in their place.

Perhaps the most confusion surrounds the *recapping* of audio gear—the replacing of capacitors. There are legitimate reasons to recap old electronics: electrolytic capacitors can bulge or burst, oil-filled caps can leak, and film caps can increase in ESR after repeated voltage spikes and self-healing. Used vintage recording consoles from the 1950s–1980s are periodically recapped by professional technicians, which means in practice desoldering several thousand capacitors and soldering in new ones. One can find numerous accounts online, at trade shows, and in magazines of before-and-after tales concerning recapping, typically positive: "the sound really opened up," but occasionally negative: "it doesn't sound as warm as it used to." Old leaking and bulging capacitors no longer provide their rated capacitance (measured in microfarads) and will have a different frequency response in filtering applications. Accordingly, *some* of the romance of vintage gear as perceived by nonexpert consumers, then, is less related to the sound of a mixing console when it was new and measured to spec than with how it sounds after it has degraded in performance. As we will see in chapters 10 and 13, this leads to numerous problems when consumers try to determine why reissues, or clones, of gear do not sound like their vintage counterparts.

Recapping is not only done to old gear: several small firms made a business model from recapping brand-new working gear made by other companies. Black Lion Audio markets their "high-grade signal path capacitors" mod where an audio interface or other rackmount gear may be sent in, and, for a cost of $300 or more, the caps are changed to impart, as the company claims, "amazing micro-dynamics and space." Despite the lack of *any* before-and-after electrical measurements or audio examples, the ignoring of warnings voiced by professional electrical engineers, and meaningless audible/electrical property descriptors (we have no idea what "micro-dynamics" might mean), this practice and market has continued for over fifteen years, attracting fervent fans on some gear forums. Other users and consumers, convinced from this discursive formation and from their own forensic analyses that their gear is not working as well as it could and that if they install the "best" caps, then they will somehow have gear that "competes" with that in the world's best studios, take it upon themselves to recap their own gear—usually with inconclusive results, but occasionally with catastrophic, irreparable failure.

Transformers/Inductors

Transformers are the other passive component that become an especially charged point in gear cultures, either because of designer decisions to use or avoid them or because of the alleged magical properties of transformers made decades ago. Whereas most components in audio gear are off-the-shelf generic parts available from neighborhood electronics stores or wholesale distributors, well-known audio companies such as Rupert Neve Designs were secretive about their transformers and had them custom-made by suppliers who worked under nondisclosure agreements. Some contemporary gear companies, including Manley and Terry Audio, hand-wind their own transformers using equipment that has not substantively changed in decades. Transformers frequently arose during our research with manufacturers—and loomed important in understanding the gendered history of audio manufacturing labor, too. Audio companies have not played up the artisanal and hand-wound nature of transformer-making, or their geographical provenance, to the extent that they emphasize the same when discussing product assembly. Gear users in broader gear cultures, however, place enormous importance on these facets.

In trawling manufacturing archives, we found that most of the factory floor employees of Reichenbach Engineering Company, who in the 1940s–1970s were a principal supplier of transformers to LA-based equipment companies, were women.[8] This is still evident in pictures from the shop floors of Carnhill, a UK-based transformer company that has bought up or absorbed several other UK-based companies including St Ives, Marinair, and Sowter, and Sweden-based Lundahl. Initially, we frequently saw women working at the stations where transformer wire-winding was done with a bobbin; the men in archival pictures appeared to work in other roles. Subsequently, though, we saw that most other key manufacturing tasks, from soldering to core insertion to molding to casing to the impregnation of cores with oil, were also completed by women. As is the case with many manufacturing companies, their public face is that of the male inventors. But behind the scenes a different story emerges: women's many forms of specialized labor, resulting from years of knowledge and experience, are essential to making not just a working transformer but a good transformer. Moreover, among the generation of male engineers in the US who came of age during the Korean War era, we have informally heard stories about their romantic fixations on "transformer winding girls," some of which resulted in relationships or marriage.

Within the audio realm, there has been little substantive innovation in transformer design/manufacturing beyond subtle tweaks to the metal alloys that are laminated together in transformer cores. Instead, much of the effort in audio transformer design in the early 2020s goes into attempting to recreate the sound of 1960s–1970s gear, which used transformers by Marinair, St Ives, and Reichenbach.[9] Twenty-first-century clones of this gear will often devote most of the marketing copy to the fact that they use, for example, Carnhill transformers that attempt to capture the sound of St Ives transformers found in 1970s Neve gear rather than those allegedly inferior 1980s transformers to which Neve switched. For Peter Montessi and A-Designs Audio, this meant an expensive and agonizing journey of commissioning and collecting transformers, since "the transformers they used back then didn't really translate too well to modern times." He explained about his process of modernizing the Quad 8 circuits: "The transformer was our biggest headache. I ended up collecting a closet full of transformers, that I had to pay for and that I now have no use for. And I'm in the poor house for 'em. But we finally found one that was acceptable for more modern use."[10]

Coil Audio, a relatively new company that produces expensive tube mic preamplifiers based on 1940s-era parts and circuit topologies but modernized for contemporary recording environments, struggled to bring products to market during the COVID-19 period due to their difficulty in finding people with the knowledge, skill, and materials to recreate their sample designs, as narrated by Steve Squire:

> We lost our 90-year old winder to the virus in April 2020 and that was a heavy blow. The tiny Chicago Mil-Spec shop he worked for immediately closed, scrapped, and everything went to the landfill—including our revised winding documents!!! In a Hail Mary last resort move, we contacted an independent Canadian winder we met at NAMM that had worked for Hammond in the 70's and 80's. . . . We sent him our very last unpotted CT-110HN and he carefully unwound it and then sent us a sample that completely exceeded our expectations—noting the lam material stack, unique winding structure, right down to the type of tape used and length. No excuses, no complaints, just a Total badass that could wind a transformer like it was 1940. He also sent a sample of the CT-41 Output and it was better too! All of this within a month! This guy

had accomplished what NONE of the major winders could—and to boot, all the materials were domestically sourced and manufactured right down to the wire. A rarity indeed.[11]

The claims here result from Coil Audio's deliberate and meticulous design process and their own sonic comparison testing of new transformers to historical ones. However, we find similar language adopted amongst amateur recordists who then extrapolate peculiar hypotheses about why vintage and contemporary gear sounds different. At its absurdist extreme, we encountered claims that transformer companies are unable to obtain the same quality of iron they once could, thus propagating a myth that all quality iron from America has already been mined or that recycled copper is more *impure* than the pure and virgin copper wires with which they imagine historical transformers were wound.[12]

The mix of terroir-style arguments about American iron purity or Canada-sourced wire with anxieties about virgin versus impure things has an uncanny resemblance to the language of fascist ethnostate playbooks in the first instance and conservative Christian masculinist fantasies in the second. These patriarchy-adjacent arguments are especially ironic considering that without skilled women in multiple countries performing artisanal labor, there would be no transformers. It is essential, though, to demarcate a big difference in transformer perceptions. Companies who make transformers acknowledge the skill and craft involved in the making of these devices, which for many companies means that women possess the primary responsibilities for this skill and craft. But consumers—including some gear manufacturers who, unlike Coil Audio or A-Designs, are inexperienced with transformer manufacture themselves—overemphasize the provenance of materials or country of origin to the expense of the skilled labor that made those transformers the marvelous objects that they are.

Opamps and ICs

It's often said that solid-state transistors made modern miniaturized electronics possible, and in the sense that computing requires chips that contain billions of transistors, this makes sense. In the case of the sound and materiality of professional audio equipment, though, the presence or absence of discrete transistors is less noteworthy than what happened when engineers figured out how to create operational amplifiers (opamps) out of a miniature, self-contained circuit populated with a handful of transistors, resistors, and sometimes diodes. One main use of an opamp is to amplify a signal, and nearly all kinds of audio equipment, from preamplifiers to equalizers to compressors to digital/analog converters, require one or more amplification stages. These amplification stages, individually and cumulatively, play a large role in determining the overall sound of gear. Two kinds of opamps matter to our discussion here: *discrete opamps* are miniature circuit boards stuffed with transistors and resistors and are typically made by an audio company to streamline their production process, while *monolithic opamps* are general-purpose, small, and affordable parts made by semiconductor firms.[13]

The first mass-market monolithic operational amplifier was the μA709, designed by Bob Widlar (1937–1991) at Fairchild Semiconductors and released in 1965 at a cost of $50.[14] The μA709 consists of fourteen bipolar transistors and fifteen resistors organized into an integrated circuit (IC) that is 1.8 mm square, and while used for audio applications, it was not *specifically* designed with audio in mind. Widlar, by all accounts, was quite a character. Besides his lifelong passion for inventing new things, he kept an ax in the corner of his office, at one point blew up the loudspeaker of Fairchild's paging system with a M-84 grenade, brought a sheep to work one day in his Mercedes after his employer tried to save money on lawn mowing, and was fond of the way of life in Puerto Vallarta, where he passed away whilst running up a hill.[15] Widlar also worked for some other "characters." Fairchild had been founded just a few years previously by the so-called traitorous eight—eight engineers who resigned from the semiconductor firm run by transistor coinventor and noted eugenicist William Shockley. According to electronics historian Bo Lojek, they left due to his management style, not his beliefs.[16]

The problem with inventions is that someone needs to actually make them. In the case of the Fairchild μA709, Fairchild could not keep up with demand, and the ballooning expansion of their business led them to create new fabrication facilities, or "fabs," around the US. One such fab was built in Shiprock, New Mexico, within the Navajo nation, and quickly overtook a nearby uranium mine to become the largest single outside employer of Native Americans (Navajo or otherwise) until the fab's closure in 1975. Most employees at Shiprock were women and, as Lisa Nakamura has noted, Fairchild's PR emphasized the value that the "nimble fingers" of Navajo women brought, even connecting their integrated circuit processes to Navajo rug-weaving and silver jewelry-making.[17] In reality, the incentive for operating in Shiprock was largely economic, as Fairchild was not obligated to pay the state minimum wage for labor done on reservations and could profit from the looser reservation regulations. The Shiprock fab followed Fairchild's first experiment with offshoring production to Hong Kong in 1961, and Hong Kong had, like the Navajo nation, a large unemployed labor pool with many potential workers who were experienced with textiles.[18] Similar conditions had prompted RCA's 1968 relocation of their primary radio electronics manufacturing facility to the newly established *maquiladoras* of Ciudad Juárez, Mexico.[19] In all three cases, manufacturers sought not just the availability of cheap labor that could be exploited without consequence, but cheap *women's* labor.

The µA709, and the LM101 that Widlar invented two years later, set the stage for a five-decade period of opamp development. The Signetics/Philips NE553X (released in 1975) and the Texas Instruments TL06X/7X/8X series (released in 1978) especially stand out, as their cost-to-performance ratio (less than $1, and capable of quality low-noise audio) led to the design of more affordable gear.[20] Initially used in consoles by Soundcraft, SSL, and others, we still find them in outboard gear and synthesizers made today—and in solar power inverters and uninterruptible power supplies. Perhaps because of their ubiquity, and since they are often found today in budget gear, just as we examined in relation to capacitors, there is a fervent gear culture interest in modding their gear by swapping TL072 and 5534s with supposedly higher-performance opamps made by Analog Devices or TI/Burr Brown. In this context of opamp-modding, the terms "high slew rate" and "fast transients" first entered the technobabble lexicon and folksonomies of gear cultures. While the former is measurable—indeed, opamp specification sheets provide slew rate information—part of the problem is the lack of an established understanding about what range of slew rates is actually desirable.[21] An opamp with a faster slew rate will have trade-offs in other specifications that may make it sound or perform worse in certain situations, or it may draw too much voltage to be a drop-in replacement. Just as we saw with recapping practices, this has not stopped wanton modding and opamp-swapping from damaging some perfectly operational—even if a bit boring-sounding—gear.

Despite significant subsequent advances in opamp design, especially on the part of US-based Analog Devices, who has been proactive in supporting small audio businesses experimenting with their products, more attention is given to historical opamp designs. If opamps are mentioned as a product's selling feature, it is probably because the product uses an API2520-style opamp, which is neither monolithic nor the result of advances in materials science. The API2520 was Saul Walker's update of the Melcor 1731 (originally introduced in the late 1960s), a 1-inch square PCB drop-in circuit produced in-house. US recording console manufacturer API made the design decision early on that their entire product line would use the same amplifier circuit—whether for mic/line preamplification, makeup gain after equalization or compression, or mixer summing busses—which simplified product maintenance. That said, there is no single API2520 object since they refined the designs over the years. Paul Wolff, a longtime owner/president of API who had transitioned the manufacturing of the API2520 to an automated process, wrote in 2000 on the rec.audio.pro newsgroup, "The changes were made because of stability, heat, current, discontinued parts (TIS Transistors), quieter input devices and technical masturbation."[22]

API has always been secretive about the 2520 and epoxied the opamp to prevent competitors from reverse engineering it. This has not stopped many engineers from trying, aided by a couple of non-epoxied prototypes that apparently "leaked" out of API. Now over a dozen boutique businesses offer their own 2520 versions, some claiming to replicate the API version from 1974.[23] These earlier opamps have quite poor specifications, as they impart DC offset and have relatively high harmonic distortion, but some consider them to be integral to the "American sound." The Jensen 990 and its updated versions by John Hardy and Fred Forssell have attracted less interest although they are better in every measurable specification, and for whatever reasons, vintage mania and cloning has not resulted in other significant 1970s-era opamps—for example, the Quad Eight AM10 or Neumann OA12—from being recreated to the same extent.

Microphone Capsules

The sound and quality of a studio condenser microphone results from the relation of several key parts: a microphone capsule that transforms changing air pressure into electrical signal; the active electronics that polarize the capsule and that amplify, impedance match, and sometimes filter the low-level signal initially captured; a headbasket, the grille of which protects the capsule from the environment and reduces the intensity of air and saliva hitting the capsule (for example, vocal plosives); and the metalwork and other material that protect the circuit from stray radio and electromagnetic frequencies and reduce mechanical shock. Of these parts, microphone capsules are often singled out as being the most crucial element of a microphone's sound. As it turns out, despite a century of experimentation, in the early 2020s, most large-diaphragm condensers use variants or copies of only three historical microphone capsule designs: the K47 and M7 designed and modified over the years by Neumann and Microtech Gefell, and the CK12 of Telefunken/AKG. All three were popularized during the post–World War II period when materials and parts costs were at their highest, and labor costs at their lowest. Making such capsules today is a challenge. Labor costs are high, and the precision of the parts requires specialized labor. As Ben Sneesby noted to us, if his company were to replicate the machining and labor workflows of yesteryear, the labor costs would amount to $1,500 per capsule.

In contrast to many newer contemporaneous mic companies that outsource most of the metalwork, capsule design, and manufacturing to Chinese companies, such as 797 or Alctron, Australian mic company BeesNeez, owned and operated by Ben and Veronica Sneesby, claims to make more of the mic parts for their unique product line than any other company. BeesNeez manufactures their own metalwork and connectors and is especially known for making their own mic capsules in-house. When we talked with Ben about their manufacturing, he noted that in making capsules, even though some of the process can be done with their precision

Figure 4.3
Royer Microphones factory, Burbank, California. Photo by Eliot Bates.

CNC machines that "maintain incredible accuracy," it still requires two and a half hours of human labor: in cutting 99.99% pure gold sheets they source from a local jeweler, in sputtering the mylar diaphragms, assembly, and capsule tensioning.[24]

Veronica Sneesby does most of the specialty diaphragm work for BeesNeez, which, as we discovered, represents a widespread gendered labor trend for microphone manufacturing. In our site visit to Royer Microphones in Burbank, California, we met Sophie Gefter, a longtime employee who was singled out for her skill with the precision tasks involved with ribbon mic manufacturing, especially ribbon element corrugation and tensioning (figure 4.3). Video factory walkthroughs of the Soyuz microphone factory in Russia depict a similar division of labor, and in the video walkthrough of the Sennheiser/Neumann headphone and microphone factory, we learn that four (unnamed, in the video) women are responsible for key stages in both microphone and headphone manufacture.[25]

We find tried-and-tested methods of microphone manufacture in most gear companies and limited experimentation in design. An exception to this is David Josephson, who patented his own headbasket designs and was the first manufacturer of mic capsules to successfully reproduce and modernize the Sony capsules used in the C37A microphone. Josephson's capsules are used in the Manley Reference Silver and Reference Gold microphones too.[26]

Vacuum Tubes

As she writes this section, Samantha is sipping tea from a Manley Labs mug—"TUBES RULE" emblazoned on one side, "DANGER! HIGH VOLTAGE" in all caps on the other. Perhaps no gear components are fetishized more than vacuum tubes: small glass bulbs that, when heated by an internal coil-based cathode, result in a controlled flow of electrons toward a plate and across a grid inside the tube. Vacuum tubes, or simply "tubes," represent the pre-transistorized audio technology era and early audio devices such as radios and amplifiers. Tubes continue to be essential for numerous condenser microphones, preamplifiers, and equalizers, and as such, vintage tube gear is highly sought after and is a big driver of the secondhand gear market. Scarcity of tube gear made in the early to mid-twentieth century, and scarcity of replacement tubes, is the cathode of tube fetishization. NOS tubes—in their boxes and in as-new condition—by RCA, Mullard, Telefunken, and Western Electric are particularly sought after and the subject of much online gassing. The scarcity of NOS Telefunken VF14 valves for Neumann U47 microphones, for example, means NOS tubes may sell for almost $8,000.[27] Tubes are loved till it hurts: toxic to produce, they contain poisonous raw materials and emit toxic gasses when heated. Many fall short of the required quality, resulting in many noisy tubes in circulation—and many disappointed consumers who spend a fortune on a prized valve only to find it is unusable. Manufacturers, too, may chance upon whole batches of unusable tubes, as Peter Montessi told us:

> I use a tube called a EF86. It's a great tube, but they're very sensitive. And you could spend anywhere from 12 dollars to 300 dollars for that tube. But it doesn't guarantee you that tube is not microphonic. So if you're spending 300 dollars on a tube and you've got to buy 2,000 tubes and out of those 2,000 tubes a

hundred tubes are good, what do you do? Float the other 2,000 down the river? You just lost a lot of money.[28]

Vacuum tube production has always featured a gendered division of labor. In early depictions of tube production, such as RCA's *Electrons on Parade* documentary (1942), women are seen undertaking every part of the valve build: chemical testing of raw materials (e.g., the tungsten for heater coils, the mica of the spacer discs), meticulously measuring the molybdenum control grid construction, fine wire coiling, sputtering the insides of tubes, nickel plate construction and micrometer inspection, construction and weighing of cathodes, tube assembly—and verification of each and every step in the process. Mullard valves were made, initially for Phillips, in Blackburn since 1938; the assembly part of the factory was split into "his" and "hers." Little has changed, and vacuum tube production is still largely undertaken by women, as EveAnna Manley pointed out following a visit to an Ei Niš factory in Serbia: "I remember walking into a room with 50 ladies just putting cathodes up heaters, there [is] a lot of hand work involved and lots of places for variations to come in, as they're still made by humans!"[29]

The fetishization of vacuum tubes is also linked to manufacturer origin. Vintage valves made in the UK (Mullard) and Germany (Telefunken) are more sought after than those made today in Slovakia, Russia, and China. The geographical and historical value placed on vacuum tubes is, however, only surpassed by the extraordinary importance placed on their sound, and the sound of tube gear in general. From warmth to richness, fullness to boldness, vacuum tubes continue to fascinate and inspire in just about every gear culture corner. Gear users spend inordinate amounts of time technobabbling about vacuum tubes, especially online. Threads on Gearslutz/Gearspace, for example, cover everything from quantity discourse, "How many U67 backup tubes do you own?", interfacial potential, "Do all tubes glow?", to blatant sexualizing: "adding a tube touch on the way in." Such threads attract hundreds of posts and thousands of views; adjacent discourse about plugins' varying ability to replicate the sound of vacuum tubes is even busier. EveAnna Manley was right: in gear cultures, tubes indeed do rule.[30]

PATHWAYS TO DESIGN

How do people become gear designers? It is well established that many of the designers responsible for the "golden age" of now canonical circuit designs came from military engineering backgrounds. This includes British ex–Royal Air Force engineers who worked at Rupert Neve Designs and perhaps, most famously, Saul Walker's designs of "encapsulated digital modules approved for military aerospace systems" for Republic Aviation following a stint at the US Naval Research Laboratory prior to becoming the lead designer for API.[31] However, this is less the case with younger gear designers. In most cases, the story begins in childhood, with taking apart broken electronics and getting electrocuted while trying to diagnose problems or putting devices back together. For gear designers such as Robin Porter, electrocution is a rite of passage:

> I played in bands and was very poor and couldn't afford amplifiers and stuff like that. My dad had a radio/television shop, so I'd get bits and pieces and knock them together. It'd catch light . . . or it wouldn't do anything, or I'd electrocute myself. Finally, I made something that worked. And then my friends cottoned on to the fact that, um, I knew a bit about this and I could fix their amplifiers. I had a friend called Jardin, who was a drummer, and he went to work for Neve in the summer holidays. He said, "There's this place I'm working and I'm going to be leaving soon because I'm going to uni. It sounds like just up your street. Do you want to take my job?" So I went along for an interview, and I thought, this is quite a cool place.[32]

Joe Meek (1929–1967) is perhaps the prototype for musician-turned-gear designer. In the words of Ted Fletcher, who worked with Meek and subsequently released a series of products inspired by Meek's designs, "I remember him borrowing microphones from AKG and using them for a session and then sending them back. Most of the time he was experimenting with things and making things actually. Modifying equipment. Just taking things to bits and putting it together again."[33] Ted Fletcher himself followed a similar pathway: "When I was a child, I was interested in mechanical things, electrical things, and at the same time, very very interested in music and appearing on stage."

A common route to design comes from music aficionados who found themselves in situations where they needed to restore vintage gear:

> I started restoring things, and my dad of course had some great classic amplifiers from his time as a session [musician] guy. When I first started restoring things my dad said, "Now listen, it's going to change it, and we have to replace capacitors at a certain point, especially paper-and-oil caps that can get leaky, but you can't put just anything in there," as a lot of top-end folks know. So I got to realize that there is *a tact* to doing this. There's *a voice* in restoration that we have to be mindful of, it's not just "a capacitor is a capacitor, a resistor is a resistor": you have to do it right. I then started doing in-the-trenches repair work, and then I started to listen, and repair, everything, from McIntosh and Bryston studio amplifiers, to tube amplifiers and valve gear, preamps, and then restoring Neve consoles and Neve modules, and making them *sound* like they did.[34]

Beyond repairing vintage gear, some designers including Hutch Hutchison honed their skills and ears further whilst doing live sound and studio work:

> I started off in typical guitar shops, fixing guitar amps. First jobs, you know? I went from that to live sound, from there to studios, producing and engineering, and developed the ears that way, and learned what people liked, how people work. I spent a lot of time in studios, and then when the time came to design, it was like—learn as you go! And I'm still learning! [I'll] continue to learn until I can't solder anymore. So you never stop learning to listen.[35]

As some designers noted to us, the options available for understanding the design of vintage gear were considerably fewer prior to the networked sharing culture that began at rec.audio.pro and later spread to message forums and other online platforms. According to Jakob Erland,

> [In the 1990s] I worked as a studio tech in one of the bigger recording studios in Denmark where I was fighting to get any information about what was inside all this high-tech audio gear they were using. And at the time nobody was willing to share any information about what pro audio really was and the whole business, especially the marketing side, was so full of smokescreens and stories about angel dust that it was unbelievable that you could sell stuff with fairy tales like that. And if you found someone who had information they wouldn't

> share it because they thought they were sitting on a gold mine. Because that kind of knowledge was so rare.
>
> I [picked] up some service manuals for some of the stuff we had back then [that] people [knew] nothing about. And I'd figure out that there wasn't much to it. It wasn't qualitatively different than any other electronics. I knew a bit of electronics because I built a Moog, a Minimoog, off the schematics. My local librarian managed somehow to find service manuals on it back in the '80s. So I took a year just trying to build it up on strip [boards], and actually got some sound out of it! I mean it's a good project; [it] basically tells you everything you'd want to know about electronics. And if you *want* an instrument, you have the motivation to *actually understand* what's going on.[36]

Reading across these formative experiences, we find commonalities. Many future designers were from working-class backgrounds and discovered gear at a time in their life where it was cost-prohibitive, which contributed to their turn to repair and design. Though not included in all the quotes above, most engineers had years of amateur experience dabbling in electronics prior to being offered a full-time job through personal connections that came through the informal networks of recording studios, bands, and gear retailers that offered repair services. On the other hand, these engineers note personal discoveries that at first seem paradoxical. Marshall Terry's discovery that "the voice" of particular pieces of gear resulted from nonarbitrary component choices led to a multiyear obsession with listening to every component within a circuit. Jakob Erland, on the other hand, found that the "gatekeeping" discourse around gear in the 1980s–1990s was exclusionary *by design*; the circulation of engineering knowledge was full of "smokescreens," "angel dust," and "fairy tales." Gatekeeping and exclusion, even if in different guises, continues today.

Circuits

Since 2000, few new circuit designs caused a widespread change in the design of professional analog gear. Rather, the period is better understood for subtle updates and modifications to circuits originally developed in the 1930s–1970s, for novel combinations of existing circuits, for new components being used in place of old ones within the realizations of circuits, and for changes in how the user interface accesses the circuit. Let's take microphones as an example. Of the hundreds of small diaphragm condenser mics on the market in 2023, most "new" ones are either transformerless and "inspired" by a circuit originally popularized by Schoeps in 1965 or transformer-coupled and "inspired" by Neumann's KM84 mic (1966–1992). Both designs have a low part count, are quick to manufacture, and can be made to work reasonably well with contemporary components. We experienced this firsthand when we both built microphones, as described in our opening example. That said, the mics we made did not sound like a Schoeps, since only the amplifier circuit was based on that design: the capsule, headbasket, and metalwork, all of which influence the sound, were of different designs, and some of the components we used differed from those Schoeps employed in their CMC5 mic. The situation is similar for studio staples such as microphone preamplifiers, equalizers, compressors, and mixing consoles, and for the circuits behind the "new" category of "mojo" devices that add saturation and distortion to audio signals: in most cases, vintage circuits dominate the design of "new" gear (see chapter 14).

The most famous British microphone preamplifier and equalizer, the Neve 1073 Channel Amp, is a clear case of the rampant confusion that surrounds vintage circuits. The 1073 was never intended to be a stand-alone product, and within its original context as part of a console was not an object with its own sound. The Neve 1073 is relatively straightforward as an abstract design, since Neve's technical drawing EH10023 provides the sole representation of the technical design and schematic. From these, we see that the 1073 is not even one circuit but consists of several BA283 and BA284 amplifier circuits. Scanning the history of Neve's many designs, we find that these amplifier circuits were modernizations of earlier ones (the BA183/184) and were followed by other "improved" amplifier circuits. By 1980, the Neve 1073 and its amplifier cards were historical circuits: newly available components allowed the design of more reliable, better-specified preamps and EQs. As Robin Porter of AMS Neve noted to us,

> Neve's philosophy has always been that you get the best out of a design from a technical point of view. So you strive for the best frequency response. You

strive for the best noise, the best distortion, the best cross-talk, all of those performance figures. And then that gets you into a ballpark where then you can start to listen to it to see whether or not you need to change anything to get to where you want to be. And once you've achieved that and once you've got that mix of the system and the components that make up that system, then that can be turned into production.[37]

The abstract 1073 is only *relatively* straightforward, though, since there are six variants alone of the BA283 amplifier cards. There were internal disagreements on the use of an additional transistor on the output of the card, so some BA283s have them, while others do not. The BA283 and BA284 were never thought, by Rupert Neve or the other designers, to be the pinnacle of amplifier designs: they accomplished certain technical tasks, including getting a microphone's output to line level, while meeting key technical specifications such as distortion and/or noise. They were built to a price point, often with inferior aluminum electrolytic capacitors "leaking like a sieve," according to former Neve engineer Geoff Tanner.[38] Because nearly all such capacitors would have burst by now and been replaced by modern equivalents, and since Neve themselves updated the components they used to realize the BA283/284 and therefore the 1073, there is no singular material object we can point to and say, "This is the real 1073." Many contemporary commentators suggest that the sound of the 1073 is due to the transformers, as we discussed before. Some aficionados claim significant differences between the transformers produced in the different factories, although ex-Neve engineers deny that there is any significant difference.[39] This partial story of the 1073 preamplifier/equalizer circuit provides a small window into engineer perspectives on circuits and the components used in their implementation. We met some gear designers who are enamored with gear from the 1950s–1970s and have dedicated their careers to either reissuing it, combining circuits from this era to make new products that were not made at the time, or subtly modifying—or in their words, "improving" upon—these designs. Other engineers we met, however, were uninterested in these approaches and instead strove to make new circuits with non-obsolete parts. Hutch Hutchison, who was a Chief/Senior Designer for

Manley Labs, Texas-based Rupert Neve Designs (a new company started by Rupert Neve, and unrelated to UK-based AMS Neve, discussed above), and Taiwan-based Fredenstein Professional Audio before joining Radial Engineering in 2018, told us,

> I don't reference other gear so much . . . a little bit in reference to what people have always liked and leaned to. In terms of topologies, component choice, I like inventing new topologies. I like doing things . . . going where nobody else has gone. That to me is fun and interesting. Trying to copy some old thing is not fun for me. That's not design. Somebody fresh out of college can do that.

Regarding his process for coming up with a new circuit, Hutch noted,

> Things start off as pencil sketches. Maybe a panel concept, maybe a circuit, but pretty soon they end up in Spice [circuit simulation software] to see if the idea will work. You can't trust Spice too much, but sometimes it's real handy. But then you have to breadboard. You have to listen, you have to tweak, and adjust by ear. And in the end, it's kind of like being a mixer or regular engineer, except it's different tools. You're still adjusting the sound. You're still making judgment calls. Except that instead of faders and knobs and plugins, I'm using resistors and capacitors and transistors, choosing things that way, but largely by ear. There's always major listening time. Anybody who designs audio gear before listening is probably gonna end up with something mediocre!

What is unstated in the interview snippets above, however, is the difference between coming up with an idea for a circuit—a schematic—and the material implementation of that idea. Most self-taught engineers at some point learned schematics and basic circuits from widely circulated pedagogical texts such as the US Army and Airforce's *Basic Theory and Application of Electron Tubes* (1952), Walter Jung's *IC Op-Amp Cookbook* (1986), or Douglas Self's *Small Signal Audio Design* (2010). Circuit layout and optimization is, however, an experientially learned skill. While software like Spice can automatically produce a circuit board from a supplied schematic, that circuit board will almost always be suboptimal: among many issues, certain components such as opamp decoupling caps need to be placed as close as possible to the opamp, and traces that use "vias" to route between circuit board layers must not introduce problems such as an increased noise floor or crosstalk. Knowledge about how to produce ideal layouts from a schematic is rarely shared: especially here, there is no substitute for the "10,000 hours" of experience that enables an engineer to produce the best layout of a design.

Conclusion: The Social Reception of Twenty-First-Century Manufacturing

Thinking about electronics writ large, the twenty-first century is typically framed as a period of global supply chains, of transnational component makers that supply a transnational wholesale market (e.g., Newark/element14/Farnell), of offshored labor and East Asian supremacy in manufacturing, and of mass-produced, automated assembly commodities. The story we tell is different: specialized artisanal labor, women contributing to gear and component manufacture, long-term familial relationships between the proprietors of local small businesses, and specialty parts agonized over by manufacturers and consumers alike. As Peter Montessi of A-Designs told us,

> If you're in the US, doesn't it make sense to support your own country that you live in instead of sourcing out? Like we're made here in Los Angeles. From my sheet metal to my paint to my transformers to my assembly to my boards, it's all here in Los Angeles. Every one of those companies that I source out to have 20, 30 employees and I've been working with them for 20 years, you know? So there's a friendship that's evolved over time . . . My assembly house, I've known him for years and he's like family, and even my transformers, Cinemag, I've been through hell and back with them. And Dave's like an older brother to me, so if something happens within our personal lives, it affects us all.[40]

"Made in America." "Made in the UK." "Made in China." What do these mean in the 2020s, and how do they articulate ideas about value—as related to nationality or nationalism—in technical objects? As we saw in chapter 3, few of the raw materials in question originate in the country in question. The largest component makers might be headquartered in the US, UK, or Japan but have moved much of their industrial capacity to Central America, Caribbean, North Africa, and East and Southeast Asia. None of the most vocal "made in America" boutique manufacturers, even those who use North American–made transformers or metalwork, currently have their printed circuit boards made in the US: PCB shops in China can produce higher-quality PCBs with a greater variety of materials for a much lower cost. This leaves, then, only three empirical things that could plausibly support such terroir-style arguments: US-based assembly labor, US-sited quality control, or the happenstance that the gear designer was in the US when they thought up the circuit.

Just as is the case with other aspects of electronics manufacturing, frontstage and backstage perspectives differ substantially here.[41] The public-facing articulations of American pride, even if not intended as such, resonate as ethnonationalist claims amongst prospective consumers. Buying American is, in effect, a kind of *ethical act* that consumers can do to fight their perception of China's increasing power on the world stage—which seeps over onto a broader laundry list of other supposed evils including communism, before transitioning into outright Sinophobia and racism against people of Asian heritage. These beliefs are certainly not coming from some of the top professional gear designers: the collaborations between Rupert Neve and the Shanghai-based company sE Electronics (founded by Siwei Zou), and Hutch Hutchison's design work for Taiwan-based Fredenstein Audio, build upon a seventy-year history of collaborative engineering projects. But the inward-facing articulations of pride, however, were some of our most striking findings regarding manufacturer-centric gear cultures: the respect shown within multicultural and mixed-gender workplaces, the treatment of small regional component suppliers "as family," the occasional relationship kindled on the transformer winding floor, or even the personal enjoyment some engineers have in the practice of soldering through-hole parts in traditional ways. While manufacturers run for-profit businesses, skilled electrical engineers would make considerably more money working in aerospace, energy, or medical sectors; their attraction to designing and making gear at a time when expensive hardware should be obsolete attests to the importance of those inward-facing articulations of pride.

5 INTERFACE

MANLEY BLUE

"It sounds the balls, but it's got to look the part too. You know what I mean? You want the supermodel with the PhD!" beams EveAnna Manley, gear designer and longtime CEO of Manley Laboratories, Inc., as she proudly leads us through a tour of her gear factory in Chino, California. The factory floor is loud, busy, and chock full of well-oiled machinery and a tight-knit team of operators manufacturing Manley's gear. EveAnna Manley is dedicated to every aspect of the gear production process. From design to sourcing parts, to build quality through customer service and relationships, her infectious passion and personality is imbued into every piece of gear that leaves the factory floor. On a machine four times his size, Andy drills into the fascia that will hold the knobs and switches, black-boxing the system: Manley tells us it's a "big old machine from the fifties and it's just tapping holes!"[1] She is especially proud of her engraving methods. Pointing out another enormous machine, Manley tells us, "This is a mechanical engraver," before showing us "an old Apple] [." Samantha asks skeptically if it is "still going," to which EveAnna proudly replies, "Yup, 5¼ floppies, which is just so crazy, and it's just a monochrome, you know, Apple] [from 1982 that's been in this building since 1994 [figure 5.1]. It doesn't have a hard drive. It's got a resident program right now that is loading up, and this is the engraving program that drives this machine. So every [one] of the historic jobs lives on a disk (holding up 5¼" floppy) and I have to load in the job." Beneath the Apple] [are drawers of floppy disks, each featuring the fonts and typeset for

Figure 5.1
Engraving machine, Apple] [, and engraving-file floppy disk at Manley Labs. Photo by Eliot Bates.

various gear interfaces. Gear including the Massive Passive Equalizer and the VoxBox preamplifier each feature more than one hundred inscriptions on their front plates, and these disks hold the information necessary for cramming all that detail onto the front fascia.

The importance of gear looks is something Manley knows all too well: her company is renowned for it. The "Manley Blue" front plates with the company and gear names engraved in white capitalized serif font are immediately recognizable to gear culture participants and fit comfortably alongside much older gear that signals to military colors with dark navy, gray, and red interfacial elements. The fact that Manley's graphics are printed with 1980s software and stored on floppy disks parallels the company's commitment to tube technologies ("Tubes Rule," remember), and ensures an inimitable design quality. Gear colors and fonts are not simply aesthetic choices; interface design is intrinsic to gear identities. Later in the book, we will connect the militaristic connotations of gear interfacial elements to the "secret weapon" discourse that permeates gear cultures. So intriguing are these elements to gear consumers that Manley clarifies the panel-dyeing process on its FAQs: "The suppliers would leave the panels in the dye tank for a shorter time and the resultant color on the panel was lighter," is the response to inquiries about color variation in some Manley gear. A further explanation as to how UV light will fade the dye over time ends, "When the dye all fades out, you are left with the color of the aluminum." Other areas of the website detail how the "fantastic conductive plastic element Bourns potentiometers" adorn their gear, and the Force preamplifier, with its "exquisitely gorgeous chassis," "ergonomic facia," and "thoughtfully organized controls" all amount to a "feast for the eyes."[2]

Knowledge of which gear interfacial elements seduce gear culture participants is one thing. Making a political statement with a gear interface is entirely another. In 2020, EveAnna Manley released a special edition of the Massive Passive stereo equalizer. Under the heading—"Boosting Equality for All!"—this rainbow-themed version of the equalizer, with its colors derived from the "original 1978 gay pride rainbow flag" and entitled "Massive Pride" was released coinciding with Gay Pride month, with profits from the equalizer to be donated to the Human Rights Campaign. With its strip of eight block colors, one each for its four bands across two channels, the Massive Pride immediately popped among the traditionally darker colors of gear interfaces—including Manley's own. Only fifty units of the EQ were made; they sold out almost instantly. So successful was the launch of this LGBTQIA+-friendly gear, that Manley repeated the event in 2021 with "Vox Aequalitatem! All Voices Equal" VoxBox, this time, presented in a pastel rainbow wash across its front fascia. The release of such products presented curveballs to gear cultures, whose demographics are largely made up of white, cisgendered, heterosexual males. Upon release of the first unit, a Gearslutz thread ensued where initially multiple gear manufacturers and designers posting under their real names praised the design and qualities of the unit. However, it did not take long for the thread to populate with homophobic rants about the "shameless marketing tactic," followed by predictable requests to "leave politics out of gear." After spiraling into toxicity, more than half the posts were deleted by a moderator, and the thread was eventually locked.

Nothing about the Massive Pride was different in sound to its Massive Passive predecessor. Indeed, the engraving looked to be almost identical on both units and no doubt derived from one of the trusty Apple] [5¼" floppy disks. Yet this gear story illustrates the power of the visual interface and the social, cultural, and political messages they convey. In these proudly pride designs, Manley dared stray—some considerable distance, in fact—from normative gear identities. Looks are not everything, of course. But in gear cultures, they most definitely are.

THEORIZING THE INTERFACE

Many affective aspects of technological desire and fetishization do not pertain to the whole technological object or even its primary intended use case (doing something with audio), but rather with the aesthetics of the physical object itself—especially its interface. But the term "interface" is a problematic and misleading term, since it tends to be colloquially used to refer to one part of a technological object (specifically its control elements that can be touched and felt), while ontologically, it implies the existence of a site or area or point of contact between two independent fascia (surfaces). In our analysis of audio technological objects, we found several other aspects of these objects and object relations that are also significant and work in tandem. Even if we think of a basic interfacing action, a user turning a knob, while the knob covering certainly has a surface that could be said to serve as an interface between the user and a circuit component (a potentiometer, rotary switch, or continuous encoder), this knob-turning experience involves material surfaces (user and knob), the act of touching the surface and experiencing a haptic response, the visual impression of the control, and even the sound of the control being turned. All this contributes to a multisensory or even synesthetic phenomenological moment.[3] Layers of semiotic meanings and a circulating techno-aesthetic discourse inflect and influence the disposition of the user prior to and during this temporally unfolding phenomenological moment.[4]

But when users, designers, and consumers talk about audio technological interfaces, they are not just talking about control elements. They are also referring to the contours of the anodized brushed aluminum of that Focusrite preamp or the Nextel finish of a Schoeps microphone, the font face and engraving/printing method of any labels or explanatory text on the front panel or mic body, the visual feedback provided by LEDs and VU meters, and even the selection and placement of jacks on the back of the device. Although none of these *need* to be touched to operate the device in the ways "scripted" by technological designers,[5] in practice, they often *are* touched: users capitalize on the wider sensory ways that they can engage with these devices.

Therefore, this chapter considers the widest array of aesthetic and morphological features that are nominally inessential to the technical functioning of the device, that pertain to the multisensory ways in which people relate to the device, and that ultimately form much of the extramusical associations people have with these devices. As we have mentioned before and will continue to argue, if it were simply a matter of achieving the sonic signature of well-regarded gear, most of the gear market today would not exist, as some affordable clones and even more affordable (or free) digital simulations get close enough to that sound for a fraction of the price. Put another way, the distinguishing *sound* of gear has little to do with the *kinds* of interfacial choices, as any particular control could be rendered in multiple ways (buttons or switches or capacitive touch for binary operations, sliders or knobs or force-sensing resistors for variable ones, or things you click on while looking at a computer screen) or removed altogether to further black-box the object.

As noted already, colloquial and theoretical conceptualizations of interfaces' actual nature disagree. In the next section, we will explore six qualitatively different kinds of interfacing that happen within studio gear. Following that, we will analyze the materiality, the stuff (brushed aluminum, cold rolled steel, bakelite) of interfacial elements, and the semiotics or associations of these specific substances separate from any specific sensory engagements with them. The final three sections iterate our phenomenological/semiotic approach by considering in turn the visual interface, the haptic interface, and the audible interface.

WHAT GETS INTERFACED

One pervasive problem with technological objects concerns the extent to which they are "black-boxed"—where the designer conceals to varying degrees their inner workings. This has historically been of interest to scholars in science and technology studies, where the trope of "opening the black box" typically refers to the revealing of human social processes, including ethical and moral choices, that made that object possible or that result from that object's circulation.[6] However, a different kind of "opening the black box" transpires within gear cultures. Instead of wanting to understand the social relations that made technology possible, this entails quite literally opening up boxes, investigating components and circuits, possibly modding them with different parts (something akin to aftermarket car parts for the hotrodding or low-rider scenes, or electronic toy circuitbending), and making these opened boxes the fulcrum of online and trade show social interaction.[7] What effects does this have on our conceptualizations of technological interfaces? *What* gets interfaced, and *where* is the interface?

Some of the pieces of vintage gear that continue to be well regarded today were considerably more black-boxed than contemporary derivatives. Maihak/Siemens/Telefunken tube amps like the V72 lacked any controls—not even a volume knob—and their position within a broadcast console further removed them from conscious attention or scrutiny. They effectively had no interface as we think of an interface today; one used a V72 whenever one needed 34 dB of gain on anything. Early compressors designed for radio broadcast similarly lacked extensive controllability or were installed in machine rooms such that any controls were set once and then left that way. In contrast, similar kinds of objects today may be centrally placed in a studio, and nearly always have considerably more controls and more types of controls. Moreover, rather than simply providing 4 dB of gain, a box of the "mic preamp" category no longer just connects mic and recorder but may require users to make a curated set of circuit-level choices (e.g., engaging or bypassing a transformer) or even encourage them to open it up and change actual components within—substituting opamps or vacuum tubes. This "de-black-boxing," as we will see later, comes to have considerable effects on the perceptions of gear, and on how gear is used and how user communities interact—which then feeds back into technological design.

Among the studio technologies that cohere gear cultures in 2020s, we have identified six different kinds of interfacing (table 5.1), one or more that may be at play with any particular device:

1. The surface of contact between user and something that appears to them as a black box
2. The surface of contact between user and a particular component or subcircuit, rather than the whole circuit with all its components
3. The skeuomorphic GUI within a computer plugin that conceals the digital controls of a hardware device (with analog and/or digital circuitry)
4. Utility modules that act as interfaces between technologies and, once set, necessitate no human involvement (e.g., DI boxes, inline mic boosters, digital interfacing standards)
5. Multisensory access to a space of operations specific to a historically specific genre of music
6. In the case of devices where the signal processing is digitally implemented, either by software running on an embedded computer or a netlist placed on a FPGA chip, the interface is the surface of contact between user and computation.

Put another way, even within one category of technological object (e.g., a microphone preamp), we find considerable divergences, ranging from those with no user interface at all (the Telefunken V72), to ones with a single gain control per channel (Antelope MP8d), to ones with a full complement of knobs and buttons (many examples), to ones that require the user to press buttons to change between alternative or optional components contained within (the switchable transformers on a Burl mic preamp or nickel/steel switchable transformers on the APA Juggernaut), to ones that the user is supposed to open up and actually swap opamps (the CAPI series of preamps) or "color" cards (the DiyRE Colour series of modules), to ones that may replicate some of the above but are marketed as the device that will get you the sound of a legendary album or producer (e.g., the Chandler TG-series

Table 5.1
Six kinds of interfacing

A	Kind of interface	B	Examples
user	panel+controls	black-boxed technology	outboard gear, microphones
user	circuit board or panel controls	circuit-level component	outboard gear with switchable transformers or user-changeable components
user	screen+computer interface for DAW plugin	black-boxed technology	Bettermaker and WesAudio "analog plugins"
technology	jacks+plugs	technology	MIDI, MADI/AVB/Dante/ Rednet, inline mic boosters, Telefunken V72
user	black box	musical genre	Motown DI
user	panel+controls	computation	digital effects units (including reverb, delay, compression, limiting, and EQ)

preamps that will help you sound like Pink Floyd or the Beatles). Like the preceding sentence, this can become bewildering to prospective consumers who may simply be wondering what they are supposed to plug their mic into, or it can become empowering to enthusiasts who, for whatever reasons, want to feel that they have had a part in the "design" process of their gear. The latter demands that the user is expected to 1) have enough "technical" expertise to be entrusted to make pseudo-design decisions (probably gained from reading web forums where this knowledge is fetishized) and 2) think/feel that this technicity is going to, for some reason, make some kind of meaningful difference (it may not). While the notion of "consumer choice" sounds fine as an unscrutinized principle, considering that audio engineering is already a domain that requires thousands of choices at every step, exactly why *should* users pretend like they're circuit co-designers, anyways?

THE MATERIAL INTERFACE

The interface concept, and multiple answers to the question of what exactly is being interfaced, suggests that interfaces are ontologically complex. This complexity extends to the array of materials used in the interfaces of contemporary audio technologies, since some have extramusical associations that matter to some designers and users alike—but that have nothing to do with sound, music, or technical functioning. When gear becomes discursive, often these substances specifically, rather than technological objects as a whole, are commented on a sort of folk material–semiotics, an opening of black (and silver, and red) boxes. But once metal alloys or plastics are formed into shapes, they also simply just exist as matter. The visitor to a major recording studio is immediately impressed by the mass of technologies within the control room. Even for those who understand what is happening within the black boxes, their primary impression of technologies can get stuck on the surfaces that call out technologies' massiveness and materiality—the *material interfaces*—those aluminum or steel faceplates; the aluminum, plastic, or bakelite knobs; or the brass microphone head baskets.

As we started to explore earlier in relation to extractivism, some of the most distinctive materials that make up the bulk of audio gear, especially milled aluminum, have few if any qualities that make them essential for their applications. Aluminum housings bring no unique audible or electrical benefit to gear, and their lightness (compared to steel) is not a major design consideration since studio gear rarely is moved. Many other materials could be used instead—wood, acrylic, magnesium, bronze, or any number of hardened plastics. Instead, following their use in war and aviation and an aggressive long-term advertising campaign by transnational bauxite and aluminum extraction and smelting companies, especially Alcoa, aluminum became affectively associated with the "cultural dream" of speed, progress, mobility, space exploration, and military domination.[8]

The allure of aluminum is experienced in the Red series of rackmount gear first released in 1993 by UK-based Focusrite—described in *Sound on Sound* magazine as the devices that "changed the aesthetics of studio outboard."[9] In contrast to the standard black or silver steel chassis, the Red series spared no expense in their red anodized aluminum case, which, based on some estimates, likely accounted for 80 percent of the materials cost for the whole unit. The faceplate had brushed anodized aluminum with rounded bevels, while the sides used custom extruded anodized aluminum hardened with a matte finish—even though, when racked, the sides of the case would be invisible. The milled silver brushed aluminum knob covers incurred another significant expense. Focusrite's economically driven transition from large-format console manufacturer to producer of outboard gear of different price points, notably, did not mean only producing "budget" objects. Instead, it entailed finding the price consumers would pay for aestheticized objects that could be appreciated inside or outside of a rack, being used to record or opened up for inspection. The resulting materiality of the Focusrite Red's several forms of aluminum reveal the metalworking tools used to make the box; the milling, extrusion, and brushing of the aluminum are just as evocative as the luster, feel, and specific gravity of aluminum as a "generic" material. In the way that the box exhibits this "interference between art and nature," we can say that the Focusrite Red preamp constitutes a pinnacle of technoaesthetics.[10]

Bakelite developed out of a nineteenth-century search for synthetic alternatives to ivory—some being known in Germany at the time as *Kunsthorn*. The eighty-year history of Bakelite was significant enough to technology historian Wiebe Bijker to comprise his main case study for his theory of a *technological frame*.[11] While we do not employ his theory in this book, he did compellingly show how the inventions of chemists only became widely adopted when they could reach multiple "relevant social groups." So which groups are these, and why are bakelite knobs used in some of the most expensive new boutique gear being manufactured today? Leo Baekeland first built upon experiments with phenol-formaldehydes to create photographic paper (Velox) before experimenting toward creating a synthetic shellac (Novolak). What we know today as bakelite is just one of several resins/plastics with that name—one that was first widely used during World War I, before being publicly marketed in the 1930s to manufacturers as supporting "attractive modern designs." Importantly, Baekeland was successful where others had not been in convincing a public that a plastic—instead of ivory, shellac, or wood—could be aesthetically pleasing. But today's aestheticization of bakelite appears to come from a different vantage point, from a contemporary nostalgia for those 1950s–1960s era products by RCA, Fairchild, EMI, Pulse Technologies, Western Electric, and Tektronix that did employ bakelite knobs.

As noted earlier, not all interfaces of contemporary gear take the form of faceplates and knobs; the PCB can be an important interface, too, especially for gear that is intended to be modded by end-users or that is sold as parts kits to be assembled by the consumer. Drip Electronics stands out in this regard. Founded by Gregory Lomayesva, a Santa Fe–based visual artist whose paintings and sculptures explore Hopi themes, Drip became known for "no expenses spared" PCBs that are beautiful artworks in themselves.[12] Their limited-edition release time frame underscores the art world association. Unlike the FR4 fiberglass PCBs ubiquitous in contemporary electronics, Drip's distinctively dyed boards are four times thicker and heavier, and their "3-ounce copper traces" and "shielded top layer solid copper" allegedly provide higher audio fidelity. Drip only sells PCBs (and accompanying instructions). But their materially massive and highly aestheticized PCBs have attracted a community of avid amateur builders, one that overlaps with computer overclockers, another gear culture willing to pay a premium for aestheticized computer motherboards.

THE VISUAL INTERFACE

Of the many memorable scenes in the 1984 mock rockumentary *This Is Spinal Tap*, the moment guitarist Nigel Tufnell points to the dials along the interface of his Marshall amplifier and declares, "The numbers all go to *eleven*. . . . Look, right across the board" stands out. "One louder" than everyone else's, Tufnell's signature amplifier draws attention to loudness by recourse to the visual aspects of interfacial elements: the sense of control he has relative to other players is projected onto panel inscriptions. While an example from guitar cultures (which have their own deep relationship to gear), similar engravings on gear fascia are sources of great pride, envy, and hilarity.

Throughout gear cultures, what the gear looks like is at least of equal importance to what it sounds like. When we consider early gear, say the 1940s EMI studio mixer, its black front panel featured a single VU meter and ivory knobs to set signal levels for each microphone input. Aside from "Main Volume Control," there were no other inscriptions on this panel or

Figure 5.2
EMI gear. Photo by Samantha Bennett.

the timber outer casing. Similarly, EMI's 1960s presence boxes featured gray or camouflage green panels with either one or two knobs. Either a single or trio of frequencies could be selected, and another knob boosted or attenuated the frequency. One EMI box's single frequency was labeled with Dymo tape and similarly stickered bantam input and output sockets (figure 5.2). As with Manley's Massive Passive, much of today's gear features many VU or LED meters, knobs, switches, and buttons extruding from colorful front panels featuring a variety of written text, information, labels, and even designer signatures. Unlike the Massive Passive, some excessive interfaces represent an erasure of engineering terminologies; gone are clearly denoted frequencies, bandwidths, and boosts, replaced with arbitrarily named controls with mystery scales. Much of this interfacial excess is unnecessary. Cluttered panels, while visually exciting, draw attention away from the sound of gear; they encourage visual engagement and exaggerate extra-audible gear elements.

Almost all contemporary rackmount gear has metering. Whether in the form of needles swinging across backlit numbered semicircles of dB indicators or rows of colorful LEDs blinking away, meters are a primary interfacial element and key visual depiction of signal presence and processing. Historical gear rarely featured any form of metering. Sound, being predominantly aural, did not necessitate much visual cueing until into the 1960s. The original Rupert Neve console featured two VU meters and a smaller meter for echo. Compare that to the modern Neve Genesys; the meters, strips of red LEDs positioned vertically atop each channel strip and master fader, can be switched to show channel input signal, monitor signal, cue mixes, effects returns, auxiliaries, and groups across stereo and 5.1 mixes. Meters can sometimes function as (yet another) means to insert women where they are absent. Rockrupel's Comp Two Compressor, for example, has images of a pinup on a swing inside each of its backlit VU meters.

As gear has developed, metering shifted away from representing the visual validation of audio signal and toward being an essential showcased interfacial element: *glow as aesthetic*. Since blue LEDs historically cost much more than the ubiquitous red, amber, and green ones, blue LEDs became especially fetishized (visible on most Manley gear). Creative positioning and LED color variety have moved beyond simple VU designs or red LED strips. For example, the Empirical Labs Distressor's "traffic light" strip of green, amber, and red LEDs denotes gain reduction. Multicolored LEDs are an interfacial norm: the more colorful the interface, the more "glow" potential when gear is powered and illuminated. The power of glow is such that the number of LEDs appearing on gear has dramatically increased.

Gear also features a host of written information and labeling. What is written on gear and why? To what extent does gear labeling align with engineering lexicon(s)? Labeling is a key aspect of gear identities, and in our examination of dozens of gear fascia, while we found some labeling conforming to standard signal processing terminologies, we also found plenty of examples of excessive inscriptions, arbitrary scales, personifications, hieroglyphs, and other unnecessary information. Gear labeling has come a long way since EMI's Dymo tape–covered presence boxes. One recent example is the Empirical Labs DocDerr. Designed by Dave Derr, this 500-series lunchbox-style channel strip is crammed so full of information it is difficult to know where to begin. Split into five sections, each one is jam-packed with LEDs, push buttons, scales, and labeling. Both the input and mix (output) level knobs feature a "1–10" scale, each labeled at single-digit intervals. The top input level section also features two high-pass filters, one labeled 70 (Hz) and the other 100 (Hz). "Push LF & MF to change HP," a blue inscription, lies between section one and two. The three middle sections, labeled "LF," "MF," and "HF," each feature a row of four traffic light LEDs and a dB scale (+-14 dB) labeled around the perimeter of each knob to denote boosting and attenuation. By pushing in the MF and HF buttons, the channel strip also functions as a compressor (although by the labeling alone, it is impossible to deduce this), with an 8-LED traffic light gain reduction meter flashing up and down the left side of the unit. The word "BAD!" is inscribed on the bottom Mix section, presumably to denote clipping at output stage. With a whopping twenty-four LEDs, five knobs, three buttons, and four logos—including the "DocDerr" play on designer Dave Derr's name—this is one of the most excessively inscribed interfaces we have encountered, and one that maximizes glow potential. While this interfacial excess might baffle, when we consider how knobs, LEDs, and ideas of technicity are the bread and butter of gear fetishization, this design perspective makes more sense.

The DocDerr is one in a long line of gear that features the name of the designer on the front panel. From AMS Neve and Rupert Neve Designs to Manley, Josephson to Terry, Neumann to GML, we see signatures-as-insignias all over gear interfaces. This functions as a means of distinguishing gear from regular audio technologies. Signatures on interfaces reinforce the gear object as intermediary between designer and end user; the signature reminds the user that in interacting with the object, they have some kind of connection to the human behind its design. Here, gear fetishization is rewarded with an albeit perceived proximity to renowned designers and recordists. We see some intriguing variations on this theme. For example, on their early gear including the RMX-16, AMS included the address and telephone number of the Burnley factory on the front fascia. The Joe Meek SC-2 optical compressor interface features two names: the signature of the designer Ted Fletcher, and Joe Meek (the recordist whose circuitry inspired the design). Fletcher said, "I knew at the time his name wouldn't be forgotten. He'd sort of live on—and it's true."[13] This immortalizing of Meek in gear at once consolidates recorded music history, perceived recordist agency, and gear mythology into a technological object. It also reinforces the posthumous canonization of Meek as technological innovator, drawing attention away from his behavior and crimes.[14]

In 2017, while organizing piles of old and discarded gear at the ANU School of Music for a long-overdue garage sale, Samantha and technical manager Matt Barnes found an Ursa Major Space Station. A diamond-in-the-rough, this early digital time-based effects processor was cast into gear obscurity until prominent mix engineer Chris Lord-Alge (CLA) revealed he liked to use it on his mix bus. After testing this gear, we decided it would be a quirky addition to the school's recording studio and patched it in, only for it to become a big hit among students. The Ursa Major's interface is one of the least intuitive we have encountered. With a white-on-black electronics diagram spanning the width of its large 3U fascia, the Space Station is adorned with ten bright red, blue, and black knobs, thirteen push buttons, a power button, a four-LED peak level meter, and a wealth of inscriptions. Most of these adhere to technical terminologies; the EQ functions, for example, feature attenuations between 0 and 10 dB, and the echo delay time knob is surrounded by stepped inscriptions in milliseconds. That said, the cluttered interface, chock full of controls and inscriptions, does match the "spaceship" aesthetic of its name. The Ursa Major, originally released in 1978, is a good example of space analogizing in studio gear cultures; the complex arrangement of labeled knobs and buttons lends itself to the utopian power and possibility of space travel.

While until the 1980s gear interfaces used clear labeling for their interfacial controls—for example, labeling frequencies with Hertz or cycles per second—much contemporary gear erases engineering terminology and replaces it with generic and broad-brush perceptive timbral descriptors. For example, the copper-and-black, art deco–inspired Kush Audio Clariphonic EQ lacks equalization-specific labeling. Instead, each of the two channels of this 1U unit features two large white potentiometers labeled "focus" and "clarity," against which nickel steel switches select between "open," "lift," "presence," "sheen," "shimmer," and "silk." This results-focused way of thinking about signal processing reinforces a quick-fix mentality: technologies themselves, rather than musicians, musicality, or recording and/or engineering expertise, will achieve your aesthetic results. In saying that, the Clariphonic EQ manual reveals all the technicalities and engineering explanations behind the labeling and an honest assessment of the philosophy behind the gear. Designer Gregory Scott proudly states, "The face of the Clariphonic—the name, front panel design, and bizarre switching matrix interface—were designed by yours truly, Gregory Scott aka 'ubk.'"[15] The Clariphonic appears to have been designed specifically for online gear culture participants—ubk was a regular poster on Gearslutz—where the most active users are gear enthusiasts who fetishize the gear itself more than its use.

Another feature of gear interfaces are logos, symbols, and hieroglyphs. On early BTR-2 tape machines, the EMI logo was pinned to the center, set in a military-style silver-winged design; the "British Tape Recorder" interfacial elements strongly allude to war aesthetics. This is no surprise, considering the origins of commercially manufactured tape recording in Nazi Germany propaganda efforts and the AEG/BASF *magnetophon*.[16] We see early examples of symbols-as-controls on Studer's 1964 J-37 multitrack tape machine. Here, the main panel of buttons features symbols of a quaver (play), left-facing arrows (rewind), right-facing arrows (fast-forward) and a three-line "wave" symbol (record). The only text-labeled button is "stop." Hieroglyphs come in handy to denote polar patterns on microphones. Instead of the words "cardioid" or "figure-of-eight," you find heart-shaped icons and infinity shapes of varying sizes. Consoles, microphones, and outboard gear use equalization hieroglyphs, too: what looks like flipped "tick" symbols and sideways-positioned tuning forks are icons that denote low- and high-pass filters and shelving equalizers. Using hieroglyphs saves space on gear interfaces, but the distilling of terminology into signs and symbols also reinforces the concealment of tacit knowledge (see chapter 13).

A large facet of gear fetishization revolves around color. Most gear continues to be gray, navy, black, green, and/or silver. This muted color palette reflects strong ties to military history and engineering traditions and is the aesthetic face of gear and gear cultures' conservatism. Since much brand-new gear conforms to this color palette, gear that deviates *does* stand out. We also see parallels between "exceptional" gear colors and those of classic cars. Doug Fearn's VT-1 and VT-2 tube preamplifiers, which retail at $3,400 and $5,900, respectively, and have a striking "Mustang red" color scheme, are some of the most fetishized in gear cultures. A mainstay at trade shows and guaranteed provoker of online gassing, Fearn's handmade, point-to-point gear is instantly recognizable by its gold inscriptions (including a vacuum tube insignia) juxtaposed against a deep, classic automotive red. Complete with a single amber incandescent power indicator light (when no "cheap" LED will do), Fearn's gear is often referred to as "classic" and "stylish": this gear ticks just about every fetish box.

After deep reds—Focusrite carved out a significant portion of the gear market with their Red and Scarlett series of preamplifiers and interfaces—ivories and variations of "British racing green" are the next most common colors; these have the potential to make a rack of gear "pop" (see, for example, outboard gear by Burl Audio and Daking). Samantha asked Ted Fletcher if this kind of differentiation was behind the choice of green for the Joe Meek SC-2 and was surprised at the response:

> I jumped in the car and . . . going into Newton Abbott there's a car accessory shop. And they were just about to close. And I walked in there and I said "I need some paint that will dry overnight, but it's got to be really tough and good paint and look nice." And the guy looked . . . he was completely disinterested and reached up behind him without looking, grabbed a can, plonked it on the counter in front of me. It was a tin of Hammerite green.[17]

The green is, in fact, "Buckingham Green," a mainstay Hammerite metal paint with both regional and regal connotations. The color anchors the SC-2 in England to aristocracy and to heritage aesthetics, regardless of the spontaneous nature of its coloring.

At BeesNeez, designer Ben Sneesby's method of coloring microphones is more personal. "Colours of the microphones! Well, we'll start with the Judas, which is a yellow-type colour. That colour actually came from our dog. His name was Judas and he was a really loyal Labrador. That was actually a Dulux colour." He went on to describe how the company worked to match the off-white of the Telefunken ELAM-251 microphone. Since the exact color was unavailable, Sneesby persevered with a specialist manufacturer: "It was custom made, we had to buy a really big batch of it because it's not the type of colour they have on the shelf." Not only Sneesby's microphones are named after his family members, the colors are personalized, too. "We have a microphone called the Oliver. So, olive green is where that colour [comes from]. Oliver is the son of a very good friend of ours who also works with us."[18]

THE HAPTIC INTERFACE

All the material interface surfaces we have discussed can be touched and, when touched, produce a sensation in the user's body. Adjustable controls, however, are the most important sites for this touch-sensation action-reaction, since they affect whether a parameter or operation is accurately adjusted and whether the user can count on their sense of touch to *know* whether the adjustment was accurate. To understand why controls have the feel that they do today, we need to trace the modern science of touch and haptics and its application to military instrumentation. But in doing so, we will go beyond the basic needs of controlling technical instrumentation and explore why controls became the site of play.

Starting with Ernst Weber's 1820 research into the psychophysiology of touch, which entailed practical experiments on and measurements of human skin, scientists have attempted to quantify the sense of touch in what David Parisi has described as a "rationalized tactility."[19] Remnants of Weber's experimental psychological method continued to be important through the 1950s in military research into instrumentation controls, which contributed to establishing potentiometer properties and knob morphologies deemed ideal for test subjects to be able to use controls for critical tasks. As Parisi noted, "Discriminatory capacities could be honed through repeated use," especially in contexts where the "'overburdened' optical and aural channels are relieved by flooding the multiple tactile channels with torrents of sensation users must be trained to interpret as data."[20] The modern concept of haptics is concerned with these "torrents of sensation" that users must "interpret" and, more broadly, pertains to the sensations of pressure, vibration, and temperature that follow the act of touching: the knob or fader or button *touching the user*.

Through archival research into potentiometers and variable resistors, we traced the history of the electrical knob in the realm of late-nineteenth-century electrical measurement technologies,[21] World War II—era weapons systems and airplane cockpits, electric power plant instrumentation, and radio station equipment. During that essential pre-musical development phase, certain normative assumptions were made concerning which bodies would interact with controls, for what purposes, and how the effectiveness of a control should be assessed. Certain phenomena such as the ease of turning a knob are far from accidental. Instead, they result from engineering psychology research concerned with quantifying parameters pertaining to accuracy and reliability—which determined the ideal knob friction (100 grams) to minimize operation times, the ideal knob width (one to three inches), and the ideal spacing of measurement marks in relation to reading time.[22] Analyzing the demographics of such studies in the 1940s–1960s, we find something related to what disabilities studies scholar Sarah Jain has noted in other domains: these researchers almost exclusively relied upon healthy male participants of normal stature, meaning that the size, look, friction, and feel of controls is directly linked to one particular construction of idealized, "average" human bodies.[23]

This led to what are termed "mil-spec" potentiometers and rotary switches, originally developed for military instrumentation purposes but then trickled down into "civilian" parts made by Bourns—designs that were subsequently copied by numerous other firms, including the Swiss firm Elma and the Japanese firm Alps. During this period, the potential kinds of interface controls reduced to a few standard elements that can be applied across numerous domains: knobs, buttons, switches, faders, and, to a lesser extent, joysticks. The interfacial components developed after World War II and in the early Cold War era by Bourns, Elma, Alps, and Grayhill are still the standards for audio equipment in the twenty-first century.

However, the instrumentation and controls of twentieth-century electrical technologies serve contradictory functions. As Rachel Plotnick has shown in the case of push buttons, button-pushing was associated with "entitled, lazy and forceful" habits of people; technologies using these were sold in America as "perfectly responsive servants" that serve as "performers of work of low social status."[24] Some of the pleasure and play associations with button-pushing are imbricated in the servile relation of technologies to their users. However, as the morphology of the button changed in the twentieth century, and especially with the development of touchscreen interfaces that lack a protruding button that requires force to use, she notes that "calls to restore forcefulness to touchscreens, or to bring back physical buttons altogether, ask not only for a different technical, more intuitive user experience but also for a more affective and sensual one."[25]

Recordists and audio engineers alike frequently discuss the nuances of interfacial controls: how pleasurable the analog knobs, and how horrible the digital controllers are that emulate them. Much noted are the feel and the satisfying "clunk" that happens when turning an Elka or Grayhill rotary switch—a haptic feel combined with the sound of the switch turning. A Penny & Giles console fader smoothly adjusts the amplitude of a channel, the user feels that they can confidently stop the fader's motion when they want. This relates to the ballistics of both faders and rotary potentiometers, which results from the kind and amount of grease that is stuffed inside the sealed control element. Users describe having confidence in an oversized toggle switch that switches a unit between two or three discrete states, or in a back-illuminated push button that provides aural and visual feedback that it has been engaged but requires sufficient force that it cannot accidentally be engaged.

All this costs money. Most of the cost of gear, besides the case, goes into the potentiometers, rotary switches, toggles, buttons, and knob coverings. A potentiometer may cost $0.25 or $25.00; the most desired rotary switches cost $80, while the same functionality could be implemented with push buttons and a simple logic IC for less than $1.00. A mass-produced plastic or aluminum knob covering for any of these should cost less than $1.00, but vintage and oversized Bakelite knobs are $10 each. The justification for using the most expensive and esoteric options is not simply reducible to any measure of accuracy or reliability. One pays for the unique haptic experience:

an experience that begins with touch, yes, but is correlated with visual and aural aspects too—a multisensory aesthetic experience.

The scientific rationalization of touch, then, led to touch's associations with servile electrical technologies that replace human servants, to the experience of using a tactile interface being aestheticized on its own accord as a site for affect and sensuality. This leads us to the way we are conceptualizing haptics in this book as the "capacity to affect or 'touch' us."[26] Whereas touch tends to imply one sense isolated from a person's broader sensorium, haptics "does not oppose the eyes with the hands, but acknowledges the sensory interdependence of the whole haptic (hand-eye-motion) system."[27] To this, we add the ear.

THE AUDIBLE INTERFACE

We are putting audibility and listening last of the sensory modalities, not because the sound of gear *should* be the least important concern, but because in many cases, interface choices have no significant effect on musical sound. There are devices with notoriously complex/crowded interfaces where the controls are never touched and one configuration serves as an engineer's so-called secret weapon, and others where the sole knob is performed like an instrument. Whether an engineer chooses to present a linear fader, a rotary potentiometer, or a continuous digital encoder to a user for a particular parameter often has an insignificant effect on musical sound—but altered haptics *may* considerably alter user perceptions of their enjoyment of using that control.

However, we can make some broader points about the relations between interfacial controls, audible phenomena, and the delineation of specific vocations. For example, one key design consideration concerns the ratio between the amount of user effort to change a control and the extent to which the signal/sound actually changes. For a stepped mastering EQ, where the selling feature of the device is the precision and the subtlety of the effect of knob adjustments, moving a rotary switch a single step, corresponding to 0.5 dB, will have an effect so subtle that it's probably more of a placebo than anything. But pressing the ratio button until it's at the "nuke" setting on the Empirical Labs Distressor (like engaging the "thrust" button of the API 2500 or the "crush" button of the Smart Research C2) is supposed to make a massive change in sound—adding significant harmonic distortion—with practically no effort, reminiscent of Plotnick's "power buttons." Due to the widespread perception amongst people who are not recordists that knob-turning encapsulates the entirety of audio arts, mixing engineers in several cultural contexts have installed "producer knobs" for studio guests—knobs that do nothing at all but become the sites for theatrical performances that allow the studio guest to make-believe that they are actually "doing" engineering and also to appease interfering musicians.[28]

These interface–vocation relations inflect broader discourses around key parts of the recording practice and expectations about how practice relates to the production of different kinds of recording subjects. The stepped control that makes adjustments so subtle they cross the placebo threshold is an essential interfacial element for the construction of twenty-first-century mastering engineering as a discipline of self-control, restraint, and subtle discernment. The drastic power button–style control is an essential interfacial element for the construction of twenty-first-century mixing engineering as a discipline of bold, dramatic, and risky choices. The disconnected "producer knob" demonstrates how producers (e.g., executive producers who run record labels, film/TV directors, or A&R representatives) who have been outside of the recording process perceive their need to exert control, and their overattributing power to the turning of that knob.

THE PLAYFUL INTERFACE

After several hours walking and talking the Javits Centre exhibition floor at the 2019 AES Convention, Eliot and Samantha have just about seen and heard enough gear to last their lifetimes. In one far corner of the floor, deep in the boomy rumble and shining out from the hoards of navy, gray, and black-boxed racks of gear, beams a bright white façade. Samantha is immediately drawn to it; we both head over to the booth. As we draw closer, primary colors pop from the much larger-than-average 5U unit, momentarily obscured by the animated gesticulations of a long-haired guy in glasses. The

Figure 5.3
Marshall Terry at his booth, AES 2019. Photo by Eliot Bates.

small-in-size but big-on-enthusiasm booth is manned by Marshall Terry, chief technical engineer at Shadow Hills Industries (figure 5.3). He proudly introduces us to his oversized white unit: the Terry Audio CEQ, a six-band stereo Pultec-inspired mixing and mastering equalizer complete with hand-wound transformers and based on Western Electric circuitry. Terry immediately agreed to an interview.

Despite its heritage influences, the CEQ is unique. While also available in black, the white display unit unashamedly undermines the militaristic-looking gear elsewhere in the atrium and appears implicitly *anti*-black box. Bright and colorful, refreshing and innovative, twenty-six Elma knobs and four switches adorn its front panel. Aside from the company insignia, "Terry," no text is on the face: no named frequencies, no boost/cut or Q scales. Instead, the broad fascia is covered with primary-color hieroglyphs of common equalizer parameters and dots depicting rotary switch steps. Terry cannot contain his excitement at the opportunity to explain the rationale behind the interface layout and color choice. "The colors are a way that we all see our equipment and bands and things," he says, justifying the lack of labeling since he wants us to "not use our eyes so much." The CEQ has a strong ludic influence: the inscribed primary-color dots and dashes appear childlike and playful, as if improvised by a toddler with large paintbrushes on a paper roll easel. The CEQ is "light-hearted, colorful, playful as well," beams Terry, as he proudly tells us he is a father and a children's music DJ: "I play music for kids under five, two days a week. I've done that for a while." Like women, children and parenthood are erased from gear cultures. Young children (and wives and girlfriends) are commonly derided as the causes that prevent hardworking dads from spending their hard-earned cash on gear. Yet Terry is up-front about his fatherhood and the obvious ways it influences his gear design.

Terry states that synesthesia, as well as influential recordings, led him to the color choices. "Even if we don't necessarily have *synesthesia,* as a major thing, there is a connection between color and vividity and certain bands of the audio spectrum," he said. Straight away, this combination of ludic design and color engraved in the interface starts to make sense; the blue, orange, yellow, red, and green markings are organized in such a way that they fill the CEQ interface with a rainbow-like visage. However, just when we thought Terry had created the most un-gear-like gear we had seen, he revealed the inspiration behind the individual color markings: "Green, to me also has a correlation to the American Army and Altec sounds of the 1960s and 70s, and the Vietnam War, which was punch and drive from Credence Clearwater Revival, so I have my own personal tastes for that that I put in for the green bass controls." Hence, the green bass markings embody the same militaristic and classic recording era values we have discussed already. "When I think of treble, I think of brassy cymbals. Brass is a yellowish kind of amber color. For the very high treble shelf and cuts, very high treble, when we think of *air*, I think of sky—that's why it's a light baby blue too, like baby blue is of sky. I'm also a big John Lennon fan, so *Imagine* is such a wonderful album cover." Terry's "baby blue" emphasis distinguishes the CEQ from the mass of navy blue gear. We are, however, only one step removed from a Beatles reference, thus reinforcing the CEQ, regardless of its appearance, as holding the same canonistic values as other gear. Terry continued in this vein, "The red cut control, I don't know, I literally felt cut, and . . . when I thought of cut I felt of blood. Affecting the midrange, we're really in a bit of the *heart* and *soul* of the music, and when I think of heart—my goodness, I think of the intro to *Dark Side of the Moon*." Terry delights in rolling out the connections between music, color, and gear, and, as the interview progresses, we realize how much of his personality is baked into this gear design. Terry's influences are in one sense conservative, yet he juxtaposes and somewhat undermines his commitment to traditional gear values with comments on social justice:

> Also, with the colors and the design and having it be white and having it not look that way [of most gear] is also how I think things feel now. Like let's use our gut, let's have fun, let's be loose, we're having more and more progressive social change, here in the world especially, the United States is getting there we're doing it more and more, and like part of that is relaxing, and stop being so critical, and coming from . . . our hearts and our chests and stuff, you know.[29]

The CEQ interface appears less engraved than decorated, as if the panel was a more artistic than practical endeavor. "I found [paint] stock at model train hobby stores, cos they love their colors too," Terry went on to say.

The CEQ exudes ludicity. Dozens of knobs and brightly colored engravings elicit an immediate play response that magnetically pulls gear user fingers toward them.

"Playing with toys," a ubiquitous analogy in gear cultures, is used to describe all manner of gear interactions (especially in online discussion) and is even used to name manufacturers, workplaces, gear retailers (such as Audio Toyshop), and gear. Recognizing the "dominance of play in interaction with technologies" since the turn of the millennium, Margarete Jahrmann noted how play constitutes such an intrinsic role in contemporary life that technological objects are built around ludicity. As such, a "new categorization of play" now exists, which consists "of the social qualities and political inscriptions of technological objects, interplayed with materiality."[30] Roger Moseley, writing about the piano's ludic potential, noted how the keyboard acts as "a field of play on which musical epistemologies have been allegorized, tested, and challenged."[31] By the end of our hour-long interview with Terry, a small group of intrigued gear enthusiasts have formed around us and the CEQ. Before turning his attention to prospective buyers, Terry offers us a final word: "I think being *light-hearted* and not *heavy*, on a variety of topics, is something that has been a part of social justice change, as a person and people. I think that's where inclusion happens, and that's really big for me."

CONCLUSION

Military aesthetics, classic cars, supermodels, and even Labrador retrievers have all informed the look of gear. The arrangements of interfacial elements—knobs, switches, buttons, faders—coupled with labeling and hieroglyphs, contribute to gear's visual, haptic, and ludic potential. The gear's interface—its layout and content, color scheme and inscriptions—exceeds its aural capacity or use value: how the gear looks and feels is essential for its social potential, and the multifaceted interface is the focus of its fetish appeal. To some extent, this explains why gear is so often analogized to the female form, personified "looks" and sex appeal.[32] So much of the meaning of gear's look is, however, constructed in the imaginations of gear culture participants, including designers. From the embodiment of the US Army to John Lennon's *Imagine*, from Ford Mustangs to royal palaces, otherworldly imaginings are projected onto gear interfaces. Gassers buy into precisely these imaginings: what it feels like to touch, to be touched by, to look at, to hear, and to *play* with the gear. But interfacing has taken two divergent paths since the 1980s. On the one hand, much recent gear is adorned with excessive and unnecessary content that replaces engineering terminologies with vague timbral descriptors. Who needs to know about bandwidth, filters, and shelves when you can add silk, shimmer, and sheen? On the other hand, interfacing is no longer just about a user's contact with a black box: the gear user's intra-action with technologies can take one of many forms, including interfacing with specific components, interfacing with computation, and interfacing with a space of activity specific to a historical era in music. The overall interfacial variety, though, works effectively in aggregate to provide fuel for the insatiable gassing of gear culture participants.

III STAGING GEAR

To some extent, the uses, meanings, aesthetics, and musical valences of gear are *designed into* technological objects—perceptible to people through the extracted and refined materials that make up the exterior and interior, and the kinds of (often excessive) interfacing presented to prospective users. However, so much of the fetishistic, technostalgic, and obdurant qualities associated with gear today exceed the materials and design; nothing about the materials and design in themselves would presume they would become central to social formations.

So where are meanings, uses, and attitudes toward gear inscribed and performed, *how* does gear become social, and *where* do gear cultures gather? Since 1995, technological objects have had the potential to be transformed into gear when staged within three media/milieux: at trade show events, in gear-focused magazines, and (from 2001) on numerous online message forums. One generic trajectory: a gear prototype is displayed at a trade show to recordist industry professionals, which leads to an early retail release being positively reviewed in a magazine; both the trade show reception and magazine reviews spill over into online spaces where gear gets gassed to the point that the object attains fetish status. There is, of course, more to it. Trade shows stage much more than just new gadget prototypes, and online gear fora discussions far exceed factual discussion of objects' technical properties and potential use scenarios.

Trade shows, magazines, and message fora also host different kinds of social formations, and membership only somewhat overlaps. The gear cultures of trade shows extend outward from those professional relationships that connect gear designers, manufacturers, parts suppliers, wholesalers, retailers, and technology media: these are tangible social networks but consist of weak ties that only get sporadically reactivated. The gear cultures of magazines consist of a periodically activated, imagined community of reader-subscribers who never actually interact with each other, with an illusion of a more palpable community insinuated through the small assortment of printed letters to the editor. These cultures are built around discourses originating from specific editorial stances. Only on online message fora do we find durable, nonseasonal gear cultures containing hundreds of thousands of registered users, considerably more lurkers, and few geographical boundaries. Despite the differences between these three—seasonal and temporary in-person social networks, imagined communities of anonymous readers, durable "always online" communities of pseudonymous users—all are milieux that, like gear manufacturers and mining operations, have depended upon the erasure of women's behind-the-scenes labor but that maintain their own hegemonic masculinities. In doing so, these gear cultures exclude women and non-binary people. Through analyzing these divergent kinds of social formations within events, print media, and online fora, we will learn more about how gear is called upon to structure social relations and how gear-centric social relations produce gear cultures.

6 EVENT

THAT TIME WE WENT TO NAMM AND MET FOXY STARDUST

Eliot and Samantha are in a queue at the most elaborately staged gear "booth" we have ever witnessed, except we haven't actually seen anything yet: we are waiting to be *let in*. To what, we don't yet know. Barrier ropes double back in a snake-like airport check-in system as the "entrance" to the booth is concealed. Here, we both feel as if we are waiting to board a blacked-out indoor rollercoaster, that we are in a cinematic conflation of *ET*, *Apollo 13*, and *The Martian*, and that something very exciting is about to happen. We are participating in constructed anticipation, held in suspense. As we wait, various signs keep us amused, in line with what Peter Blesgraaf noted about the design of queuing systems at theme parks: "the inclusion of entertainment for people in the queue, such as decoration, interactive elements, and active entertainment with actors."[1] It's all there: "Please turn your mobile devices to spaceship mode," "Prepare for enriched sonics when traveling through the onboard magnetic field," and another, *Spinal Tap*-esque notice, "Experience analog sound at the speed of light." At the entrance, a large video screen beams down a woman dressed in a space suit, apparently in a spaceship, explaining what it is we've all been waiting for: the launch of the Universal Audio LUNA DAW. The "broadcast" cuts out with intermittent white noise emulating signal interference. As we reach the end of the queue, a UAD rep dressed head-to-toe in a white space suit adorned with NASA-like insignias greets us: "Hello! My name is Major Tom!" he enthuses, before introducing a video projected in widescreen onto a wall. At only a couple of minutes long, the video is a semiotician's dream. Super close-up images of a Studer analog tape machine path locking a 2" reel to the heads and the sound of its "play" button being pressed follow extreme close-up images of an EMI TG-series console. We zoom in on Neve 1073 preamplifiers, controls for the oscillators, filter sections, and envelopes on a Moog synthesizer before resting on a super close-up image of a Lexicon 224 and its faders, red metering, and buttons. Scant technical information accompanies these glossy images; the video is simply comprised of extreme close-ups of analog equipment—and one heritagized early digital signal processor. Who needs the sound, technical specifications, or purpose when you can just *look* at gear? Old gear. *Analog* gear.

Around the corner, we are finally in the booth. Another video is projected onto the wall. We need to grab a set of AKG headphones and plug them into the station. There is no time to do this, watch the video, and then take the headphones out before someone in a NASA-style outfit hurries us along to the next part of the exhibit. And that's exactly what this is: an exhibit of gear as though it is already an artwork or a museum piece. "This is going to be an audio transmission with media," states a UAD spaceman rep. We move on to the next exhibit (figure 6.1). This time, glass cabinets featuring all versions of the Apollo interface adorn the room, positioned in freestanding plexiglass cabinets as if jewelry (see chapter 13). Again, we take our headphones and plug them into a station, listening to information about brand-new gear spoken of in the past tense—as though we are from space in the future, looking at a product designed on Earth in the past. Another listening station. This time, we hear audio recorded with and without the Studer A-800, which does indeed sound impressively Studer-y as UAD has successfully modeled tape compression, and with and without the Neve BCM-10 console. The Neve does not quite convince us as much as the Studer, but we move on. Around another corner, a gigantic video is surrounded by four Quantegy 2" tape reels in each corner with the word "warmth" hanging above it, spelled out Tracey Emin–style in bright red neon light. Next, we listen to a Moog plugin on the LUNA. It sounds like a Moog. The "Ravel" grand piano sounds like a very good soft piano, modeled on a Steinway.

Figure 6.1
UAD Apollo museum display, NAMM 2020. Photo by Eliot Bates.

Finally, we reach the end of this exhibit. As we turn a corner, a woman dressed in a silver spacesuit, blonde wig, and alien-style sunglasses greets us. She introduces herself as Foxy Stardust, and Samantha poses with her for a photo (figure 6.2). We are now in the main exhibit room, completely concealed from the rest of the NAMM floor. The low-end cacophony still envelops it as we attempt to engage with a UAD rep—yes, still dressed up in a NASA space suit—who explains the LUNA to us. After all that analog imagery and plexiglass vitrines, and listening to old gear dressed up as new, we arrive at the product: a basic DAW GUI with an edit and mix window similar to Pro Tools or Logic. After a couple of technical questions about functionality, the rep realizes LUNA perhaps is not for us. "If you're happy with Pro Tools, stick to Pro Tools. We're not trying to convert people," he says as he signs off on our "LUNA Test Pilot Flight Plan": an A4 sheet with various DAW instructions presented in the style of a classified government document. He stamps it with an insignia—"LUNA Universal Audio Mission Control"—before sending us off to a bar to collect a LUNA T-shirt, LUNA patches, and a couple of neon orange shot glasses.

Figure 6.2
Samantha with Foxy Stardust. Photo by Eliot Bates.

GEAR EVENTS

New gadgets are invented every day. But products that manufacturers are unable to bring to market, or which are unclearly framed, will not circulate and substantively change music and engineering practices. What *kind* of thing is it, how might it be useful, and how should people *feel* about it? What was LUNA? Was it the future of "the analog"? Was it the promise of music in worlds yet to be explored, or a nostalgia for the utopian naïveté of the 1970s space race? Or was it just another DAW, a rehashing of technologies we've already had for decades? If the LUNA product launch were to succeed, the last question would ring less prescient for most booth attendees, and the new, untested digital product would have a chance of captivating their interest. As we saw, Universal Audio based the entire aesthetic, marketing, and promotion of the DAW on its affiliations to 1960s–1970s analog hardware technologies and the canonized albums made with them, all to sell a mid-twenty-first-century piece of digital editing/mixing software to a generation of musicians born two or more decades later. Even launching a software platform that will never itself be regarded as gear relied upon gear and the widely held values within its cultures.

The AES convention, at least since the 1950s, has been a principal staging ground for prototypes and new products, and a showcase for audio and technology research. Harald Bode introduced what many regard as the "first" modular synthesizer at the 1960 AES show; at the time, it was unclear whether the synthesizer was a musical instrument or if it was a new kind of studio gear.[2] Inspired by this, a young Bob Moog soon after launched the first commercial line of related products. But the show is not only a place for showing products, and many pre-prototype developments in audio are discussed in the adjacent technical program. The large attendance at AES (15,000+ annually) and even larger attendance at the music instrument and audio technology trade shows NAMM (45,000–115,000 annually) and Musikmesse Frankfurt (50,000+) make them essential sites for the staging of gear. We regard each of these trade show floors as possessing its own gear culture, since each show is organized around technical objects that structure most of the human social interaction that unfolds there, and since there are durable social relations within each trade show that might not extend to other sites. As we found, many people in the conjoined recorded music and audio equipment industries have long-distance friendships that are reanimated at the annual trade show.

Beyond matters of technological categorization or function, trade shows are places where technologies are concretized, where meanings/uses of and attitudes toward technologies are first made durable or inscribed, and where social bonds are reaffirmed.[3] However, this does not just happen through conversations. Trade show floors are deafeningly loud (with an ambient noise floor exceeding 85 dB), entail constant collisions between booths and people and technical objects, are visually and aurally overstimulating, and reek of attendee bodies. The floors, flat and concrete, feel heavy underneath the feet (it doesn't take long before the aches set in); some gear manufacturers attempt to keep gear enthusiasts longer at booths by rolling out relief-inducing fluffy carpets. The relentless sensory overload operates simultaneously at different time scales: the extended synesthetic instant of the actual demo, the rhythm of the product pitch, the suspended time of the unstructured visitor walk-through, the pervasive overstimulation and temporal disorientation that characterizes everyone's overall trade show experience, and the informal anticipation of—and reflection on—the trade show that spans months in online forums and at exhibitor headquarters. Our analysis will elucidate points within this spatial, temporal, and multisensory cacophony where gadgets become gear.

A product as basic as a cable *could* be regarded as just some wire with insulation around it, but in the context of the 2020 NAMM show, the Whirlwind cable booth featured a poster with a woman's eye, two meters tall, with blue eyeshadow and red mascara gazing down at you, multicolored lighting in the tubes that seduced the attendee toward different makes of cables, and young booth-worker women ready to provide product brochures (figure 6.3). It worked: way more attendees perused the lit tubes of cables than the more straightforward product presentations of competing companies. Just a couple years previously at the 2017 AES convention, Redco, a Connecticut-based maker of affordable custom cables and reseller of more expensive bulk supplies, apparently encouraged one of their employees to create an artwork, "Cable Gal," a mannequin wearing a skimpy beaded dress made of several multicolored XLR cable snakes (figure 6.4). Cable Gal "greeted" visitors to the Redco booth, despite being unable to speak about the products.[4]

For cables to become "sexy," they apparently need to be associated in multiple ways with women's bodies—whether the presence of saleswomen, as a seductively lit display being gazed upon by an all-seeing gargantuan female eye, or in the form of clothing that fits the female form. Redco's Cable Gal had an additional "implied identity" by allusion to her "ethnic" braids and cable-woven patterned dress. While these tactics are not found at all booths, "booth babes," most of whom are migrant workers of East Asian and Eastern European citizenship,[5] were a fixture of AES in the 1990s and early 2000s and still are hired to perform such roles at NAMM. This continues to cause confusion for some attendees; plugin designer and gear manufacturer Steven Slate took to the online forum Gearslutz during NAMM2020 to decry the sexual harassment of one of his female product presenters, who is a full

Figure 6.3
Whirlwind Cable Eye, NAMM2020. Photo by Eliot Bates.

Figure 6.4
Redco Cable Gal, AES 2017. Photo by Eliot Bates.

employee and user/presenter of Slate's technologies—in other words, who was *not* at NAMM to serve as an objectified gendered body. Many product brochures over the years have featured white, cis-gendered, often tall, slim, and blonde women conforming to Western standards of "beauty," sometimes scantily clad, in proximity to but not actually using the technological objects. Chapter 10 begins with one of the most extreme examples of this we have ever encountered. However, in the context of NAMM or AES or Musikmesse, which typically require a professional affiliation to even attend, such optics are not intended so much for public consumption as they are for professional trade show attendees—including people working at other show booths. This multiyear attempt to make cables sexy may not have succeeded in convincing general consumers of this association, but at least within a network of audio professionals helped maintain the idea that *anything* related to audio—even its accessories, components, or materials that otherwise might be ignored—are potential objects of fetishization.

Collectively, we have attended music technology trade shows in various locations since 2004, including many of the large US-based shows, smaller regional shows in London, Paris, and Berlin, and one-off trade shows and gear events held in Istanbul and in the US and UK. In the sections that follow, we provide brief historical overviews of two major trade shows, AES and NAMM, before presenting select findings from our ethnographic fieldnotes taken at several of these events and at one-off gear-related events such as "80 Years of Recording at Abbey Road." We had intended to conduct ethnographic observations at Musikmesse-Frankfurt in 2020 and 2021, but these events were canceled due to the COVID-19 pandemic; as of the publication of this book, Musikmesse will no longer happen in its original form.[6] The focus of our ethnographic analysis is twofold: how gear is staged and the kinds of human socialization that trade shows enable. Since gear mediates social relations at trade shows, these two concerns are wholly entangled with each other.

AUDIO ENGINEERING SOCIETY (AES)

The Audio Engineering Society (AES) was founded in New York City in 1948 to represent the growing needs of trades that diverged from radio technology and its societies or from the acoustical societies that had become more narrowly focused on architectural acoustics.[7] The first AES convention in 1949 had 3,500 attendees, and the society expanded over the next two decades to include overseas chapters in Japan and Europe. The largest gathering continues to be the annual fall US convention (typically located in New York but sometimes in Los Angeles, San Francisco, or Las Vegas), but ever since the Cologne 1971 summer convention, the AES Europe convention has been similarly structured, albeit at a smaller scale. AES also hosts several committees: its standards committee is responsible for developing digital audio carriers such as AES/EBU and standards such as AES42 for digital microphones. AES has numerous regional chapters, student groups, and field-specific sections that organize meetups or focused events, which we will not discuss here; they are important in relation to specific audio trades but did not appear to inflect the gear cultures we are examining.[8]

As illustrated in chapter 1, the AES trade show takes over the main floor of the Javits Center in midtown New York City and is filled with gear: new gear, vintage gear, stages for showcasing gear in music-making contexts, gear-related T-shirts and tote bags, and workshops on gear use. AES is not *just* about gear, though, serving as a space for exhibiting less-fetishized technical objects such as acoustic panels, surface-mount components, mic stands, and hardware/software intended to support an increasing number of trades related to audio. The technical program is downstairs between a dozen conference rooms; these are not filled with physical gear, but many papers and posters invoke gear in multiple ways. Some of the in-between hallways and unmarked spaces are occasionally used for workshops and demos, including the microphone build session hosted by MicParts and the Audio Builder's Workshop (the main outreach activity of the AES Boston chapter). AES occupies the southern wing of Javits; in the northern wing, the NAB (National Association of Broadcasters) show caters to the broadcast video/lighting/sound equipment market. In between is a food court with mediocre and overpriced food but, crucially, the only space in the whole center with places to sit, which is routinely commandeered by folks like ourselves trying to get some relative quiet to conduct interviews, by audio education school groups comparing notes after a tiring stretch of trade show trawling, and by manufacturers trying to feel out potential business partnerships. The several days of AES/NAB is the nexus of a wider flurry of activity around professional equipment that spills over into nearby studios, bars, and restaurants.

While AES has come to promote their convention almost solely on the promise of gear, it strikes an uneasy balance between unabashed gear fest, engineering conference, and networking opportunity for its professional members. The uneasiness comes from the ways in which gear is called upon to socially mediate relations—even during networking. As one well-known microphone designer noted in an AES-adjacent event that Eliot attended in 2018, "Few of us probably did well socially in middle school," and AES was the first place he had ever been in his life where it was "ok to be a geek." Being a "geek," in this context, goes beyond a shared interest in technologies and their workings: it means that memories of social relations between people are materialized in specific objects. It also means that technologies may come to symbolize expectations or promises around social relations; the gear is a point of communication. Just seeing a circuit with some distinctive and unusual properties can trigger emotion-filled memories of trade show friendships past and of expert gear designers who are no longer with us. This phenomenon is not just audio-specific, as similar sentiments run through Bo Lojek's social-technological history of semiconductors.[9] However, it stands out in the context of a trade show and conference where recorded musical sound, and audio generally, continue to be framed as the privileged medium where emotions are conveyed between artists and audiences. Many contemporary gear designers such as Geoff Daking, Hutch Hutchison, and Robin Porter got their professional start as performers in local or touring bands, later switching focus to gear repair and then design. But at AES, it is gear that mediates professional/social relations and produces the individual subjectivity of some self-described "geeks."

It was a major challenge to conduct trade show interviews over the ambient noise floor (between 78 and 85 dB) and the hubbub coming from the MWTM (Mix with the Masters) stage.[10] Hubbub is not *our* term for it; most exhibitors we talked with were unhappy with the MWTM stage imposition, since it did little to generate any interest in new technologies, it glorified a previous generation of privileged major record-label audio engineers who are no longer representative of people working in audio trades, and it consumed a lot of space, energy, and attention. The stage was straightforward: the "master" was behind an SSL AWS console and manipulating a reconstructed Avid ProTools DAW session of a song they wanted to discuss, and the audience got to see the "master" mute and unmute parts of a mix of a mainstream rock/pop song where most signal processing had already been rendered to the tracks. The nearly $100,000 console was inessential for demonstration purposes but had a performative value in indicating that *masterful* mixing was to be anticipated in the space; after all, no one else at the Javits Center was behind a $100,000 SSL console at that moment. Additionally, the SSL console at MWTM functioned to link commercially successful recordings to expensive gear. The sound from the stage was certainly audible but competed with the ambient noise floor, meaning only bold mix changes were perceptible. The stage featured, in effect, forty-five-minute to one-hour-long "taster sessions" for what people might expect if they were lucky enough to attend the week-long sessions in the south of France at La Fabrique studios—lucky to be one of the "chosen few" who had been "carefully selected" to attend, and "lucky" to have an extra €4000 (plus airfare and food) to spend. For the first five years of MWTM seminars, only one of the forty "masters," Young Guru, was a Black producer, and as of 2022, Sylvia Massy has been the only woman "master" to run a week-long seminar.

One such "master" taster session was run by mix engineer and Los Angeles–based celebrity Chris Lord-Alge. Known in gear cultures as an acronym, CLA took over the MWTM booth at AES 2019 to talk us through mixing a live album for US pop star P!nk. The session alternated between a deafeningly loud bombardment of sound and video clips through enormous Barefoot monitors as CLA performed exaggerated movements on the SSL console and gave a hard MWTM sell. "Come and spend a week in France with CLA!" bellowed CLA, referencing himself in the third person throughout the whole session. Samantha attended his taster session with colleagues and was instantly reminded of another gear event: a demonstration of the SSL K Series console at Peter Gabriel's Real World Studios in Bath in 2012, also featuring CLA. In both gear events, CLA used the same time-tested formula. His extroverted, engaging, and charismatic persona shines through as he gleefully sells his own agency packaged up as mix retreats, gear endorsements, and even signature plugins. Boisterous, loud, bombastic, and egotistical, CLA commands the audience's attention but always pauses in his sessions to lower his voice before either speaking about or video-calling his mother, who beams with pride as CLA shows her his adoring audience and tells her of his latest achievements. Peppering his talks with nostalgic stories of 1980s and 1990s recording and production excesses, CLA effortlessly leverages his professional stature to canonize gear (in both cases, an SSL console). As one of the most recognizable names in gear cultures, he has become an expert in packaging technological agency for sale to gear culture participants—through a heavy dose of nostalgia. Gear does not, however, always play ball in these scenarios. In the Real World Studios event, just as CLA had begun to swing his chair up and down the SSL console to performatively recreate a live mix of Green Day's "Holiday" in front of dozens of professional audio attendees (figure 6.5), smoke billowed from the back of the console. In rushed the in-house engineers with fire extinguishers, and the event concluded early.

Figure 6.5
CLA on the SSL console at Real World Studios, Bath, 2012. Photo by Samantha Bennett.

We recognize the tireless work undertaken by current AES President, Dr. Leslie Gaston-Bird, in her drive to diversify convention speakers and the organization as a whole. A good example of this is her securing legendary turntablist Grandmaster Flash (nee Joseph Saddler) as the 2019 AES keynote speaker. Grandmaster Flash helped popularize turntablism and sample-based music production, originally in hip-hop but with an influence extending to many subsequent styles of electronic music—electro, breakbeat, and all derivative genres.[11] He demonstrated his singular backspin technique for creating a continuous rhythm part out of 4-bar or 2-bar long sections of pre-recorded music on vinyl records and anchored his practice within the Bronx borough.[12] This unusual (for AES, not for popular music history events) lecture/demo was introduced by New York–based engineer Paul "Willie Green" Womack, who has also been involved for several years with trying to increase the representation of Black engineers, producers, and musicians within the milieu of AES. Unlike the typically sedate AES keynote talks, this time, there was an intensely emotional vibe in the room, as Gaston-Bird and Womack had facilitated AES entry for a significant number of aspiring young local audio professionals working in hip-hop. Grandmaster Flash was legendary, and the whole event was tightly organized and bursting with funky breaks.

More surprising than his appearance at AES was the fact that it took until 2019 for AES to recognize the contributions that hip-hop, electronic and dance music, let alone pioneering artist/producers such as Grandmaster Flash, have made to audio in general. This reflects the exclusionary nature of AES's activities that keep the space appealing to older white cis-gendered middle-class men, thereby perpetuating the historical exclusions of everyone else.[13] Black music recordists, musicians, and genres have had an immeasurably great influence on all recorded music, especially with regard to introducing new production styles and approaches towards using technology,[14] but electronic music and hip-hop have creatively and commercially flourished *despite* AES. Grandmaster Flash did record at some commercial studios, but his performative innovations, and those of the Belleville Three who helped create Detroit techno in the 1980s, happened more often outside multiroom studio environments. The repurposing of affordable technology that may have not been designed with creative production in mind (e.g., turntables,

Auto-Tune), or discovering the unique voices of synthesizers that had failed to catch on for their original use cases (e.g., Roland TR808 and SH101, Yamaha DX100), are only now starting to be properly recognized for the veritable technological innovations they are. AES has instead privileged the gear cultures of classical music engineering and major-label rock/pop production, instead of DIY/maker cultures, technology-hacking, dance music, or production undergrounds.

AES 2019 saw a broadening of the scope of audio-related accomplishments that deserve society-wide recognition; the recognition of Black music(s) and its associated technologies and production was long overdue. However, on the show floor and in a few panel sessions, we saw evidence of the broader audio industry's continuing fragmentation. The old rock studio/classical hall paradigm is increasingly at odds with fields such as immersive audio, audiovisual postproduction, game music/sound, podcasting and vlogging (a featured topic at this AES conference), virtual reality, foley, location nature-recording, AI audio, and musical information retrieval. Since these vocations all require technical skill and specialized equipment, this calls into question what an "audio engineer" is. The audio engineer is no longer a white middle-aged guy sitting at an SSL console (as in the case of MWTM), as music is mixed in many ways. The audio engineer includes not just the electrical engineers who work with analog circuits but also DSP coders and software engineers. An audio engineer may not even work in a music-related vocation at all. Gear at AES, while a huge draw, is in tension with the changing role of audio recordists and professional audio production domains.

Our fieldnotes from AES2017 also highlighted a contrast between the east wing of the trade show floor, where the plugin and software manufacturers demoed their technologies with club and electronic music examples, and the rest of the trade show, which appeared to celebrate classical music recording, classic rock aesthetics, and associated canons of the golden age of record production. The show spatially demarcated "front-end" analog, by which we mean devices used for recording such as microphones and preamplifiers, from digital mixing and audio processing. This had not been the case at AES shows in London that Samantha visited in 2010–2011, or at AES

Paris, which Eliot visited in 2016. While much software strives to emulate classic analog technologies, many novel approaches toward working with audio—for example, transient sculpting, intonation adjustment, loudness maximization through clipping, spectral repair—were also demonstrated. These technologies attracted a different demographic of trade show attendee: younger engineers/musicians with a more pronounced gender and racial/ethnic diversity. Perhaps for the first time ever at AES, there was a queue for the women's restroom—despite the disproportionate space still consumed by men and the legacy of endless analog.

NORTH AMERICAN MUSIC MERCHANTS (NAMM)

The NAMM (National Association of Music Merchants) began life in 1901 as the National Association of Piano Dealers of America, representing a burgeoning American industry. Since the early decades, it has attracted celebrities; Charlie Chaplin helped demonstrate piano sales techniques, and Thomas Edison attended in 1918–1919. From then, the association formally expanded to encompass all music instrument related trades, including manufacturing, retail, and instrument pedagogy. As Trevor Pinch has discussed, due to the entrenched dealer networks, when electronic instruments including organs and synthesizers came to market, they tended to be sold by piano dealers.[15] Hence, since at least the 1960s, NAMM has been a place where you would find electronics related to music. By the 1980s, NAMM had permanently located their main annual show in Southern California. With the increasing importance of home and project studio recording, and the relative convenience of the Anaheim Convention Center site for exhibitors from East Asia in comparison to AES's New York location, audio equipment exhibition at NAMM has significantly expanded. Today, a separate show floor houses more than a hundred pro audio booths, not counting similar displays scattered among the synthesizer and digital audio (DAWs, plugins, and related equipment) exhibition areas on the main show floor. The importance of audio to NAMM is perhaps best indicated by their establishment of the TEC Awards ceremony in 2011, a red-carpet affair that "recognizes the individuals, companies and technical innovations behind the sound of recordings, live performances, films, television, video games and multimedia."[16] NAMM's global report, which ranks US sales revenues in twenty-three product types representative of its membership, puts pro audio second (at $853 million) and microphones fifth (at $627 million)—considerably above pianos, which was NAMM's original concern. Gear is, therefore, a billion-dollar industry.

NAMM, as an industry consortium, is concerned primarily with protecting the interests of its members, which in practice means a fiscally conservative, pro-small business outlook aligned with larger consortia such as the National Association of Manufacturers. While this has included lobbying Congress to increase funding for music teaching in K–12 and supporting the RESTART Act to provide COVID-related relief to small businesses, it has also recently entailed opposition to the Lacey Act and CITES, which restrict trade in endangered and illegally harvested wood.[17] The 2020 "Global Report" by NAMM, which includes overviews and risk assessment profiles of several dozen import/export markets, positively notes about Canada the projected growth of their pipelines and mining operations while sounding a warning for the United Kingdom about "the growing need for companies to have, and market, their sustainability credentials."[18] However, unlike some other trade associations, NAMM does not appear to have provided resources for its member companies regarding what such sustainability credentials might look like. NAMM provides no guidance on how companies might attain carbon-neutral targets, has never provided clarity regarding conflict minerals and supply chain ethical issues, and has not even ventured into the typical greenwashing of corporate sustainability initiatives we find in computers and larger component manufacturers. As such, the industry remains unaccountable for its contribution to the environmental and human costs of the production of its goods.

A few weeks before the global COVID-19 pandemic, in January 2020, we and 115,000 others experienced the best-known public face of NAMM: the annual show at the Anaheim Convention Center. If AES's default was loud, then NAMM was "one louder." Despite the din, individual booths were a case study in contrasts. Some of the more innovative designers of analog circuits—for example, Grant Carpenter (designer of the Gordon mic preamps), Jakob Erland (designer of the Gyraf line of tube gear), and JZ Microphones (a Latvian company)—had spartan displays with few pieces of gear and appeared to be eager to talk about audio and gear design. Some of the clone manufacturers, however, had fancy and expensive booths but little to say about the specifics of the technologies they displayed. Oktava Microphones, a former Soviet state-owned business, floundered at political humor with the slogan "Make Oktava Great Again," just next to Heil Microphones, who had come up with dozens of custom variants of their handheld vocal mic, ranging from camouflage to an American flag to shocking pink.

NAMM is not just a place where gear is displayed, it is a place to see—and to be seen. The structure and layout of NAMM reinforces existing hierarchies and social classes of people involved with instruments and music technology retail. On our visit, we noticed a clear status differential between musicians performing on the larger and more featured stages, those on smaller stages struggling to keep a small audience's attention, the many artists hired to demo gear for various booths whose often virtuosic efforts were barely noticed by attendees, and aspiring musicians who had enough connections (school, work, friends, or family) to get a NAMM badge but no official role. Exhibiting at NAMM is expensive: at least $2,600 for the smallest 5'x10' foot booth, increasing to tens of thousands of dollars for larger spaces, not including the considerable expense some of the firms renting the larger spaces made to create custom displays and to hire musicians and booth workers. As is the case at AES, a whole sub-network is continually operating in parallel to the main schedule of events. We regularly heard about or were invited to "secret" and "private" parties held either in VIP-access-only areas of the show or offsite at hotels, restaurants, and LA-based businesses.

80 YEARS OF RECORDING AT ABBEY ROAD STUDIOS

Not all gear events happen at trade shows. "80 Years of Recording at Abbey Road Studios," a 2012 event, featured Brian Kehew and Kevin Ryan talking about their then newly released book *Recording the Beatles*. At the first of these talks, around one hundred people—including Samantha, one of the only female attendees, who attended with Evangelist Studios owner,

Lewis Durham—gathered in Abbey Road's Studio Two to listen to Kehew and Ryan's gear stories intertwined with well-trodden anecdotes of Beatles recordings, references to the studio "atmosphere," and tales of busy recordists, squirreling away behind closed doors making iconic records (figure 6.6). Here in Studio Two we sat, the odd ones out as young people in a sea of retirees and the odd female partner, on the "famous red chairs," listening to Kehew and Ryan's tall tales, all the while encircled, surrounded—*overwhelmed*, even—by Abbey Road's enormous gear collection.

The event was an intriguing blend of "behind-the-scenes" revelation, history lesson, nostalgia, recording workplace celebration, gear exhibition, and merchandising opportunity all wrapped up with affirmations that Abbey Road is the "best studio in the world."[19] Kehew and Ryan, perched on two stools, begin a PowerPoint presentation that takes us through the history of the studios and particular pieces of gear. Kehew and Ryan are also promoting their book *Recording the Beatles: The Studio Equipment and Techniques Used to Create their Classic Albums*, a five-kilo coffee table tome limited to one thousand copies. At over five hundred pages long, the book features extraordinary gear details, alongside hundreds of photographs and diagrams, painstakingly collected and methodically organized into an authoritative account of the history of Abbey Road's gear: its TG series consoles, presence boxes, microphones, and BTR tape machines. The book details "individual systems down to the very last ohm, breaking down each recording session to the final tape edit."[20] A labor of the love of gear, Kehew and Ryan's opus concentrates on gear made by EMI engineers for use in Abbey Road studios and on Beatles records, although the same gear would have been used to produce recordings by many other musicians and artists over an extended period. The overall tone of Kehew and Ryan's technical specification-oriented book is clear, factual, and detailed. Their talk, however, was punctuated with romanticism, employing liberal amounts of nostalgia, religious analogy, mythology, and secret revelations as they meander through Abbey Road's gear, recordings, and artist highlights. Beginning by stating we were in "a mini museum," Ryan went on to suggest that the chairs on which we sat just might have been chairs used by John, Paul, George, or Ringo. This is a good example of how gear events play on intangible mythology—the aura, ideas, and imaginaries

Figure 6.6
Lewis Durham in Studio Two. Photo by Samantha Bennett.

of gear and its affiliated personnel and workplaces—as a means of discoursing gear. But more is going on. This talk also taught attendees how to socialize around gear: how to speak about it, how to connect it to recorded music history, which engineers, artists, and recordings matter, what kinds of work the gear was used for, and in which studio the gear was used. The gear talk downplays the technical specifications detailed in the book and ramps up the technostalgia in a way that would leave even the least-savvy gear enthusiast hankering after an Altec compressor.

Following the talk, event attendees were invited to wander the perimeter of Studio Two and observe some of the gear. Each piece of gear in this "museum" is adorned with museum-like placards. A BT-3 tape machine—one of twenty made, and only three that survive—is described as "a special machine." The TG12345 Mk II console was "used on legendary albums." The EMI RM-1B ribbon microphone is described as "incredibly rare, only one of two in existence" and is attributed to Alan Blumlein, "pioneer of many recording advances." Rather than technical specifications, the museum placards situate gear through era descriptors, personnel, and historical recording contexts. Retrospective and sentimental, the gear was presented less as tools of the trade and more as representative of a specific era, locality, and sound-recording ideology. Bewildered by well-made gear, which was designed to be used in recording and production workflows presented as display objects, Durham concluded that gear did not belong in a museum.

At the end of the event, Samantha and Lewis browsed the merchandise. Close-up shots of console channel strips with LEDs aglow adorn mouse mats, mugs, coasters, and notebooks, and there are plenty of Abbey Road–branded T-shirts and tote bags. The most intriguing item is a set of prints produced and signed by Director of Engineering Peter Cobbin, featuring close-up black-and-white images of TG console channel strips that claim to put the viewer "within touching distance" of the knobs and controls. These signed prints, priced at £200 each (£500 for the set), attracted much interest. Cobbin's gear prints exemplify how, in gear cultures, imagery of gear can be fetishized almost on par with the gear itself. Where it is difficult—or even impossible—to own gear, pictures of gear are the next best thing, whether that is via online circulation, retail catalogs, or in expensive merchandising of limited-edition prints. To put the console "within touching distance" also invokes haptic ideation, which is central to gear fetishization; if it is not possible to touch the gear, imagining touching is the next best thing. "80 Years of Abbey Road" celebrated the tacit knowledge of its elite studio personnel, the scarcity of the gear produced by EMI engineers, and the canon of its recordings, reinforcing Abbey Road's place as one of the world's most recognizable studios. By foregrounding heritagized recording equipment, it signaled to broader gear cultures the origins of much of their ideology.

CONCLUSION

In a keynote at the 2021 Audio+ symposium in Victoria, British Columbia, Annelise Noronha, after discussing her acclaimed career in professional production, songwriting, and tonmeister roles, brought the discussion to a critique of the AES and its meetings. As someone dedicated to teaching a new generation of studio workers, especially aspiring women, trans* and non-binary engineers,[21] in her assessment, the organization did not contribute to facilitating networking among up-and-coming audio professionals, especially those from underrepresented backgrounds. Like in many milieux, the Diversity and Inclusion committee provides a space for some underrepresented members to meet others who share this designation, but those underrepresented members are the ones tasked with undertaking this additional, essential labor beyond their work as audio professionals. Moreover, conceptualizations of "technical" inform what can be published in the *Journal of the Audio Engineering Society*, reinforcing a definition of "scientific" that precludes sociocultural research (regardless of the rigor or repeatability of the methods). Our contribution to Noronha's critique pertains to documenting the ways in which the spatial organization, including the gendering, sexualization, and racialization of the milieu (which marks white male bodies as being full gear culture participants worthy of respect where everyone else is marked as ancillary and accorded less respect), works hand in hand with gear staging to (re)produce the ideas, meanings, and extramusical aspects of audio technologies.[22]

The kinds of gear staging, gear fetishization, and analog (tech)nostalgia we observed at 2010s and 2020s gear events are not new, although they adopted different aesthetic forms. At their first AES Convention in San Francisco in 2004, Eliot met self-described "gear pimp" Fletcher, at the time the owner of Boston-based boutique retailer Mercenary Audio, who proudly discussed some of his trade show antics from the 1990s. Fletcher had recounted the same for *Mix* magazine, where he mentioned hot-rodding a Harley-Davidson golf cart, which he drove around the trade show floor at AES 1996 carrying a banner proclaiming "Analog's Back And It's Pissed." In his most infamous stunt at AES1997 Los Angeles,

> Fletcher created a piece of alleged artwork—a battered Alesis ADAT and a Mackie mixer impaled on a 4-foot spike—entitled "Shit on a Stick" and displayed it in his booth. It was his artistic interpretation of the meatballing of pro audio equipment, as the age of low-cost mass manufacturing kicked into high gear. Alesis didn't agree with his aesthetic vision. AES officials threatened to close his booth down and eject him from the show if he didn't take it down. After some bluster about First Amendment rights and possible retaliation, Fletcher dismantled the display.[23]

From Foxy Stardust's UAD LUNA space launch to Redco's Cable Gal, from the Mix with the Masters stage to Kehew and Ryan's Abbey Road celebrations to Fletcher's "Shit on a Stick," gear culture events could never be accused of being solemn, "engineering" affairs. They are spaces for play, and that play, regardless of how dramatic, is a performed tactic of connoisseurship. These performances underscore the social/technical value of boutique audio gear and its retailers/distributors at a time of pro audio store attrition, and in the face of competition from more affordable and portable software-based tools. More accurately, they are *fraternal* spaces for *men* to play, the staging grounds for male anxieties about the potential of analog loss and for obdurate re-inscriptions and performances of nostalgia for "the good old days" of the record industry, and its large-format, major-label, multiroom studio economy.

7 PRINT

> Samantha,
>
> Let's be honest. You're a gear geek . . . just like us!
>
> You love gear, music, performance, self-expression, and being surrounded with creative energy. We know the feeling!
>
> Imagine spending your time surrounded by the latest and greatest gear, hearing right from manufacturers about hot new releases before anybody else! Imagine being surrounded with like-minded gear geeks and musicians who eat, sleep, and breathe music performance and production. Imagine visiting the world's largest music instrument and pro audio retail store, a world-class recording studio, or taking lessons on a new instrument—on your lunch break. At Sweetwater, you can turn your passion for gear into a meaningful career in the music industry.
>
> Rock on,
> Jordan Applegate
> Sweetwater Director of Recruiting

May 2022. As Samantha checks through emails before working on this chapter, she clicks on a circular from Sweetwater, one of the biggest music instrument and audio technology retailers in the world. Being "just like us," they address her, a subscriber of their mailing list, as part of their gear culture. If gear cultures are Formula One racing cars, Sweetwater are McLaren, driving everyone round the bend with their unrelenting gear obsessiveness. Most days, they email to ask if she has "checked out the freshest gear," whether it is a #NewGearDay for her and if so, if she has "shown the world" evidence of the new gear. Sweetwater reminds her that they have the "hottest gear" and plenty of "gear picks." They extend an invitation to join their "GearFest ONLINE: A monumental celebration of musicians & gear we love" and encourage her to browse their "We *heart* gear SALE" and check out the "clearance gear" on offer.

Sweetwater's website sweeps its visitors up into a gear frenzy: "gear experts" are on hand to provide advice because "buying gear is complicated." "Trending gear" and "top new gear" are positioned front and center on the page, and there are "plenty of ways to score great gear"—with financing, of course. Visitors can "browse used gear," enter a "monthly gear giveaway" competition, and ultimately "get the gear you want" on a special Sweetwater credit card. Sweetwater operates both as an online retailer and as a mediator of gear cultures. What sets this Fort Wayne, Indiana–based company apart from other retailers is how gear manifests in one of the largest and widely circulated gear print publications: *The Sweetwater ProGear Select Catalog*. The two-kilogram catalog contains "just a fraction of our music gear selection."

Since the 1990s, the Sweetwater catalog has been a staple print publication that documents gear for US-based gear-interested people. Over its several hundred glossy pages, the pictures and evocative prose juxtapose dreams about owning the most expensive and sought-after gear with knowledge about technologies available at a variety of price points. Every gear-buying studio Eliot has visited in the US since the 1990s has a stack of these catalogs lying around somewhere, and colloquially, the catalog has been known as premium bathroom reading for every individual who had ever ordered a gadget from Sweetwater in the past.

SITUATING PRINT MEDIA

The Sweetwater catalog demonstrates the continuing power of print media to mediate and reinforce gear ideas, and how print media dovetails with online gear cultures. This is not new: Paul Théberge documented the synergy between print media and user clubs in a pre-1995 music technology culture. In building on his work, we investigate the following macroscale questions. How does gear manifest in print in the twenty-first century? What has shifted in terms of gear representations? How does print media assert itself as an authority in gear cultures? How do print media and advertising uphold and reinforce fetishization as underpinning gear cultures?

We primarily analyze the content within the two most prominent twenty-first-century magazines, *Tape Op* and *Sound on Sound*, and the weighty *Sweetwater ProGear Select Catalog*. These two magazines represent the range of attitudes and approaches taken within contemporary trade publications, and as we will briefly discuss, both attempted to expand beyond a magazine format to encompass events and to host online message forums akin to regional user groups.[1] *Sound on Sound* magazine was founded in 1985 by Ian and Paul Gilby; the former remains the CEO. Its current executive editor is Dave Lockwood. The publication is based in Cambridge, UK, and circulates globally. In addition to the print publication, they host a popular YouTube channel with over 2,000 videos, mostly consisting of trade show reports, recording-studio tours, and tutorial podcast episodes. *Tape Op: The Creative Music Recording Magazine* began in 1996 as a hand-photocopied and hand-stapled zine celebrating indie/punk music and independent engineering and production. Larry Crane, the magazine's founder, has been an active musician and sought-after recordist; based in Portland, Oregon, he runs Jackpot Studios.[2] From early on, the magazine hosted DIY articles—how to mod microphones, how to make a plate reverb, affordable acoustic treatments—alongside reviews of budget and high-end gear, interspersed with reader letters, album reviews, and interviews with recordists and gear designers. After a few years of funding the magazine with credit card debt, in 1999 Larry partnered with Sacramento, California–based recordist John Baccigallupi, who contributed publishing experience and graphic design acumen. As of 2023 they have 38,000 subscribers.

We refer to three distinct eras in this chapter:

- The pre-gear cultures period prior to 1995. *Sound on Sound* was established prior to the formation of the gear cultures we feature in this book. We build upon Paul Théberge's documentation of this period.

- The gear culture buildup period between 1995 and 2010 when magazines experimented with hosting online gear forums and events. Here we expand on Samantha Bennett's work on the music technology press.
- The maintenance period from 2011 to the present defined by fully established print, online, and trade show gear cultures. In contrast with the buildup period, magazines stopped organizing their own events, and magazine-specific fora declined to dormancy. Alex Annetts's extensive yet (academically) undercited work discusses this period and links the press's gear fetishization with masculine performativity.

Print media functions as a conduit between gear manufacturers, gear retailers (including Sweetwater, who advertises in magazines), and gear users, and this mechanism is intertwined with fetishization and GAS. "That's what keep [sic] the industry running," admitted Paul White, former editor of *Sound of Sound*, in a 2013 interview with Annetts.[3] Despite the blurring of amateur and professional domains made possible in part by decreases in cost of recording equipment in the maintenance period, print media continues to reinforce amateur/professional binaries through discourse and imagery. Print media reinforces fetishization and GAS, yet at times attempts to push back against these dominant consumerist-fetishistic trends.

At a micro level, we conducted discourse and image analysis of three specific genres of gear discourses: reviews, advertising, and gear designer interviews. Here, we focused on several sub-questions. Which types of gear are reviewed and why? What contrasts between gear reviews exist across publications? Which themes are present in twenty-first-century gear advertising? Who is interviewed in gear print media and why?

In *Any Sound You Can Imagine*, Paul Théberge recognized a "cottage industry" of "double production," in that the proliferation of new digital technologies concomitantly developed a symbiotic music technology press. Within this "cottage industry," one domain (manufacturers) produced technologies, while the other (the press) produced consumers. Théberge grounded his tech-deterministic stance on the music technology press in the ways sound (re)production devices became the focus of technology magazines in the early twentieth century, and how technical characteristics became key subject matter in "hobbyist magazines" during the 1970s–1980s.[4] Alex Annetts built on Théberge's work, theorizing that music technology print media in the maintenance period "served to establish exclusive, mediated forms of communication" within communities. Similar to Théberge, Annetts suggested that the so-called democratization of technology, a common trope for framing technological change, is undermined by "notions of masculinity and gear fetishism" and is deeply flawed because it ignores the exclusionary nature of the discourse and practices around audio technologies.[5] *For whom* is technology democratized? Certainly not for women, nonwhite people, LGBTQIA+ people, or those living with certain disabilities—and not for those working on the ground or underground in extractive industries, who will never see a first-world studio. Thus, "democratization" is only a potential outcome for those individuals already part of the hegemony of masculine gear cultures.

Despite the proliferation—*not* democratization—of audio technologies, a pro/amateur demarcation continues to be intrinsic to the industry, even after one hundred years. Whereas in the pre-gear cultures area print publications made a clear demarcation between the professional audio industry and amateur/semiprofessionals, subsequently that line became blurred.[6] Print media have exhibited a confused attempt to hang on to this demarcation, even as gear manufacturer advertising has changed. Tomaz de Carvalho, too, analyzed how within "80 Years of Abbey Road," recording fora, and print media including *Canadian Musician*, *Recording Magazine*, *Sound on Sound*, and *Tape Op*, the discourse is pinned upon "pros" guiding "amateurs" to "master" their own gear and home studio setups. Applying Foucauldian discourse analysis to articles, de Carvalho noted, "Home recording 'pros' and recording professionals make sure to remind other home recordists about their individual properties, positions and roles within the discourse. 'Pros' and professionals have the authority to function as gatekeepers, limiting what is done, by whom and how."[7] While print media serves other purposes, we are most interested in how the pro/amateur dichotomy reifies audio technology knowledge that allegedly only professionals possess, justifies the need for an audio technology press, reinforces GAS, maintains space for gatekeepers in the masculine hegemony of the audio technology industry, and excludes so-called amateurs and all of the aforementioned groups who do not comply with the masculine hegemon of gear cultures.

Methodologically, this chapter is informed by a corpus analysis of gear advertisements appearing in print media between 2017 and 2022, drawn from fifty issues of *Tape Op* and *Sound on Sound*. Advertisements were coded according to presentation of technology or technologies, colors, words and word pairs, and any particular or special features. Patterns in the advertisements, including recurring themes and keywords, imagery, and aesthetics, were pooled and compared with earlier findings. Additionally, we analyzed a corpus of eighty "Behind the Gear" columns, each interview focusing on a gear designer or manufacturer, coding each for keywords and main themes. We traced patterns in the ways gear is mediated through print media to gain a better understanding of its role and mechanism in broader gear cultures.

Across all periods, tech-utopianism is perpetuated through the idea that technology is, in some instances, solely responsible for successful recordings. According to print media, gear is the answer to just about everything: to sound, whether that is "clean" and "pure" or "colored" with "mojo." "Power" and "control" are still key facets of what it means to be technologically inclined in the professional audio world, and much gear discourse in print media hinges on a single GAS-oriented parameter: you've either got it, or you haven't.

REVIEWS

As well as interviews with some of the best-known major-label recordists, producers, and recording-active musicians, *Tape Op* was originally known for providing useful and pragmatic reviews of the "hidden gems" of budget gear but broadened its scope to include everything from sub-$100 gadgets to $100,000 mixing consoles. We first read the 2010s decade of microphone reviews to gain a sense of a magazine-specific style and then conducted a detailed analysis of all microphone-specific gear reviews published between 2014 and 2015, since all of the main framing devices, rhetorical styles, review environments, and individual author writing style variations were well represented during this period.[8] The products in question ranged from $199 (sE Electronics X1R ribbon mic) to $2,999 (Earthworks PM40 PianoMic system).

Reviews vary considerably in terms of the use of professional audio terminology or description of specific design or component choices. One review attributed the "astonishingly low self noise" of a Microphone Parts mic to the use of "Wima capacitors and Vishay Dale resistors," while another described an Ashman Acoustic microphone containing a "small-diaphragm capsule mounted flush on the face of a 4 cm Perspex (acrylic) sphere" that contributed to the mic's "flat response up to around 5 kHz, and then a gentle rise in frequency response that peaks with almost a 5 dB boost around 15 kHz." Other reviews, though, lack these kinds of details, instead restricting their discussion to impressions of the external mic body and accessories. For example, the grille of the sE Electronics ribbon mic seemed "a little flimsy" to a reviewer, the large and imposing presence of the Monoprice Lollipop-Style condensers either made singers react with "wow" or intimidation, while the reviewer of the Telefunken M80 focused on its "nice leather bag, a well-designed mount, and a Telefunken mic cable with a 90-degree female connector." While single reviews may not cover component parts, technical specs, *and* mechanical construction, a reader of multiple reviews would be made to feel that these aspects might or should matter at least *some* of the time.

Most of *Tape Op*'s gear reviews draw upon one or more real-world recording sessions where musicians other than the reviewer are performing in the studio; in 30 percent of reviews, the actual artists are named, while in others, we only find out the nature of the instrumentation and musical style—for example, a Christmas single, a Latin jazz horn section, or "brooding indie pop." Occasionally, we read the results of "studio tests" or an ad hoc recording made in the reviewer's living room. Some of the reviews, whether professional sessions or ad hoc tests, provide select details about where the microphone was positioned and its orientation relative to an instrument—useful tips for any reader's future experimentation with their own microphones.

None of the reviews we read were "bad": all indicated that the microphone in question would be good on certain sources, in certain environments, and for certain desired aesthetic outcomes. This makes sense, since even the cheapest microphones on the market today for the most part will work reliably assuming they are placed correctly, in an appropriate recording environment, with a quality source sound. However, this did not stop reviewers from comparing the microphones with historical or currently manufactured microphones, and from detailing every piece of gear in the signal chain they used to make their test recordings. Every review mentioned at least one other comparison microphone, while one mentioned five other products—two of which were not the same type of microphone. The tendency to present each review product in relation to other brand names and well-known products extends beyond the musical domain—for example when the Microphone Parts kit for modifying a cheap MXL mic was described as "the guts of a Porsche 911 with the body of a VW Beetle."

The last example hints at the most important aspect of *Tape Op* gear reviews, and where gear inscriptions happen (or not). Here, a liberal nontechnical lexicon featuring vocabularies, metaphors, analogies, and similes is employed to describe multiple gear phenomena. This may include the sound that resulted from the objects, the reviewer's embodied and subjective experience while conducting tests for the review, or the relation of the products to others of the same type. Where the sound of the Peluso P-84 SK on drums was apparently "like a baseball bat to the chest," on some sources, the ADK Microphones Z-67 was "gentle and forgiving," in contrast to the AEA N22, which "begs to be used up close." Describing a frequency band of a microphone as "harsh" *always* implies something bad, whereas the inherent violence and sports valence of the baseball-bat-to-the-chest metaphor indicated something *good*, as did the "gentle and forgiving" euphemism. Why *not* invoke sports, especially when the MicRehab Miktek CV4 is "worth the price of admission, because it will get you in the same league as the über-expensive vintage mic superstars"? We never understood how a microphone inside a studio could cause the user to be "stopped in their tracks," unless that was perhaps related to the reviewer who "jumped out of my seat" when encountering a microphone that "stands out from the pack." Microphones do not just come in "packs," especially when they are "trying a little harder to not follow the herd." Some microphones apparently make more sense to the prospective buyer who already owns an "arsenal of fantastic mics," especially the Earthworks PM40, whose aluminum briefcase suggests to others that "I've got an upscale rifle here." Conveying a truism but persisting with weaponry metaphors, the reviewer of the Audio-Technica AT5045 admitted that "there is no single 'magic bullet' piece of gear." However, much of the mystique around that MicRehab microphone relates to the modded mic's "proprietary secret recipe" of changed components. We most took issue with one simile made in 20 percent of the reviews: that listening to a microphone in a session "was like being in the room." In our professional and pedagogical experiences of using a variety of microphone types across a breadth of recording and production workflows, we have yet to hear a microphone that sounds like "being in the room."

Plenty of shiny new gear glosses the pages of *Tape Op*, and in the latter pages, a small advert states, "Tape Op is made possible by its advertisers. Please support them and tell them you saw their ad in Tape Op." The symbiotic relationship between gear manufacturers and the music technology press, as theorized by Théberge in the pre-gear cultures era, is now explicitly placed among the gear advertisements in a leading audio technology print media.

For comparison, we analyzed a corpus of gear reviews printed in *Sound on Sound* between 2017 and 2021. These reviews, approximately 30 of which are printed each issue (350 per year), are particularly intriguing due to the summary sections: a highlighted box with two columns, "pros" and "cons," summarizes the benefits and limitations of the gear. Each review was coded for its gear type, positive and negative comments, and keywords including word pairs. Five categories dominated the types of gear being reviewed for the five-year period: signal processors, plugins, microphones, sample libraries, and synthesizers—all of which feature in the high double figures (40, 50, 60) per year. This demonstrates how hardware, including gear, remains prominent amongst the twenty-first-century software-centric audio technology industry. Hardware outboard signal processors and microphones are two of the most reviewed gear categories, with a balance of original technologies, clones, and reissues from established and bespoke manufacturers across both high-end and budget price ranges.

Let's take four hardware signal processor reviews that exhibit several commonalities. The reviews are overwhelmingly positive. Key take-away points are summarized in the "pros and cons" box, and rarely does anything of note feature in the "cons" box: in most, there are simply "none whatsoever." Similar keywords are used to describe gear's benefits, including "classy," "build quality," "classic," "vintage," "versatile," and "elegant," as well as "competitive" and "good price." Where there are "cons," these are usually brief, arbitrary, and, at best, moot points—for example, "There's no mains power switch on the front panel" (a recurring con for signal processors) or, simply, "expensive," which could be said for most gear. Nothing in the "cons" box critiques the sound or build quality; the manufacturer/reviewer relationship is unthreatened while the magazine maintains an, albeit flimsy, façade of impartiality to the reader. Most *Sound on Sound* reviews are positive in tone and rarely situated in any context beyond the functionality of the technology itself; reviewers rarely detail how the gear performs in an audio recording workflow. Moreover, the technical specification is narrated and often padded out with basic descriptions of audio engineering theory or functionality, which in some situations, particularly pedagogical, can be very useful. The reviews we surveyed included the following outboard signal processors from 2017 to 2021:

- 2017: Grove Hill Audio Liverpool, Trident 80B & Deca Dent 500, Locomotive Audio Model 14B
- 2018: SSL Fusion, Retro Instruments Revolver, API 529, Neve 500-series
- 2019: Eventide H9000, Roger Mayer RM58, Pulse Technologies EQP-1A
- 2020: Manley Laboratories ELOP+, Drawmer 1976, AMS RMX16 Reissue
- 2021: RND 5254, Fredenstein Artistic Plus, Manley Laboratories Nu Mu, Trident Hi-Lo

What we found supports our earlier findings on fetishization, GAS, and hegemonic masculine formations in online gear cultures: both media encourage fantasizing, gassing, and gear ownership as the ultimate goal.[9] As discussed in the *Tape Op* examples, gear's looks are of equal if not greater importance to gear's sound, described with a lexicon of feminine word pairings including "elegant styling," "classy looking," and "looks and feels."

These pairings serve two key functions. They connote high quality and mark connoisseurship, reinforcing the amateur/professional boundary that print media work so hard to maintain: buy this gear and you will join the "pro" ranks *and* satisfy your GAS needs. And while not overtly sexual, these word pairs signal to the dominant male readership that they *should* be looking at the gear; that gear discourse, even if not explicitly sexualized, is at the very least feminized; and that the expense incurred is a fair trade-off for the ultimate gear acquisition.

Heritagization discourse also surfaces in the review pages. The words "vintage," "classic," and "original" are littered throughout, particularly when discussing microphones and signal processors, and reviewers work to explicitly link new gear to gear of the past. Reviews also recapitulate advertising themes. A focus on the "bespoke," "build quality," and general "performance" of gear forms a continuum of precisely the kinds of themes found in audio technology advertisements since the start of the buildup period in the 1990s.

In its reviews, *Tape Op* balances technical specifications with real-world contexts along with some metaphorical lexicon. While it overlaps somewhat with gear culture themes, there is still a significant contextual basis of real-world audio production application and praxis that underpins the reviews. By contrast, *Sound on Sound* replicates much of what we will see in online gear cultures. The types of gear being reviewed, the hierarchical discourse, and overall tone serve as important pro/amateur boundary maintenance work.

ADVERTISING

A flick through *The Sweetwater ProGear Select Catalog* reveals gear in abundance. On page 275, the studio section starts with preamplifiers—Manley preamplifiers, no less. The featured Manley FORCE listing—"'Clean,' 'quiet' and 'professional'"—abuts a photo of the preamplifier, its traffic-light style LED meters flickering out from the page. Later in the catalog, a full-page Manley advertisement jumps out—"**investment grade** audio jewelry" (bold type reflecting original advert)—between images of the CORE reference channel strip, Nu-Mu compressor, Voxbox, and Stereo

Variable Mu (the name of the latter device EveAnna Manley strategically trademarked). The word "jewelry" perches between two microphones in shock mounts: the Reference Cardioid and the shimmering Reference Gold—apparently a "holy grail" microphone (page 455). This advertising listing is loaded up with gear culture values. While "its sound will never go out of style," more important than sound is justifying the price point. Much more than gear for studio workflows, this is sonic *jewelry*. Rare and precious, of a higher "grade" than "professional," it represents an investment that will retain its value. Like expensive jewelry, this gear is timeless, with lasting value: it is divine gear. As Annelies Moors noted, "Gold is a fetish par excellence," and rather than representing value, as currency does, it *is* value.[10] Indeed, gold is "an important source of economic security . . . a means to store wealth . . . worn and displayed on the body," and it "produces a particular public presence . . . claims to status, notions of the self, forms of identification and socialities."[11] There is something else in this clever analogy: jewelry is often displayed on special occasions. Since gear attributions depend upon connoisseurship, the more expensive and elusive the gear, the higher its fetish potential. As a Gearslutz moderator, EveAnna Manley is aware of this and catches gear users in a feedback loop; the discourse is captured and played back into the advertising. Manley knows full well that her customers may not all be recording engineers, but all *are* purveyors and collectors of the very *finest* gear. This is evident in the design and construction of the gear, too. Clearly visible through the grille, the microphone diaphragm at once connotes build quality, robustness, and fragility. As with rings, the diaphragm acts as a bezel-set jewel in the golden band of its casing. All these themes in the Sweetwater catalog are commensurate with long-standing trends in marketing professional audio technologies: little has changed in the structural and functional aspects of print media advertisements.

We also analyzed advertisements in *Tape Op* and *Sound on Sound* magazines between 2017 and 2021 for gear imagery, the size of advertisement, text and fonts, and the relation of key themes to those found in previous scholarship. Whereas earlier work assessed many types of music technology equipment, we focused solely on studio gear. Within ads ranging from budget devices to high-end gear, many recapitulate common gear culture themes and convey the object's fetish potential. This is not surprising, seeing as "the messages of the marketplace (advertising) must reflect the symbolic breadth of the person-object relation."[12]

One novel aspect was a clear political shift toward technological supremacy rhetoric. Despite the efforts to maintain it, the pro/amateur boundary has eroded significantly in the maintenance period, and, as such, manufacturers go further to categorize and distinguish their gear as reigning supreme. Take an advertisement by AMS Neve (*Tape Op* #124, 2018): "For first class recordings, it has to be Neve—no question." This headline features three separate assertions of technological supremacy: "first class," "it has to be Neve," and "no question." Here, Neve positions themselves as a superior gear brand through an authoritarian tone; these assertions are not for question or discussion. Neve gear is now marketed as a barometer of audio professionalism; if the recording domain does *not* feature Neve gear, it *cannot* be professional.

Without wishing for this chapter to eat itself, an advertisement for gear retailer Sweetwater (*Tape Op* #124 2018) conveys a similar tone: "The Pro Audio Authority for over 30 years." Interestingly, the advertisement depicts both professional and budget gear in the context of the same 2010s studio setup, which is indicative of how Apple computers, AVID's Pro Tools, and budget keyboard equipment sit alongside gear including the Rupert Neve Designs Shelford Channel Strip, Blue, and Mojave microphones. Here, the home studio setup, the reality of many current studios that feature mostly budget technologies, is "professionalized" by the Neve channel strip and microphones.

Against images of a starry night sky, a Cloud Microphones advertisement (*Tape Op* #124, 2018) claims their microphone DIs and preamplifiers will "elevate your sound." The stars-and-stripes flag is prominent in the bottom right; coupled with the night sky and colorized to a gray, black, and blue color scheme, this imagery connotes the NASA space agency. Cloud taps into a continuum of space-age themes and technological utopianism as seen in pre-gear cultures professional audio technology advertisements—for example, the "moon walk" Fostex advertisement from 1990.

An ad for the Warm Audio WA-67 (*Sound on Sound*, January 2021), a clone of the Neumann U67 microphone, centers the gear on the page. Under the heading "Paul. Mick. Bonzo. Kurt," the advert conjures up the posthumous rock canon and plays off displaced agency. Whereas many gear ads during the buildup period used recordist testimonials, and pre-gear culture advertising depicted technologies adjacent to popular musicians, such agency ascriptions here have been replaced with a mythologized version. Were these musicians so good because of the microphone? (Which could not have possibly been the Warm Audio version, of course). This is a huge leap, but it does not stop there. References to a "classic '67-style circuit," "custom all-brass K67-style capsule," "boutique capacitors," "premium 7-pin cable from Gotham, Switzerland," and "custom large-core transformer from Lundahl, Sweden" foreground locality, build quality, and bespoke components. This is clearly aimed at the Gearslutz market: gassers who, priced out of an original Neumann, can at least manifest their connoisseurship in a clone while simultaneously brushing against The Beatles.

Quantities of analog gear are fetishized, too. Retailer KMR's advertisement (*Sound on Sound*, January 2022), with its close-up shot of a gear rack, represents idealized amounts of gear. "When you push the quality of your recordings to the limit, there's still nothing quite like the sound of analogue outboard," they exclaim. But what is "the limit" here, and once one has transcended the "professional" boundary, where does one arrive? What *is* clear is the precise type of gear that will achieve this quality: the rack consists of a Retro Instruments 176 Limiting Amplifier, a Chandler EMI TG12413 Zener Limiter, an AMS Neve 1073 8-Track Preamplifier and a Lydkraft TUBE-TECH compressor. Together, this coveted gear would cost more than $17,000 USD, not including the rack. In the lower left of the advert, over twelve additional professional audio manufacturer logos are printed. KMR's logo, an analog tape reel, anchors the whole presentation in a "heritage" era, despite the ad ostensibly selling twenty-first-century gear.

There is, however, a simplicity found in some high-end manufacturer advertisements, representative of a shift from pre-gear cultures. For example, ads by Royer and Mojave microphones simply include an image, the model number, and review quotes with no testimonial, tagline, or added text.

D. W. Fearn's coveted outboard equipment features an image of two stereo equalizers with a one-word pun for a tagline: "unequaled." Overstayer Audio distills its message into "analog equipment for recording and mixing built in Los Angeles," managing to include build quality, locality, and "analog" in the same short phrase. Hammer Audio microphones are simply "hand-made in the UK." And Josephson microphones use an image, model number, and one quote from Steve Albini, who commissioned Josephson to make the "ultimate" snare drum microphone. Advertising space for high-end manufacturers is also much smaller now than it was in pre-gear cultures. The few whole-page adverts are reserved for bigger software companies, such as AVID and Waves; gear advertisements are smaller in size, one-fourth or one-eighth of a page.

INTERVIEWS

Despite gear taking center stage in professional audio discourse, perceptions of agency are key to validating gear as belonging to the professional audio domain. This is achieved especially in the interview format by associating gear with the right kinds of people. Online, recordist interviews have been less popular than threads about gear. However, all print media features interviews in some form. But who is doing the interviewing, who is being interviewed, and how does this contribute to present gear understandings? We analyzed more than eighty columns of "Behind the Gear," a feature published in most issues of *Tape Op* from 2000 to 2018, coding frequently occurring keywords and themes. *Sound on Sound* published interviews in each issue; these are uncategorized and vary widely. Since the interview corpus is so large, we limited the scope to 2017–2022, looking at how interviews have changed in the maintenance period.

At this stage, it is important to declare that we interviewed *Tape Op* editor Larry Crane for this project, but we were declined interviews with the editor and staff members of *Sound of Sound*, despite attempting to conduct them at AES conventions in 2017 and 2019 (they ignored Samantha outright). These interactions with *Sound on Sound* staff appear to reinforce previous findings about the magazine's responses when asked about culture and gender-oriented details;[13] our interactions only amplified our perception that the magazine excludes women and gender non-binary people. An insidious playing down of women's technical expertise pervades the tone of the publication. Additionally, we note that Crane's interview with us was extremely generous in time and scope. We found Crane to be open, affable, and supportive of this project, and keen to detail the importance of women in his life. As gear culture participants ourselves, we have sought to eliminate our biases as much as possible; however, we encountered limitations in this milieu.

With *Tape Op*, founder and editor Larry Crane has always sought to situate any discussion of gear within the context of recording and production workflows. Conscious of the pro/amateur binary in other print media, Crane actively sought to break down this divide with features and reviews that granted the reader more agency. In fact, a lot of the content was, and still is, reader generated, with numerous features authored by industry participants at all levels of recording. Additionally, most *Tape Op* articles are interviews, which, as Alex Annetts noted, results in a "dialogical rather than didactic" interaction between knowledge holder and reader.[14] Part of this direction was, as Crane reflects, around the recognition that not all great music is made in elite recording workplaces:

> I certainly sought to validate alternative ways of recording in the beginning. Say [the] four-track cassette recorder is not any less valid than a huge, fancy, multi-million dollar studio . . . that was something that I found offensive in other magazines and books and stuff . . . well that's nice and all, but you're going to work in a real studio when you get serious. But I had friends that were releasing records recorded on less than high-quality equipment, and then records that were getting critical acclaim, or sometimes lots of sales. Any sort of recording scenario can be valid, and sometimes it suits the music. Sometimes it might not even suit the music in an aesthetic way that makes it any better, but the music's still valid.[15]

This focus on workplace and agency separates *Tape Op* from its counterparts and functions "in a less exclusive capacity" than other print media. In saying that, *Tape Op* "permits the articulation of a variety of discourses regardless of whether they perpetuate or rupture the dominant tropes of masculinity and gear fetishism."[16]

Tape Op's "Behind the Gear" column takes a regular interview format with a gear designer or manufacturer. Over the years, these have ranged from prominent names including Rupert Neve and George Massenberg (each featured in the column more than once) to boutique designers such as Ted Fletcher, EveAnna Manley, and David Josephson. These interviews almost always show images of the designers themselves and pictures of the gear in varying forms: inside the box showing components/circuitry or finished product(s). Less obvious is the relationship of these interviews with advertising. Interviewees appear to have been engaged out of a genuine desire to reveal aspects of the gear that might otherwise be concealed, as well as to champion audio technology developers who do not receive the same kinds of attention and acclaim as recordists. After reviewing and coding these "Behind the Gear" columns with frequently occurring keywords, we found interesting commonalities between gear designers that diverge prominently from the fetish discourse so rife in gear cultures.

Words related to the construction and build of gear and its components are central to the discourse (figure 7.1). Locations of manufacturing ("British," "Germany," and "Latvia") commonly feature.[17] Words related to equipment testing and gear's sound (for example, "listening," "sound," "test," and "metering") are more frequent than gear culture discourse stalwarts (for

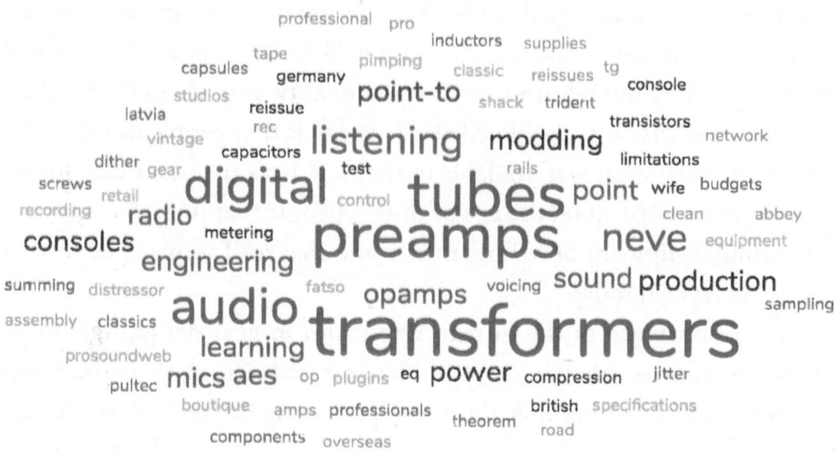

Figure 7.1
Tape Op "Behind the Gear" keywords.

example, "classic," "vintage," and, of course, "gear"). The biggest disparity between designer/manufacturer discourse and that of gear culture participants concerns fetish potential; designers rarely used terms such as "wife" and "pimping." Gear designer interviews serve to somewhat counterbalance the fetishization happening elsewhere in *Tape Op*, since many designers do not set out to design "gear" per se, nor do they fetishize it, but rather apply their talents toward designing useful audio technologies. Lacking a grounding in fetish discourse, gear designer interviews instead focus on three key areas, broadly categorizable as learning, human sensoriums, and tradition.

Gear designers emphasize sharing and champion lifelong learning as a rite of passage. John Hardy (#15, preamplifier designer) echoed many of our interviewees when he stated, "You blow things up, you learn from that. Electrocute yourself, you learn from that. Little by little you learn . . . Hopefully you don't keep electrocuting yourself every day for the rest of your life." John LaGrou (#32, Millennia Media) corroborates this when he states, "Over the years, I've refined a test methodology for new circuits . . . Trial and error and listening tests are as essential to audio product development as they are to fine audio recording. It's all about listening critically, and one of the key elements to listen for in mic preamps is dynamic stability." The testing process is far more methodical and detailed than the ABX testing that goes on in Gearslutz threads, since designers listen and employ trial and error to improve sound quality. Skipper Wise (#40, Blue Microphones) agrees, "What we did is learn from the past quality, and what I learned from a lot of things that are considered classic today is that people did their best at the time with what was available to them. What's the best I can do, and I believe that's what stood the test of time—people that took that approach in building equipment or writing music, and that stuff seems to be timeless and is still revered today."

The importance of the human sensorium in gear designing and use is another recurrent theme. As John Hardy stated, "It may even be a great design but there's something about the packaging or the look or there's a vibe there. You know, we're dealing with human beings here." Gear design is about more than components, circuitry, and sound, and gear designers are more concerned with the human—as opposed to the fetish—potential of gear, including its look and vibe. Malcolm Toft suggests that end users being involved in the design process helped make older gear great: "I would say that the history of Trident (mixing consoles) is based on the fact that we were recording engineers, we weren't really electronic engineers. The whole thing we designed primarily using our ears." An iterative listening process made up for the lack of professional electrical engineering experience, and this testing-learning-building-using gear pattern came through in many other recordist-designer testimonies.

The "Behind the Gear" column does, however, reveal gear designer ties to tradition and the building of "classic" gear, especially via canonization and heritagization discourse. Fletcher (#34, Mercenary Audio), owner of one of the first US-based boutique audio stores, said of the move into the "vintage" market in the early 1990s, "The real vintage explosion, yeah. We kind of saw it coming and, frankly, we kind of made it happen." Being "classic" is a key driver behind the design of some contemporary gear: "If you're competing with something that's classic you have to be classic," stated Dave Derr (#33, Empirical Labs). Doug Fearn spoke of how "classic" gear exemplars provide benchmarks and points of diversion in the design of new gear: "As nice as those Pultecs were, they really were a minimal cost design, and they really cut a lot of corners to hit a price point, I would imagine. I wasn't under those restraints quite as much so I was able to do some things that wouldn't have been cost effective for Pultec to do. The inductors that I use in there are custom made for me by Jensen. We worked for a long time tweaking that design and getting them just right. They are extremely consistent from unit to unit and the matching is absolutely precise." Through its "Behind the Gear" column and recordist interviews, *Tape Op* consistently links gear to professional recording and electrical engineering practice, and to the people and processes involved in gear design.

We find a very different frame in *Sound on Sound* interviews. Since its inception, *Sound on Sound* has featured interviews with recordists affiliated with "classic" recordings. For pre-gear cultures, most interviews focused on either current or past rock and pop records, and interviews were skewed toward replicating mythologies and revealing "secrets." Former *Sound on*

Table 7.1 *Sound on Sound* magazine, 2017–2021: Number of interviews per year, by gender and area of expertise		
Year	**Gender split**	**Expertise**
2021	59 male, 16 female	21% of women interviewees Change of editorial from White to Inglis
2020	58 male, 6 female	9% women-authored articles A broader range of audio interviewees: FOH, EDM, podcast producers, etc.
2019	54 male, 2 female	3.5% women interviewees Slightly more visibility of non-white male producers Still very low numbers of women interviewees: Catherine Marks, Laura Escudé "Leader" op-eds and mix reviews make up bulk of the "interview" pages
2018	45 male, 5 female	10% women-authored articles Large proportion (11) of mix reviews "Leader" op-eds by editor Paul White More EDM-focused
2017	44 male, 1 female	2% women-authored articles—Jayne Drake only woman writer Mix reviews—by SoS journalists "Leader" op-eds by editor Paul White Opinion pieces by SoS journalists Features bizarre article by David Ricard: "Sounding Off, Sound Bright"

Sound editor David Mellor reflected on the magazine's fetish culture from the outset:

> I think the role of the magazine was that it just displayed it for you, you opened the pages and it was sexy, it was desirable and you want it and you look at the pictures and read the text and think "that guy's had access to that piece of equipment and I haven't" so there was this real feeling of envy. It made you feel bad, like you couldn't compete. The unspoken sub-text was that you needed this equipment.[18]

The "guys" that Mellor refers to are the journalists and recordists who were interviewed for the magazine before 2009. We wanted to assess how—or if—the culture has changed since then. Here, we took a different tack. Given that *Sound on Sound* has no column comparable to *Tape Op*'s "Behind the Gear," we instead analyzed the kinds of topics discussed in all interviews between 2017 and 2021, and the expertise and gender of the interviewee. Until the end of 2020, issues generally contained between four and six interviews, meaning a total of forty-five to sixty-four per year.

All *Sound on Sound* interviews are categorized under "people," are conducted by in-house staff, and feature similar trajectories. If the subject is a recordist, their perspective is sought if it may "reveal" something about how a record was made. Interviews are revelatory in tone and often feature references to secrets, secret weapons (gear objects or gear settings), and secret locations—studios or other recording locations beyond what might be deduced from liner notes or Discogs credits. Except for Sue Sillettoe, a mainstay pre-gear cultures *Sound on Sound* interviewer, all were middle-aged men. What we see in table 7.1, however, is a very recent swing away from the male-dominated culture of *Sound on Sound*'s first thirty-five years toward a more diverse set of interviewees, topics, and emergent professional audio domains. Whereas for the magazine's duration its focus was almost entirely on male recordists of well-known commercial hit records, since 2021, it has diversified to include producers of podcast and game audio—and many more female interviewees. Up through 2017, fewer than 5 percent of interviewees were women, but that number jumped to 21 percent in 2021. Interestingly, this marked shift coincides with a change of editor-in-chief from Paul White, who editorialized the magazine for decades, to Ian Inglis. That is not to say the problematic tone of the magazine has disappeared altogether. Many interviews still perpetuate fetishistic gear culture discourse, routinely ascribe agency to technological objects, and play into secrets and heritage discourses.

CONCLUSION

While the most extreme modes of boundary maintenance will be discussed in our analysis of online gear cultures, even retail and professional audio mailing lists have ramped up the rhetoric. A sales representative for Sweetwater Sound once cold-called Eliot, and with far too much inflection in his voice asked, "Are you READY to take your mixes TO THE NEXT LEVEL?" (Perhaps Eliot *was*, but they didn't immediately need any Sweetwater gear to do so). Prospective customers are supposed to be "blown away" by a new "secret weapon" to add to their studio "arsenal"; we will discuss the many implications of this in chapter 12 but here wish to underscore the role of such discourse in prepping prospective gear culture participants for the imaginary war that is to come—the primary brand identity of Shadow Hills and others. The #NewGearDay announcement, in this light, shows Sweetwater's attempts to enforce a convergence between their traditional print media (the catalog), their advertising work, trade show style events, and online gear cultures.

We recognize a continuum between the types of advertising themes, gear reviews, and interviews in all three periods: the pre-gear culture era (pre-1995), the buildup period (1996–2010), and the maintenance period (2011–present). *Tape Op* has more consistently contextualized gear in workflows and emphasized human engagement over fetishistic scenarios. *Tape Op* champions the gear designers and manufacturers by highlighting their expertise, and the secrets discourse that is printed is more factually informed than based on myth perpetuation. *Tape Op* was, in the 2000s, more than a magazine; the TOMB (Tape Op Message Board) provided a Gearslutz alternative and created a context for subscribers and fans of the indie artists typically championed by the magazine to socially interact rather than "passively" consuming within an "imagined community."[19] TOMB's decline probably had to do with the general lack of any promotion or advertising specifically done for the forum; it is more remarkable that a loosely moderated site with no publicity ended up attracting over 17,000 members and hosting over 700,000 posts. *Tape Op* also organized conventions from 2003 to 2008 that were designed to articulate the same concerns as the magazine. However, *Sound on Sound* has followed a different trajectory. Only since 2021 has its fetish culture—described so well by David Mellor—begun to shift, and we cannot predict if this shift will take hold. In the meantime, gassing is the norm, and the magazine's focus and overall tone is on the gear, its ownership, and its secrets.

8 ONLINE

MIC GUTS: THE NEUMANN U67 REISSUE

How did a request for images and videos pertaining to the guts of a Neumann U67 reissue microphone result in one of Gearslutz's most discoursed threads? We may never learn whether the pseudonymous OP was trolling, joking, or serious.[1] Neumann has never participated as a company on the forum in question, but that did not stop the thread from precipitating 845 responses over a four-month period, and 1,045 responses over a two-year period. With over 91,000 thread views, and one of the highest Google search placements of any discussion thread related to this reissued $7,000 Neumann microphone, this thread has considerable impact on widespread perceptions of whether the reissue is "worth it" and of similar quality to the highly coveted, "vintage" Neumann U67.

Unfortunately, but not atypically, this popular thread, like thousands of similarly framed threads, falls far short of being a useful resource for anyone genuinely interested in the mic. Yet this example serves as an exemplar as to how gear cultures form, operate, and gas online. In over 1,000 responses, not one participant ever mentioned using the microphone in a music production workflow context, or that over 500 posts passed before a single sound sample of the reissued mic was posted (albeit a flawed "shootout"). But if music, musical sound, and working with musicians is not being talked about, what *is* being discussed, how did the discussion last so long, and what does this mean for online gear cultures? Popcorn emojis at the ready, we took a deep dive into the thread's dynamics and analyzed the social implications for a group of passionate gear enthusiasts brought together by the inner workings of a Neumann U67.

We found several social functions in the first twenty-one posts. The most common kind of posts are those that challenge or insult the OP, another poster, or forum users in general. This might be accomplished by questioning the premise of the thread itself, demonstrating the use of technical vocabularies (either accurately or not), bragging that the poster has actually purchased the object (or more than one of it), claiming superior understanding of the economics of audio technology manufacturing (accurately or not), or claiming the ability to hear minute differences that other posters cannot. We also find the performance of connoisseurship throughout the thread. Ten of the first twenty-one posts contain such challenges or insults, and therefore were speech acts more concerned with jockeying for status within the online gear culture than with the aural qualities (real, perceived, or imagined), the use value, or workflow context(s) of an as-of-yet unreleased microphone (table 8.1). As with most online gear threads, the focus was extra-audible aspects of the gear.

These posts exemplify the various engagements and performativities involved in social formations around gear. Not until post 125 could a Gearslutz member claim they had actually received the microphone; their first action was to open up the mic and then take and post pictures of the innards—just as requested by the OP. This suggests that one primary motivation for purchasing gear is to obtain social capital within an online community as opposed to using it to record vocals, for instance. In post 127, a grateful user compared the picture of the disassembled mic in post 125 to a Marylin Monroe centerfold "for the slut that lives inside me." No one claimed to find this response odd. In fact, as we have hinted already and will discuss further, this particular message forum has been structured around the sexualization of gear since its inception in 2002, and a microphone is more likely than any other kind of technological object to be discussed in terms of sexiness—and mic-desiring in terms of sluttiness. With such a concern for visual aesthetics and optical experiences, and what we term technological voyeurism, it's no wonder then that so many posts in the thread were complaining about what kind of box or case the mic came in—a matter that

Table 8.1

Categorizing first 21 posts of an "epic" gear thread on Gearslutz

P#	Main topic	social function
P1	OP wants to see mic internal pics	information request
P2	insulting OP, asserting status over OP	jockeying for social status
P3	popcorn emoji	signaling this will be a long and funny thread
P4	off-topic response to P2	
P5	off-topic response to P4	
P6	I want to hear how it sounds	indicating OP's thread premise is flawed
P7	The mic is not *too* expensive	demonstrating connoisseurship
P8	What's the appropriate price?	thought experiment
P9	dismissing the premise of P8	showing discernment, reasserting social status
P10	I went to NAMM and bought 2	economic capital→ social capital
P11	Technical discussion of mic parts + construction	correcting OP, asserting superior knowledge
P12	off-topic response to P4	
P13	insulting OP and prior thread antics	social status / posturing
P14	complaining there's no wooden box	rambling, staking claim in thread
P15	quality costs, but we're in a disposable society	critique of society and forum users
P16	response to P14, Neumann sells boxes	opportunity to assert possession of even rarer mic
P17	more complaints about box, it looks "boring"	complaining, asserting place in thread
P18	generically insulting posters who complain about Sennheiser owning Neumann	posturing
P19	affirming P18's discernment by listing quality Sennheiser mics	asserting knowledge
P20	complaining about mic price vs. perceived component cost	staking claim in thread
P21	insulting P20 for not having used pro mics	posturing

would not be a major concern at a busy professional recording studio where mics such as this might be left up permanently on a sturdy Latch Lake stand. Boxing and unboxing the mic, however, does relate to the normative state of mics owned by many Gearslutz users—unused and in their boxes, unless they are being brought out to be disassembled and photographed or subject to pseudo-scientific shootouts with other lonely mics.

Did the OP ever intend to purchase the microphone? Or was this thread a stunt? Later in the thread, several prolific contributors challenged the OP on their intention and funds, the implication being that participation in this thread necessitated more than $7,000 in disposable capital. Therefore, the economic aspects and expectations being discussed, the sexualization of gear, and technical (or pseudo-technical) discourse all work together: they constitute the dominant strategies of users for maintaining or increasing social status within this predominantly pseudonymous and almost entirely male gear forum. Some of the challenges inherent in analyzing a thread such as this—we have undertaken extensive analyses of dozens of other popular threads spanning 2005–2020—stem from the numerous paradoxes contained within. What initially appears to be earnest technological-engineering concerns and a passion for the design and materiality of gear turns out, after careful analysis, to be little more than an effluvium of technobabble.

THEORIZING ONLINE GEAR CULTURES

Our Neumann U67 example demonstrates several salient features about twenty-first-century online gear cultures. A *problematic object* often resides at the center of gear discussions, rather than the center being musical works, music production workflows, musicians, or music audiences. Problematic objects mediate social relations in very specific ways, since they offer numerous opportunities for the human participants in gear cultures: to demonstrate their command of locally specific technical languages and forensic practices, to attempt to maintain or increase social status by demonstrating the correct taste and aesthetic dispositions, and to attempt to increase social capital by publicly performing their private object purchases.[2]

However, while the Neumann U67 thread might appear to be about microscale minutiae—a microphone that has not yet become available for purchase—problematic objects like this are the glue that holds together a city-sized social formation and around which a whole platform's infrastructure is organized.[3] That said, as it would be for any cultural formation, the complex and multiple nature of individual experiences and collective goals show the need for more nuanced local and comparative analyses. The overlap between the active users of multiple forums means that the bounds of any online gear culture exceed any one forum, subreddit, or social media group/feed. Not all forums or social media groups inculcate to the same extent a distinct cultural identity that individuals can aspire or subscribe to. In other forums, the discussion might be a straightforward, factual, experience-driven discussion about microphones. The sheer quantity of content produced by online gear cultures—hypertexts, meme images, emojis, animated gifs, vlogger gear reviews, and gear shootouts hosted briefly on file-sharing services—presents unprecedented opportunities for large-scale research, but the ad hoc nature of their production and organization, and their frequent ephemerality, pose significant challenges for analysis.

Our analysis here is selective, focused on elucidating several complex questions, many of which are discussed in greater detail later: What conceptualizations, materializations, or performances of gear dominate gear-focused online interactions, and why is discussion of music, musicians, and audiences marginalized on general gear forums? What cultural work does the Gearslutz style of technological discourse do, especially around gear sexualization and key archetypes of gearsluts and gear pimps? Who participates in or lurks around online gear discussions? Who is actively excluded from forum participation? Why do the majority of active participants post either pseudonymously or anonymously? How do posters attempt to gain social capital through forum participation? Finally, what effects do online gear interactions have on professional audio engineering, the broader audio equipment industry, and its various trades writ large, in terms of pedagogical possibilities and product design?

Our data begins with Eliot's participant observation experiences in online gear fora, where they began being an active user in 2002.[4] While this participation was not, at the time, being done solely as research, it provided a long-term "insider" understanding and perspective of cultural norms and cultural differences across fora. It also helped us identify paradigmatic discussion threads that had become part of forum lore. Starting in 2017, we undertook more systematic research using a variety of methods. We generated quantitative data on forum activity over time, keyword incidence, and measures of user engagement. Attending meetups of forum users held at trade shows helped us compare (typically pseudonymous) online activity with the fuller lives of individual people. Our work on the distinctive modes of sexualized discourse found on one forum, Gearslutz, necessitated crafting a novel multimode methodology that attended to words, images, and fetishized material objects.[5]

Because the history of audio engineering online and audio-specific forums is largely undocumented, we begin with a brief history of online gear. In the first period, the 1990s, media were newsgroups and email lists, and manufacturers created static websites to advertise their products. In the 2000s, activity moved to the medium of message forums. Hundreds of different fora experimented with moderation styles and topical focus. In some, the focus was on musicians, recordings, and charismatic moderator personalities (e.g., Mixerman, Fletcher, Steve Albini), while others focused on distinct kinds of technological objects (e.g. microphones, field recording equipment, or a single DAW platform). A handful covered gear in general and were organized around different price tiers of gear, analogous to social class. By the 2010s, experiments in forum focus receded, and Gearslutz became the hegemonic gear forum catering to recording studio–specific gear and, to an extent, location and live sound equipment and electronic/software instruments too.

A BRIEF HISTORY OF ONLINE GEAR

Online gear cultures were in a nascent state by March 1992 with the topical Usenet newsgroup rec.audio.pro.[6] Initially accessible only to individuals affiliated with research universities or large tech companies such as DEC and AT&T, newsgroups were accessed through command-line interfaces for

UNIX mainframe computers.[7] Rec.audio.pro, like many other newsgroups, gained many US-based *netizens* after September 1993—the "September that never ended"—when AOL began offering Usenet access to its dial-up modem-subscribing households.[8] This influx of new networked users allegedly did not understand the *netiquette* (social norms and Usenet-specific etiquette) that had been established over fourteen years.[9] The first crowdsourced information resource specific to audio technology was the Harmony Central Effects Database, a school project of Scott Lehman that launched in November 1994, which merged articles about guitar and bass pedals, rackmount effects processors, and MIDI-enabled devices with user reviews of the same, later adding a user bulletin board.[10] Lehman initially spread the word about Harmony Central on rec.music.makers, an early example of platform bifurcation and cross-pollination.

Gear manufacturers started utilizing websites in 1996. E-commerce was not yet widely adopted, so websites presented information to support their mail-order or phone-order businesses, as in print magazines.[11] By the early 2000s, some gear companies, especially makers of DAWs, had created user forums, including "The Duck" (the nickname for the "Digidesign User Conference" for ProTools users) and equivalents catering to users of Cockos' Reaper, Steinberg's Cubase/Nuendo, and most other DAWs. Notably, these forums expanded beyond tech support and became, to varying extents, hangouts for users to discuss gear, audio technologies, and music production in general.

Not all fora and listservs were started by active companies. For example, since 1995, the Ampex Mailing List, dedicated to the repair of vintage reel-to-reel tape machines made by Ampex (from 1948 to1983), has been the primary site for sharing knowledge about that niche. It also has served as a meeting place of sorts: Marshall Terry (whom we met in chapter 5) singled out his interactions with renowned microphone designer David Josephson—who hosts the Ampex Mailing List—as facilitating Terry's own career development. There are hundreds of similar specialty listservs, mailing lists, and Yahoo/Google Groups. That said, by any measure of user engagement (e.g., number of registered users, total number of posts, posts/user, or average posts/thread), product-specific fora, whether hosted by the company or by third parties, are marginal milieux for online gear cultures. They also do not represent a major break with prior modes of sociality around gear, such as the local meetings of technology-specific user groups and the radio-mediated sociality of pre-2000s ham radio "tribes."[12]

From 1999 to 2006, the second period of online gear, hundreds of new forums were launched. Multiple experiments with designing online gear engagement marked a clear expansion of interest in gear and in structuring social life around it. Forums launched during this time included magazine-related ones (e.g., *Recording* magazine's recording.org, *Sound on Sound* forum, *Tape Op* Message Board, and Mix Online), genre-specific production forums (e.g., Hydrogen Audio, Future Producers, and AudioSEX), specialty electrical engineering forums (e.g., diyAudio and GroupDIY), and personality-driven forums (e.g., The Womb Forums, which coalesced around engineer/author Mixerman, and Electrical Audio, which coalesced around Chicago-based engineer, rock musician, and poker aficionado Steve Albini). Other forum experiments were built around narrow categories of technological objects—for example, the 3d Audio mic and gear shootout forum, John Sayers's acoustics treatment and studio design forum, the Taperssection "concert taping" and remote recording forum, and the Lathe Trolls vinyl disc cutting forum.

Although all these had brief periods of intense activity, and some continue to exist less actively today, the most popular and influential forums have been what we term *general gear forums*: ProSoundWeb and Gearslutz (renamed Gearspace in 2021 after the period of our primary research). ProSoundWeb (pre-2006) was the first of these to inculcate a broad-reaching gear culture formation. Specifically, within ProSoundWeb's Rec Pit section, the conversational range was much wider than had been the case with first-period gear communities: a discussion of a particular technological object might transition into a discussion of musicianship and applied technique, studio humor, and lots of off-topic material such as politics, marital relations, or personal experiences of drug and alcohol addiction. Most users went to Harmony Central to learn about and discuss, say, a guitar pedal and left when their interest in that topic was exhausted. However, users would initially come to ProSoundWeb to learn about a technological object and stay there to talk about practically anything and everything—in the process, learning something about other like-minded, gear-interested people. The Rec Pit was the origin of numerous memes that continue to resurface within gear cultures, and argot/cant specific to that gear culture (e.g., Alsihad as a euphemism for ProTools) continues to serve as markers of belonging.

ProSoundWeb was an evolving experiment in what online gear forums could or should be.[13] Proclaiming to be a "professional audio center" under the motto "independence, open communication, industry knowledge," in September 2000, they first launched a message board, which in 2002 they converted to a threaded message forum utilizing the phpbb 2.0 software (one of the earliest open-source forum packages).[14] The first Rec Pit subforums were named after individual forum moderators, each of whom had a distinctive online persona and correspondingly attracted different participants. John Klett's subforum, for example, explored gear repair and design. At the time, Klett was well known to employees of larger US studios as a sought-after gear repairman. Nika Aldrich's subforum attempted to demystify the science and practical application of digital audio technologies. Jason Hall (posting as j.hall), an engineer/producer from Kansas City, focused on indie rock recording. These fora explored gear within sensible contexts and set a precedent for two styles of gear sociability still found today: a penchant for debating with experts, transpiring alongside the valorization of the forum's power users.

One of the most popular subforums was hosted by an engineer who went by the pseudonym of Mixerman. "Zen and the Art of Mixing and Production" focused on the practical aspects of studio work. Mixerman initiated a series of "global" production collaborations called CAPE (composers, artists, producers, engineers) where forum users were assigned to often unlikely teams to write, record, produce, and mix a song in a fixed time frame. Eliot participated in several CAPE sessions in 2003–2008, contributing oud, percussion, and keyboard parts to a melodic/industrial metal production called "Burn in the Pyres" with contributions from a Kurdish violinist, Swedish mixing engineer, and German keyboardist/guitarist—none whom had met in person or previously collaborated.[15] Mixerman's subforum, and the Rec Pit more broadly, became best known to a wider readership when Mixerman began publishing a series of forum posts that appeared to be a diary of a major-label American rock recording project that went horribly wrong. *The Daily Adventures of Mixerman: Bitch Slap or Bust* was ultimately published as a book.[16] The pseudonymous diaries assembled an avid fan community, inclusive of many full-time working engineers, producers, and studio musicians, that continues twenty years later.

The Rec Pit was an international forum that attracted several thousand users and some 170,000 posts over two and a half-years—an unprecedented amount of user engagement compared to prior milieux. However, in April 2004, ProSoundWeb management shut it down in their first of several forum restructurings. After the relaunch, much of The Rec Pit milieu followed Mixerman and his friends to the MARSH (Musicians', Artists' & Recordmakers' Superfine Hangout), but ProSoundWeb management continued to tinker with the structure. Two years later, their attempts to charge for user access, and internal conflicts over rules about pseudonymity versus real name user requirements, led to ProSoundWeb's demise. In 2006, Mixerman, who later revealed himself to be Eric Sarafin, used the notoriety of the diaries to start The Womb Forums. Until its closure in 2017, The Womb Forums was a place for working and aspiring engineers, mainly those serving niche genres or local band scenes, to discuss techniques—both gear- and people-focused.[17] At its peak, The Womb attracted 10,723 users, fewer than a thousand of whom were regularly active, but this small user base still created over 384,000 posts.[18]

The second period of online gear (1999–2006) demonstrated the extensive reach of online gear cultures and differing focal points of gear discourse, where forum moderators and users alike explored the relation of gear to life—and how gear could become a *way of life*. While most registered users of newsgroups and early bulletin boards posted with their real names, pseudonymity on the Rec Pit provided the necessary cover so that experienced professionals could write music industry satires and critically discuss their lines of work. As we will see in the next section, the widening of topical possibilities also extended to the discussion of politically controversial matters, contributing additional motivations for pseudonymous online participation. Arguably the most positive outcome of gear fora in this period, especially The Rec Pit and The Womb Forums, concerns the pedagogical function that they sometimes served. For example, the drum/guitar miking and rock/metal tutorials posted by Slipperman, the pseudonym for Big Blue Meanie studio owner and New Jersey–based heavy metal engineer Tim Gilles (1961–2021), became legendary in online communities—both for providing a detailed and replicable process of what would typically be regarded as trade secrets and for Gilles's sardonic and self-deprecating wit.[19] With the collapse of the major studio economy,[20] the customary industry internships and apprenticeships no longer existed to the same extent, and quite a few working engineers today, especially some who lived in countries that lacked major studios or degree programs in audio technology, got their start in audio at that time by trawling forums and asking questions of professional engineers.[21] Moreover, Brad Blackwood's subforum at ProSoundWeb was the first place where there was extensive discussion of the working practices of mastering engineers; arguably, the expansion of interest in being a mastering engineer is directly attributable to that forum and subsequent similar ones.

The widening of topical possibilities did not necessarily widen the demographics of participants. If anything, the broader topic discussions narrowed the engagement as new gendered lexicons were established among male participants. Fewer than 1 percent of the active posters in any of the forums we analyzed *publicly* identified as women, although we have met a few cis and trans women audio engineers who chose to try to pass as heteronormative men in order to participate. Even if the international reach of the forums increased, the English-only policies of many forums, and criticism by self-appointed "grammar police" when encountering poor sentence constructions or usage mistakes, excluded those who did not already have considerable command of English, audio-related technical English, *and* the rapidly evolving argot specific to the forums.[22]

GEARSLUTZ/GEARSPACE

As of June 2022, with 415,000 users and 13.8 million posts,[23] Gearslutz[24] is by far the most active forum related to recording, engineering, mastering, and studio-based gear cultures—and later expanded its remit to cover instrumental gear for electronic music, hip-hop, and rock. Gearslutz, the largest single studio-gear-interested community, proved a rich ground for our research. More than just a forum, several of its moderators produce trade show video content and maintain an active presence on trade show floors, which adds a celebrity aura to some of the prolific users. The broader Gearslutz business, which proclaims itself to be "the no. 1 website for pro audio," therefore has come to have a disproportionate influence on distribution and retail markets, on designers and manufacturers, and in assembling user communities for many different kinds of audio technologies ranging from studio gear (outboard and transducers) to acoustic treatments (absorbers, diffusers) to electronic instruments (synths, samplers) to software plugins.

Founded in March 2002 by Julian Standen, an American expat in Britain, and Meg Lee Chin, an industrial musician who had lived in the US, UK, and Taiwan, the site was infrequently visited in its first two years. But with the instability around the frequent restructuring of ProSoundWeb's forums, Gearslutz became a new home for some former RecPit members. From the onset, Gearslutz was structured around gear rather than creative or technical practices such as musicianship, engineering, and production. It is a for-profit commercial site, supported by advertisements that manufacturers or retailers buy directly from Gearslutz; based on page impressions it has been estimated that by 2017 the site was earning upwards of $250,000 annually from advertising.[25] Many new companies and products were "born" via user discussion on Gearslutz—for example, Kush Audio, Louder than Liftoff, Michael Joly's mic mods, Foote Control Systems, Steven Slate Audio, GIK Acoustics, and numerous "cloners" of classic circuits including Warm Audio and Stam Audio. Several site moderators launched their own brands—for example, Lindell Audio—or online retail outlets, such as Front End Audio and Zen Pro Audio. Even established brands that predated Gearslutz experimented with workshopping new products on the site; the Empirical Labs Arouser, a plugin related to their very popular Distressor hardware compressor, changed in form and functionality through a sprawling 5,279-post thread. In other words, what happens on Gearslutz is not just gear chat, but manifests in the transformation of gear itself—and the manufacturing of vast quantities of it.

Leaving aside the site's name for the moment, the bulk of hardware studio gear discussion happens in five top-level subforums. Three of the original subforums continue to be amongst the most active: they discursively divide gear into "high-" and "low-" end, with a "middle" category defined instead by quantity: "So Much Gear, So Little Time." This mirrors 100+ years of tiered marketing of consumer audio, pianos, and other musical instruments. Without a clear dividing line between market segments, the "high end" subforum became perceived, wrongly it turns out, as the place where the "real professionals" could be found.[26] "Low-End Theory" became the place where amateurs and do-it-yourself aficionados shared notes on good gear that could be bought on the cheap. The third subforum was for "anything else"—as long as participants subscribed to the notion that *excess* amounts of gear were good. Many users focus their energies on only one subforum; a small subset of power users contribute thousands of posts and have the chance to become respected within that subforum but not necessarily on Gearslutz at large. A fourth subforum, "New Product Alert," is intended for discussion of recently released individual products. All manner of products can be promoted here, from studio gear to instruments to guitar pedals to acoustic treatments to software to entrepreneurial services. Since few products will be of interest to all, this subforum has less of a cohesive social identity to it than the preceding three. We witnessed a vitriolic back-and-forth on a product launch thread for a $799 clone of a $15,000 microphone, for example, which attracted users on smaller budgets, convinced they have the chance to get the same sound as the big studios. These posts are then responded to with insults from high-end gear users who complain about double-blind listening tests and component differences in the clone.

Despite hundreds of thousands of threads within these four subforums alone, there are a comparatively small number of premises for starting a new thread. First, some threads begin by discussing one new piece of gear: its merits, its looks, its interface, its technical features, sometimes its sound, and its relation to existing gear on the market. Second, many threads begin with a question concerning how a particular sound was achieved on a historical recording, where the social exercise becomes fishing for the "magical" or "secret" piece of gear that made it sound so good. Third, a popular thread type is the "shootout" or "A versus B" thread, which might be initiated by a "newbie" (the local term for a beginner gear user) who wants to know "which preamp should I buy, A or B" or might start with a studio session where two or more mics are used on the same application and users partake in a guessing-game activity. Fourth is the "gear porn" genre, which are all the threads that euphemize gear as sex, demand that users "show us your racks," or pose questions about which gear is sexiest, or that consider what qualities define a true gear slut—including excessive gear acquisition. We will discuss this fourth genre in a moment, but in the meantime, consider that all four thread genres start with the assumption that *technologies themselves* are responsible for the kinds of outcomes normally attributed to people; gear has been delegated agency. Whereas thousands of engineer and producer interviews in *Tape Op* document that working professionals know that the most important factors in a recorded part are the musician, a quality instrument (or in the case of singers, no voice-related impediments), a conducive recording environment, and high-quality performances, Gearspace represents the world reversed.[27]

Now for the site name. Although Meg Lee Chin, who set up the technical back end and hosting software for her ex-boyfriend Julian Standen, did not care for the name Gearslutz, Standen regarded it as "edgy" and insisted on keeping it.[28] We wish to clarify that the use of the word "slut," by itself, does not lead to only one potential outcome. Within a feminist or LGBTQIA+ space, sexualized language may have multiple connotations and social functions, and within the industrial music subculture in which Chin had participated, as a member of the band Pigface, "extreme" language and imagery can often be a focal point for social critique, reclamation, and individual catharsis. However, within less than a year of operation, on Gearslutz it was commonplace for users to refer to each other as "true sluts," to refer to gear as "sexy," to call gear dealers "gear pimps," and to "slut shame" other users for not following through with purchasing gear that they "lusted" over. Indeed, the entire site quickly adopted a discourse revolving around framing gear through euphemized misogyny. Threads dedicated to showing pictures of women *next to* gear (one assumption being that women did not know how to use or have any interest in using gear) or holding guitars—pinups preferred—visually contrasted with "show us your racks" threads where male Gearslutz users were never depicted next to their stacks of expensive, boutique gear.[29] Quickly, that discourse became the bedrock of the site, whether in the fifteenth response to a new product announcement or in a thread on the "Geekslutz" subforum where gear technicians, semiprofessional designers, and dabblers might describe a vintage Marinair transformer as "sexy."

Sexualizing gear became a principal technique that forum regulars utilized to preserve a particular social order—in other words, to drive out women, LGBTQIA+ individuals, and any men not perceived to be "true sluts." In spaces demarcated around hegemonic masculinities, "slut" is widely understood as derogatory, critical, and highly gendered.[30] Referring to a woman who has multiple casual sexual partners as a "slut" contributes to notions that consent to sexual activity is a given and thus perpetuates rape culture.[31] Therefore, the "word play" on Gearslutz takes on much darker meanings strongly aligned with power, addiction, and the control of women. Furthermore, the discrepancy between the objectification of women and erasure of men in images underscores a widespread sense of the ugliness of male bodies. The hegemonic masculine formation here, then, does nothing to affirm any potential beauty or worth in the bodies of the male users: worth is only a possible outcome of gassing.

It is indisputable that Standen, a very active site user, was aware of all of this, and aware that the large Gearslutz logo was broadly perceived to be NSFW (not safe for work); for years, the site could be accessed through a mirror—gearsz.com. In the site's FAQ, the question "I am a little edgy about the site name. Do you have a SFW (safe for work) version?" is answered "Yes! You can use **Gearsz.com** to navigate the site, although all this will do is change the logo. . . . This is good if you are worried about who may be looking over your shoulder and judging!" Strikingly, the concern here is not whether the logo, and sexualized and misogynistic discourse more broadly, is problematic to prospective users who find it offensive, or how the name serves to further normalize the widespread violence against women,[32] but whether the nearly universally male users of the site are likely to get "caught in the act"—at work or at home—for looking at a site that *intentionally* blurs the boundaries between audio and pornography.

That said, starting in 2020, eighteen years after the site launched, Standen seemed forthcoming with the discrepancy between his stated values that the site be "welcoming and inclusive to all" and the reality of social interactional norms. In one podcast, after Jules discussed the brief-lived politics subforum, which turned into a "cesspit" due to matters like "right-wingers saying 'here's another jihadi beheading video,'" the interviewer incongruously suggested that Gearslutz "seems like it's in a place where it's like beautifully welcoming of anyone in our community." Jules responded, "I think it's a little bit *lamentable* that we could have more *women* on the forum, and . . . I think we have more than we think we do, because they just don't identify themselves as being female with their picture or their avatar. I'd like to have more *conspicuous* female contributors. And when people do contribute, I'd like the guys on the site to behave better" (emphasis in original).[33] Ostensibly, this sentiment contributed to the decision in March 2021, after years of calls, to change the site's name to Gearspace. So what happened in the three years since? The frequency of *some* sexualized discourse decreased; the word "slut" is only found in 0.06% of threads versus 0.12% before, but the word "sexy" is still popular, being used in 0.09% of threads versus 0.1% or more in previous years. However, some of the decrease is attributable to self-censoring. In a recent thread on the decline of the NAMM trade show, over a dozen responses lamented the end of the trade show "booth babe" era without employing blatant sexual language: the interactional norm on the site has not changed one bit.[34]

Gearslutz inculcates a cultural formation where the generation of cultural capital and social status is partly tied to the ability to rehash the terminological conventions specific to the community. This overlaps with studio gear discourse, although some terms index something different than they would in a professional setting.[35] To perform this discourse and connoisseurship means a user must memorize copious product names and possess the "ability" to correctly differentiate between very similar names such as U87 and U87ai, and KM84 and KM184. One must be able to discuss, albeit in an idiosyncratic way, particular components within these devices that are purported to be responsible for their "sound," whether it might be a particular NOS vacuum tube, German-made electrolytic capacitor, or nickel microphone diaphragm—whether or not that part in fact matters in the way that is claimed. There is a whole vocabulary of what we term "technobabble" that relates in some way to actual electrical engineering discourse but is used differently on Gearslutz, meaning that users drop references to "slew rate," "transient response," "intermodulation distortion," and "femtosecond jitter," even when they may have no idea what it means, whether it matters, and if it does, what degree of the property is meaningful and why. There is also a local lexicon of nontechnical aesthetic descriptors for audio—warm, fat or "phat," soundstage, air—found in studio settings, too, that may have different connotations online. These terminologies and aesthetic preferences converge on the domains of contemporary "audiophiles" (historically termed "hi-fi enthusiasts"), although only to an extent: when audiophile "cable freaks" attempt to discuss esoteric and expensive cables with Gearslutz regulars, they are routinely ridiculed and prodded into continuing to post, becoming regarded as a social media lolcow.[36] Most importantly, the preponderance of technobabble creates a milieu where it becomes hard for any user to recognize actual technical knowledge.

To an outsider to this gear culture, the content moderation policies might seem odd and ad hoc, but they have not changed substantively in twenty-two years, even as moderators came and went. This is impossible to observe in action if one treats Gearslutz as a static text, but clear to see if one is a registered user, checks the latest posts daily, and screenshots threads pre- and post-moderation (although as researchers, we realize our analytical methods differ from the ways some users would engage with the forum). The forum guidelines suggest that "political" and "religious" threads are off-limits, which has been consistently enforced through recourse to a very specific reading of what "political" or "religious" might mean. Threads by evangelical Christians discussing how God showed them how to mix have never been censured or removed, for example. One particularly charged topic consists of Mac vs PC "wars," meaning threads where posters level insults based simply on the computer operating system that each other uses—these *are* moderator-censored or removed if the thread has become too messy. This is another example of the "battles" or conquests staged in gear cultures and thriving of war metaphors.[37] It is permitted—even encouraged—to post pinups of women and comment on the appearance of professional female recordists or musicians, but discussing the exclusion of women from professional audio is considered "political," and such threads will be swiftly removed in their entirety. Posters are free to try to dissect the gear that led to the sound of the classic Motown albums, but posts that talk about the 100+-year systemic exclusion of Black people from myriad audio-related trades are removed for being "political." A Black Lives Matter–inspired thread in June 2020 inquiring simply about whether there were any Black-owned pro audio businesses was removed in three hours when a regularly racist poster complained to the moderators that it was "political."[38] It is apparently fine for posters to quip that using certain kinds of low-end gear will make you "gay" or "ghey," but if a poster talks about how being queer inflects their engineering inevitably, that will result in homophobic tirades that are *not* quickly removed; once, Eliot observed this escalate to the doxing of a trans* person which was left up for days.[39] Conspiracy theories popular among Q-Anon followers (chemtrails, covert government 5G chip implants) were left up for more than a month, as was a veritable death threat from a vociferous anti-vaxxer against specific forum users during the initial coronavirus peak. Not even the death from COVID-19 of Sam Clayton Jr.,[40] who was unusual in being a Gearslutz regular who actually had led a widely respected career (as a live sound engineer and reggae recordist), quelled the brawl. We have witnessed moderation on Gearslutz to be selective and inconsistent, and nonaligned to contemporaneous equality-motivated movements. Moreover, within this conservative cisgendered-heterosexual masculine hegemony, deeply problematic posts remain online and are even enthusiastically quoted—posts that in other fora would result in the user being perma-banned.

All this might seem surprising for a platform that claims to be about professional audio and that obtains media and press permits to cover major audio-related and music instrument–related trade shows. It is less surprising if we realize Gearslutz is just another articulation of the manosphere, comparable with loosely moderated redpill and incel-themed subreddits.[41] The difference, however, is that the Gearslutz manosphere is not constructed simply through discursive statements but requires gear to function. Gear, gear talk, and gear imagery mediate social relations in every subforum. We are not the only ones to note the mismatch between the male posturing forum conduct and how professionals normally discuss similar issues. As Larry Crane (who, recall, had prior forum moderation experience) noted,

> I don't think their experience is broad enough to make the decisions, but then they want to have an opinion and yack off. I don't know why. Notice it's mostly men, or 99.999% men that are doing that kind of stuff. I've never seen very many women. I've seen [London-based mastering engineer] Mandy Parnell . . . show up on a forum and discuss a little bit of mastering ideas, or some equipment or something. Maybe I've seen her on there, but it's very rare, and you're gonna see it's . . . like all these guys trying to do this one-up thing.[42]

OTHER FORA

Gearslutz is not the only gear-focused forum to bear new products and companies. From the maker culture—and electrical engineering–focused forum GroupDIY, launched in June 2004, several product designers launched successful lines of boutique gear—for example, Gyraf Audio, Serpent Audio, Goly/PCB Grinder, Total Audio Concepts, and Hairball Audio. Several businesses set up storefronts facilitating a broader DIY/maker culture through providing difficult-to-source components, project PCBs, metal cases, and kits that come with gear-assembly instructions—for example, those by AML, CAPI-gear, The Don Classics, and Studio 939.[43] Some of these are not-for-profit businesses, if not legally, then at least in principle. While the gendered demographics do not apparently differ on GroupDIY, the interactional style and social capital concerns do. When we asked Jakob Erland, a forum admin, why he felt this was, he noted, "I like the way people address each other. It's kind of like your local bar where you know people will behave."[44] Most forum power users have bona fide electrical engineering expertise and a low tolerance for pseudo-science and the performative technobabble of other forums. The collective goal is to design and make things that work well, and power users appear to be most motivated to contribute to a post if it advances the state of knowledge about audio or shares existing knowledge more widely. In short, legitimate expertise is recognized and rewarded.

One hypothesis we had was that GroupDIY avoided the toxic analogizing, digressions, and culture found on Gearslutz and amateur production fora such as AudioSEX or Future Producers due to its craft/maker ethos.

However, GroupDIY is only one of many places where DIY audio is a central focus. For example, Gearspace's Geek Zone subforum, formerly known as Geekslutz, is comparatively more constructive and expertise-driven than Gearspace's general-interest fora. Even there, though, Geek Zone has hosted copious audio myths and snake oil audio fixes. The balance between engineering talk and technobabble—skewed in some threads toward the latter—makes those threads more prone to caustic arguments. Therefore, our initial hypothesis was not borne out; simply being organized around craft and design did not lead to a preordained outcome. The significant amount of expertise amongst GroupDIY's moderators and power users, and their lesser interest in personally profiting off their designs or participation, both contribute to a different discursive style.

All gear fora, including GroupDIY, are masculine domains. However, there are fewer rewards for the performance of masculinity than on Gearslutz/Gearspace. GroupDIY lacks the combative and lengthy off-topic discussions and diatribes, or the "gear porn" pile-ons characteristic of Gearslutz. Perhaps because of this, GroupDIY has had less engagement: over eighteen years, 80,897 members produced 814,305 posts organized into 59,713 topics—far fewer posts per thread than on Gearslutz and fewer than one-third of the average posts per user compared to personality-driven forums. Any thread is considerably more likely to contain a useful resource, however, because posting from expertise and sharing engineering knowledge are socially valued more than winning an argument.

CONCLUSION

Message forum participation has declined across the board as Facebook groups, Instagram and Twitter feeds, and private Discord channels became the primary sites of online activity. While there is plenty of pro audio and gear-related activity on these, and search and web traffic statistics indicate that Gearslutz and audio message forum interest "peaked" in 2011, to this date no platform has surpassed Gearslutz/Gearspace in terms of reach and continued "engagement."[45] This reflects a generational distinction: of all the forum power users for whom we have been able to glean demographic information, the considerable majority are Boomers or Generation X: people who "came of age" on computer-mediated communications during the era of listservs, newsgroups, and message fora. These "web 1.0" platforms provide more durability for conversations spanning multiple years and do not utilize algorithms to recommend topics of interest. In one sense, they are "slower" and less trend-based media,[46] they provide users more of a sense of control over which threads they want to view or participate in, they enforce looser character count limits, and they do not demote active threads for not getting enough "likes." In another sense, they are socially conservative media that present the illusion of free speech while subjugating individual experience and contrasting perspectives to the inertia of an ill-informed, gear-fetishistic groupthink.

Professional media creators working in audio, photography, video, and other fields will often note some variant of the expression "you can't polish a turd," and no amount of gear or recordist expertise can make up for a poor performance. On audio forums, however, this notion is reversed. Due to their technological agency, gear *can* polish turds—but only if one uses the best microphone through the most expensive signal chain and if one discusses the gear with the appropriate local technobabble argot and sexualized language.

IV ENCOUNTERING GEAR

Until the advent of digital recording technologies at the turn of the 1980s, gear was housed in concealed spaces. Gear imagery in professional recording studios, particularly shots of well-known musicians sitting next to lesser-known recordists who leaned over large-format mixing consoles, would have rarely circulated outside the occasional photo in a music magazine. In the 1980s, cheaper digital technologies and consumer-model multitrack recorders proliferated amongst a broader cross-section of musicians and amateur recordists, leading to nascent technology-centric communities. Commensurate to the rise of the contemporary music technology press in the 1990s, a cottage industry of behind-the-scenes documentaries, such as the *Classic Albums* series (1992–present), and the growth of postsecondary audio and music technology education, exposed the professional audio technology domain and its gear to broader audiences. Print media, videos, and education laid important groundwork for twenty-first-century gear cultures.

To understand the multiple valences of how gear culture participants encounter gear, we will examine the paradoxes of human/gear agency, "shootouts" and other social performances of testing, and the practices of gear organization and ornamentation in studio spaces. In our analysis of the agency configurations around gear, we distinguish between two concepts, control and agency, to examine the first paradox: increasing the number of *controllable* parameters may result in a decrease in perceived *agency* for the user. The second paradox concerns the juxtaposition of a "free speech" sentiment within gear cultures, with the widespread attributions of agency to gear, including displaced agency, which diminish the possibility of personal agency for a gear culture participant. One place this can be seen is in teleological analyses of recordings from "the golden age," which entail irresolvable paradoxes concerning what actors and factors—human and technological—were responsible for that "classic album." This also can be seen in *displaced agency* configurations, where a technological object becomes allegedly imbued with the personal agency of a famous recording engineer or musician.

Testing is another prevalent form of gear encounter. The audio technology and trade media, trade shows, websites, and online fora often use testing to inspire gear culture participants to the amateur reviewing and evaluation of gear. Whether that means Samantha partaking in a blindfolded "magic trick" event at NAMM, or a ninja-themed technology shootout site in Aotearoa, or the competitive and ludic nature of interminable online "my-EQ-is-better-than-your-EQ" forum threads, testing is a main locus of gear sociability, and one way gear culture participants make sense of their gear.

Finally, fetishization goes beyond individual objects and becomes central to organizational norms in gear cultures. Gear users love nothing more than sharing and looking at vast quantities of gear organized and arranged in specific ways. Gear is depicted racked and stacked, adorned and aglow with blinking fairy lights, and perhaps set off with a bubbling lava lamp atop an animated meter bridge. From "Kurt Cobain's microphone" to studio cats, gear is encountered through agency configurations, often flawed ideas of testing, and conservative organizational norms.

9 AGENCY

THE STEVE ALBINI/KURT COBAIN MIC

"It's irreplaceable," declared renowned alternative music recordist Steve Albini about his Electrovoice PL20 dynamic microphone. "They stopped being microphones and now they're artifacts," he described two of his Lomo 19A9 tube condensers. The PL20 is a common choice for a range of purposes in professional studio recording and radio broadcasting. A slight variation on the ubiquitous RE20, the PL20 retails for around $350 USD on the secondhand market. The less common and now scarce Lomo 19A9 is a 1960s Russian valve microphone, again used for a range of professional instrument recording (usually in stereo pairs), and retails for around $1,500 USD secondhand. Albini's microphones were, however, not just any microphones; Albini used them in 1993 to record Kurt Cobain's vocal performances on Nirvana's multiplatinum-selling album *In Utero*. In 2017, Albini decided to auction the microphones in two separate lots on Reverb .com. Around one month later, the PL20 sold for $11,750 and the pair of Lomos for $9,300. Even high-profile gear events like this one are never quite enough to quell the insatiable appetites of gear culture participants, and lively debate ensued.

Albini's "Kurt Cobain" microphones exemplify one way that technological objects become imbued with agency. Unless intentionally modified, the technical and operational aspects of gear—for example, a microphone's construction and its capacity to behave as a transducer—are the same regardless of who is doing the operating or whose acoustic sound is being transduced into electrical signal. In the PL20/Lomo example, however, such agency is displaced, and gear's valuation relates to that displaced agency. We say "displaced," because the microphones here stand in for the creative artistry of Cobain and/or Albini, yet Cobain and Albini are irrelevant to the technical and operational aspects of the microphones. It would be impossible to recreate the identical set of circumstances in which these microphones were used, these were only three among many used on the sessions, no one buying the microphones today would be able to record the late Kurt Cobain (or Nirvana, for that matter), and the broader workplace, recordist, and gear combination is also unique. The excess valuation and personification of the mic—this PL20 sold for more than thirty times its normal cost—is only possible since Kurt Cobain sang through this microphone, since Steve Albini recorded him doing so, and because the result of that performance and recording is heard on a canonized recording. Such excessive value articulates gear heritagization and valorization, but the gear itself comes with the promise that the microphones are at least *part* of what made *In Utero* a multiplatinum-selling album. It is as if Cobain's artistry and Albini's recording prowess are now baked into the microphones, which carry the potential of recreating such a successful combination.

Albini's microphone sale elicited a broad range of responses. From an(other) opportunity to celebrate and commemorate the work of a posthumously canonized artist, to a purported moment of "democratization," many reports in the music press pitched the sale as an opportunity for fans to "own a piece of rock history." The sale also provoked outrage, particularly in online gear cultures, where some participants resented the excess value placed on these otherwise relatively accessible microphones: "We don't enjoy history traded as currency," wrote one Gearslutz user in a lengthy thread. The thread ignited with noble tales of users selling similar microphones with history and provenance at market price without informing the buyers, with accusations of opportunism and a "cash grab," and with exasperated cries of "how *could* he" (sell them). Therefore, many gear culture participants intrinsically know that agency should not *solely* be ascribed to technological objects, and when it comes to gear, not all agency is equal.

AGENCY AND CONTROL

Agency is the ability to act or to produce an effect. Some rationalist writings about agency restrict agency to humans a priori. We instead follow a twenty-six-year tradition of understanding agency *relationally*—and of not conflating agency with a much more limited idea of "human conscious intent."[1] Therefore, we append our initial definition of agency to consider how widespread attitudes toward action, effects, and causality lead to routine practices and material assemblages that subsequently (re)produce persistent effects in the world. Sociotechnical systems are notoriously complex to analyze, inspiring descriptors like the "mangle of practice."[2] Many participants in such systems—for example, technological users and nonusers—do not really understand how such systems or technologies actually work, and gear cultures are no different.[3] Users may not even recognize if they are in control of a technology. Yet part of the miracle of modern electrical technologies is that users *can* operate them, at least to an extent, even if they have no idea how they work or what these technologies can and cannot do.

One challenge for us in writing *Gear Cultures* has been disentangling the mismatch between our own experience and expertise and those of some gear cultures participants. We understand how this gear works, we have built some of it ourselves, and we use it to teach audio engineering practices or to make commercially circulating recordings. Clearly when some professional tracking and mixing engineers perform skilled embodied operations with and around gear, these are "felt, learned but not or no longer rationalized,"[4] and those moments of flow are an example of the "dance of agency"—the ongoing give-and-take between human and technological agency.[5] In other words, certain kinds of work benefit more from flow than from rational conscious intent. However, our research brought us in constant contact with the magical beliefs of many kinds of gear culture participants, whether non-using collectors or working professionals. Following Bortolotti, we recognize that even though such beliefs are irrational and ill-grounded, in many cases for the believer, they are examples of "epistemic innocence," meaning that they may have temporary positive psychological effects in boosting an individual's perception of their own agency—even if the facts suggest otherwise.[6] Our interest, therefore, is to attend to the widest set of "agency configurations" that are operative here.[7]

The agency concept is insufficient by itself to explain the beliefs and practices about technologies and their use; interface designers do not consciously try to design in agency but instead design in *controllability* in divergent ways. Within the same class of object, say microphone preamps, one preamp may have been designed to provide nearly no user control (the Telefunken V-72), while another provides extensive parameterized control (Dave Hill's Europa). The same control parameter can be presented to users via different kinds of interfacial elements, each with unique consequences. A parameter presented via a switch or button prioritizes precision and recallability but limits the number of options for that parameter, while a parameter presented via a rotary potentiometer or fader provides the illusion of infinite tweakability but precludes precise recall.[8] Control labeling may emphasize engineering, musical, or metaphorical aspects—or controls may lack labels altogether.

All objects result from design decisions about which parameters to expose or obscure or hide from the user, which parameters should feature precision recallability and which should promise infinite variability, and how use choices will be described. Correlating controls with the explanations made about them in user manuals, in manufacturer blurbs, and in reviews, we find a wide variety of outcomes, roughly arranged around three types of scripts.[9] An object may be designed to give a sense that the user is in musical control—if I use this box, it will make everything sound better or have more mojo—but unaware of its circuit-level components and technical aspects. Of course, gear users do not *need* to know the component-level construction of gear to use it to great effect, any more than someone needs to know the technicality of a kettle's element to make a cup of tea. An object may appeal to the user's technical competence and knowledge but obfuscate musical-aesthetic outcomes. Other objects are designed to be mysterious sites of play where the user's ears ideally guide their use of the object, or where they can delight in the act of pressing a "nuke" button during a mixdown. Through these three script types, we see that musical, technical, and ludic aspects are in tension: designing an object to control one aspect will de-emphasize or preclude the

Table 9.1
Possible perceived and/or operative agency/control relations—basic

	Agency	Control	Agency	Control	Agency	Control	Agency	Control
Human user	x		x			x		x
Tech object		x	x			x	x	

others. No single design strategy has yet to emerge as an accepted best practice, though, or as the script for the preferred user experience.

In our research on gear use, we found that, contra the extensive literature that defines "affordances" as *properties* of an object,[10] knobs *do not* always afford turning. Madeline Akrich's concept of scripts examines this very problem. Her discussion focuses on how designers attempt to script idealized uses of objects, yet the objects may be refashioned or reconceived by end users to some extent.[11] In a similar manner, Christian Pentzold and Andreas Bischof note it is more useful to understand affordances as *processes*.[12] Counterintuitively, when examining the agency configurations around a particular object, there is no obvious a priori relation between agency and control; users may feel (or be externally perceived by others as feeling) a sense of agency even when they are not in control of an object, or they may feel a lack of agency when they would be apparently in control (table 9.1). Things are further complicated when considering how some technologies are scripted to serve as a surrogate for having an actual experienced engineer in the room, leading to situations we describe as being characterized by *displaced agency*.

AGENCY IN GEAR CULTURES

The post-2000 used and new market for gear is buoyed by the widespread myth that the sound of recordings is primarily a result of technological objects—more than the strategic use of them, the musical performances that were fed into them, the skills of the individuals who made them, or any other number of sociomusical factors contributing to the overall sound of a recording. In gear cultures, technical objects ascribed with agency as such are asserted to have a greater importance on music/sound-recording processes than the user. This belief reduces the sense that a famous engineer's sound was most defined by their skills and cumulative expertise applied to a complex recording workflow, rather than by the brute use of a tool configured with one "magical" set of parameter values. In this configuration, technological objects, rather than music, become imbued with the aura typically associated with an artwork and its "unique existence at the place where it happens to be."[13] Gear inflated in importance in this manner tends to be "vintage" gear. The EMI TG 12413 limiter ceases to be merely a box for reducing the dynamic range of a signal; it becomes synonymous with Abbey Road Studios, with the "British invasion," and with the now-legendary engineers and producers who in 1966 worked on The Beatles album *Revolver*. Owning that box, the promise goes, allows you to own a piece of that musical, technical, and cultural legacy.

This relates to widespread perceptions that by turning GAS into actual acquisition, a user/consumer will be able to participate in some space of activity related to "the golden age" of audio—and hopefully leverage the acquisition for the temporary gain of cultural capital. By extension, by *not* turning GAS into acquisition, by *not* owning this auratic technology, one is excluded from this space of operations. In other words, this is a cultural norm of ascribing technological agency and discounting user agency. But, within this configuration, there is *some* user agency: the ability to act on GAS, to acquire or not to acquire. Acquisitions, when made, lock the user into a constrained sound palette or set of operations, despite that during the "golden age," many engineers were trying to overcome those very same constraints, particularly regarding noise reduction. From a capitalist standpoint, it is awfully convenient that the main path for user perceptions of their own agency results from marketized transactions for luxury goods that stand in for knowledge or experience. The social performance of those acquisitions, whether in the form of message forum posts or the museum display of hoarded objects, demonstrates how acquisition-centric agency should be understood as a defining feature of gear sociability, too.

Participants within those gear cultures defined around this agency configuration, then, are scarcely conceivable as independent agents acting on their own free will. After all, within these gear cultures, individual skill sets and experience are secondary to object ownership, and individual choice is secondary to replicating technologies and their settings from the golden age of audio. Rather, gear culture participants who perform their conspicuous consumption of luxury goods hope to buy their way into participating in a gear culture rather than having to confront their irrational magical beliefs—about music writing, composing, arranging, and performance, and about performance practice and technical expertise and the social-collaborative nature of studio work—which would disrupt their sense of epistemic innocence.[14]

From the technological object's vantage point, so much has been invested in its acquisition, placement, and shootout testing, that the object has been blown up and exaggerated to a caricature of itself: the Trump baby balloon of the studio. On the one hand, there's a reductionism at play whereby musical writing, arranging, composing, performance practice, engineering, production, programming, and even vinyl cutting become reduced in significance compared with semiautonomous "technological effects." On the other hand, the discourse around these objects vacillates between a form of constructed knowledge that features stylistic allusions to facts and maths and science, and colloquial aesthetic attributes that get mapped to specific named parameters for newly released gear. "I need more presence" leads to the desire to turn a presence knob rather than going through a process of critical listening that would reveal that the perceived lack of presence was actually a specific frequency and/or amplitude imbalance.

DISPLACED AGENCY AND PERSONIFIED TECHNOLOGIES

A widespread myth avers that using a specific piece of gear will enable one to achieve the same sound that characterizes legendary albums and, by extension, legendary engineers—what we term *displaced agency*.[15] Such perceptions are capitalized on by manufacturers. Take, for example, the reissue of the AMS RMX-16 digital reverberator. "A classic—reimagined. The world's first microprocessor-controlled, full-bandwidth digital reverberator has characterised an enormous number of seminal recordings from the 1980s onwards," reads AMS Neve's strapline. The first problem, though, is the inability to obtain *exactly* the same technological object that was used in the original instance of any one of these seminal recordings—excepting the Kurt Cobain mics or other occasions when that exact piece of legendary gear is sold on the used market for grossly inflated prices. This is due to the auratic effect we discussed earlier. Next best would be a well-maintained vintage model of the same object, followed by, in decreasing order of desirability, a contemporary reissue by the same company, a "clone" made by a boutique company (e.g., Chandler, Retro), a Chinese-manufactured clone (e.g., Warm Audio), or, if all else fails, a software plugin emulation of the piece of vintage gear (UAD, Softube, Brainworx). The same displaced agency functions at all levels, and we see it feed back into design and manufacturer blurbs, such as in the AMS RMX-16 reissue, all the way down to plugins. For example, the software company Waves Abbey Road Reverb Plates claim to model "the reverb sounds on Beatles, Pink Floyd, Radiohead recordings."

Not only do plugin companies emulate or model or take impulse responses of vintage gear, they also model the specific units and equipment settings of famous recordists. This appears to have begun in 2007–2008 with the software company Waves releasing the CLA (Chris Lord-Alge) presets for their (modeled) SSL plugins. We have seen—and definitely heard—CLA demonstrate these at multiple trade show events where he is a gear culture draw. Since 2008, a whole subgenre of *personified technologies* has emerged, where the line is blurred between whether the company is selling you a technological object or selling you some part of that person.[16] Here, agency questions become more complicated. Plugin sale residuals do create a new income stream for Andrew Scheps, for CLA, and for the other handful of engineers who have been packaged in such a form. That said, such products market the idea that you no longer need to have this legendary engineer in your studio with you to make your mix sound like it was recorded or mixed by him. As of 2023, it's always been a "him"; no women or non-binary people have been commodified as personified professional audio technologies in this manner.[17] As a result, the idea of the Scheps or CLA "sound," which developed over the decades when both worked in professional audio, has been reduced to the basic parameterized control of a few skeuomorphically rendered knobs.

These examples do, however, raise questions around who or what has agency in gear cultures. To what extent is agency real or perceived among gear culture participants? And what are the implications of the personification of Kurt Cobain through an Electrovoice PL20 microphone, or Andrew Scheps through a plugin?

In displaced agency configurations, agency exceeds a dyadic relation between the user and the technological object. Sticking with the plugin examples, this involves, at a minimum, four entities: 1) the plugin user, 2) the plugin, 3) the engineer being personified via the plugin, and 4) the gear owned and/or used by that engineer that was modeled by the plugin "manufacturer." These four entities are not freely associated, however. Setting aside empirical evaluations about whether the plugins come close to achieving their marketed aims, let us assume for the sake of argument that the plugins *are* successful in *perfectly* replicating the sound of the gear owned by the legendary engineer. If this is the case, then where does that leave questions about the agency of the user? At best, anything they do will be mimicking the legendary engineer and they will never attain their own sound: doing so has been deliberately scripted *out* of the plugin by the designer. Moreover, if replicating gear objects or gear settings produces an outcome that is *the same* as having had that engineer or musician in the studio, this negates any sense that specific skilled people including musicians were responsible for the creation of legendary albums and discounts the sense that the legendary engineer ever had any personal agency at all. If we assume, alternately, that the plugin falls short, then not only will the user be restricted to trying to mimic the legendary engineer, but they will also fail at doing even this activity. User agency, within displaced agency configurations, therefore, is limited to mimicry and conspicuous commodity consumption.

AGENCY FOR WHOM, FOR WHAT?

Thinking through agency questions beyond the scope of the individual consumer, we run into unresolvable contradictions. On the one hand, companies, marketers, journalists, and academics announce that the latest product, whether a piece of gear or a plugin, will "democratize" recording or make engineering more "accessible." On the other hand, it appears that most of the gear being bought does not get used for music-creative work and making recordings that circulate; gear appears to function more consistently as social mediators within the gear culture, as will be examined in chapters 10 and 11. Regarding accessibility, professional audio technologies continue to be designed around assumptions that consumers are "fully abled" regarding hands, vision, and hearing. Ability is defined based on normative assumptions about Euro-American normative early adult male bodies and their size/strengths, just as was the case with US and UK instrumentation research during World War II and the immediate postwar era: for the many people with physical impairments, blindness or color blindness, or profound hearing loss, these technologies may be less usable or wholly unusable.

Of course, that's not even taking to task the problematic assumption that a $1,000 box that only can perform a narrow range of functions or DAW system requiring a monthly subscription fee are financially accessible to the 1.3 billion citizens of the forty-six least-developed countries or the three billion additional people living elsewhere in poverty.[18] Even within a privileged first-world context, the promise that owning the means of production will allow anyone to compete with world-class studios belies the reality that all commercially successful popular music depends upon songs and creative inspiration that are realized by teams of often geographically dispersed, skilled individuals—not by isolated engineers or a string of gear purchases.[19]

To buy or not to buy? Gear cultures limit the range of potential human agency to the field of consumer choice, which is as much true for gear designers as it is for general gear buyers. Gear designers and gear mediators are consumers too, purchasing components, circuit boards, interfacial elements, solder, and cases; some of their choices are pragmatic, while others pertain to fetishism at the component level. Gear user concepts—the technobabble and performed connoisseurship—certainly play back into these designer choices. Gear mediators, such as magazine equipment reviewers or web forum moderators, are also consumers, as noted by Théberge.[20] When reviewers and moderators stage their own positive encounters with gear—"I bought the review unit!"—prospective consumers frequently model their behavior.

In presenting prior versions of this research, we fielded questions about our analysis of agency. We could not possibly believe that these technological objects actually *have* agency, right? Isn't agency a specifically human matter, a quality of political efficacy? Is this just another case of the popular Marshall McLuhan adage that "the medium is the message"? But these questions miss the mark in many ways. They are the wrong questions to ask.

Yes, systems of belief, myth, and symbolism are ubiquitous in gear cultures, but they are second-order effects. Gear cultures are first and foremost social formations around masses of gear that result from the substantial material reorganization of the world. They arise at the time when these material reorganizations would seem to be less necessary than ever before. The agency configurations that depend upon technologies' magical effects cannot be explained away *only* as those delusional beliefs that relate to epistemic innocence. Neither can they be explained through a game of pretending as if technologies have agency. They structure social relations in hegemonic cultural formations and are an obligatory passage point for participating in gear cultures.

CONCLUSION

Gear imbued with agency, whether the sonic signature of a renowned recordist or a posthumously canonized artist, or a container of fetishized components, exaggerates its social potential in gear cultures. The promise of a particular sonic signature artificially inflates the technological potential of gear as it increases its fetish appeal. In gear cultures, agency is less related to the use value of gear or its potential in a professional recording scenario than to the social engagement potential during gear acquisition. However, branding gear with a famous recordist fetishizes it via a particular recordist's process or technique. This undermines the many factors contributing to successful audio production, including writing, arranging, performing, and recording. The packaging of alleged recordist agency into gear attempts to commodify the uncommodifiable: the years of experience acquired through iterative professional recording practice. Gear that embodies such agency claims to allow gear culture participants a perceptually closer proximity to the golden age of sound recording and its concomitant recordists and workplaces. This displaced agency is present throughout gear cultures and beyond. Our fourfold model of displaced agency contributes toward analyzing broader music technology cultures where software, particularly plugins, have more recently come to stand in for missing professional recordists. Ultimately, attributed and displaced agency play an important role in gear cultures in structuring social formations and increasing fetish value and gassing potential; however, such agency configurations preclude any possibility for personal agency outside of consumer transactions. We doubt very much that Kurt Cobain would approve.

10 TESTING

TESTING AS DRAMATURGY

There's a normative routine for browsing the trade show floors at AES and NAMM. You go to a booth. They have a new technology on exhibit. Would you like to listen to a comparison? Sure, why not. You don some headphones after a fleeting anxious thought about contagious diseases of the ear or hair. The test begins with the engineer excitedly discussing their new gizmo while you're trying to listen and trying to filter out the 86dB ambient noise level of the trade show floor. It's impossible to hear anything, so you take off the headphones, nodding, improvising something like "Oh yes, I definitely hear a difference, good work. What was this called again?" before taking a business card and wandering off to another booth. The act is complete. The test, although unsuccessful at demonstrating the gizmo, did serve its primary function in constructing the subjects of the play: the excited engineer, the curious customer, the authoritative testing apparatus, the business card, and the technological object: not-yet-concretized but yearning for a future where it is regarded as distinctive and useful.

In NAMM 2020, however, we experienced a test that took the staging and dramatization a bit farther. We had just entered the audio wing of the NAMM show and intended to do a quick walk-through in preparation for taking photos and conducting interviews. We had only been in there a minute, long enough to reach a booth with a few headphone stations set up, where we were approached by a suntanned man named Jeff wearing a sleeveless leather vest, no shirt who asked us, "Do you think you have good ears? Do you want to try your chance at winning $100?" Jeff gesticulated to a transparent lanyard around his neck, dangling in front of his bare chest, containing the $100 bill. We had no clue what was going on, and a quick scan of the booth of the generically named 3D Audio (unrelated to Lyn Fuston's company called 3D Audio, which we discuss later in this chapter) suggested that this was a new microphone company. Regretfully, Samantha agreed to give it a go, although we still did not know what was going to actually be tested. Jeff and his assistant were not quite ready to do the test, so Eliot went over to the listening stations to listen to the 3D effect the "Barnaby microphones" were intended to impart on a peculiar set of sound sources: the University of Georgia military band, some classical guitar, the California wildfires, piano, and some foley/sound design for film.

Samantha was called over to the raised carpeted stage by Jeff's assistant, a tall blonde woman we estimate was in her twenties, who spoke slowly in a voice that sounded like she was imitating early science fiction android speech and who (we discovered later) was not familiar with the technologies and appeared to only be there for the staging. One male booth employee pulled out an expensive DSLR to begin filming the stage show, and, almost spontaneously, a crowd of onlookers gathered to watch the spectacle unraveling in this gear booth. What in the world was going on? The assistant put a blindfold and a pair of noise-canceling headphones on Samantha, and Jeff explained that Samantha would win $100 if she could tell repeatedly whether she was listening to Jeff jangling keys in real-life or a recording made with the Barnaby mic of the same keys being jangled.[1] From Eliot's spectator vantage point, it looked like Samantha was the unwitting audience member who was about to be sawed in half at a magic show. From Samantha's perspective, the whole experience felt in part like being a stooge at a comedy show or like being an unwitting participant in a stranger's kink; the combination of Jeff's leather waistcoat, bare chest, blindfold and strong cologne, and the camera and the assistant, all served to sexualize this gear-testing stage.

This test was reminiscent of Edison's infamous "tone tests," which were major public events where the promise of wax cylinders was demonstrated.[2] However, this booth purported to be exhibiting *professional* audio equipment, not a consumer playback device. With 3D Audio, such public stagings were updated with an eye for kink and fetish—one whip away from role-play. The microphone was scarcely of importance in a staging that foregrounded the unwitting audience member (in this case, Samantha), Jeff's leather-clad body, and the filming of the encounter as if it were an amateur public porn shoot. The lack of disclosure or consent about the blindfolding and filming was particularly unnerving but perhaps was intended to make the otherwise pointless staging stand out in a room full of comparatively boring show booths.

Nothing about this test is related either to the kinds of tests that scientists and researchers do to enhance our understanding of human hearing or the tests engineers and manufacturers need to do to perfect their products. However, it does relate to a popular nonprofessional domain of testing that is as old as audio technology itself. As we will explore, even in less kinky examples, popular tests are always performative and involve staging. Tests exist not only to demonstrate expertise and proficiency but to (re)define what counts as expertise and to re/shape the conduct of practice. The performativity of and around testing is socially constitutive and is a key means through which technologies become social within a gear culture.

TESTING AS INTRA-ACTION

Testing is a primary mode of *intra-action* between people and audio technological objects.[3] Whether it's a designer using a scope to measure power supply ripple, a beta-tester putting an object through its paces in a studio setting, an end user conducting a shootout between several similar objects, or a group of forum users poring over the methodology of an informal ABX test, seemingly everyone can be involved with testing in some way. Yet, surveying much of the published literature on audio technology, one might conclude that testing was a fringe activity or only happened if it was in service to some subsequent and important goal: expanding engineering or psychoacoustic knowledge, developing a new product, or making a technological choice in a studio in service to an artistic goal.

Testing scenarios are "manifestations of practices, ideas, values, and institutions" and can often become an end in and of themselves.[4] Though we might colloquially say that technologies are being tested, listening tests and analysis visualizations are just as much about employing technologies to test and validate human ears, eyes, knowledge, skill, and social bonds. Testing always involves at least four base entities: the testing human, the audio technology, the testing apparatus, and data emerging from the test. In the case of informal online testing scenarios, additional entities include active audiences, those who ask for follow-up testing or become amateur testers themselves; their commentary on the test may gain a life of its own. Gear tests generate significant and intense discourse, particularly online but also in print media and, as our opening example suggested, in bizarre trade show scenarios. Therefore tests, like technologies themselves, produce their own associated milieu.

Testing is not simply a revealing of the world "as it is"; rather, as Karen Barad notes, it is a "[site] for making meaning."[5] Therefore, testing as a space of activity is especially revealing of the intra-active nature of technology use in general: the test itself produces the phenomenon that is being measured and concurrently produces the subject and objects—even unwitting and blindfolded ones. One example of this we observed in action concerned technical developments in analog-to-digital and digital-to-analog converters, a recurring technological problem because the implementation of sampling theory in clocking circuits results in a kind of distortion known as jitter. When we talked at AES with Dave Hill (1954–2023), a recordist and the lead designer of the Crane Song line of boutique audio equipment, about his new Quantum converter technology, he enthusiastically obliged a detailed story about buying extremely esoteric measurement equipment that could measure femtosecond-scale duration jitter—and learning himself how to hear what femtosecond jitter sounded like (a femtosecond is 1/1,000,000,000 of a microsecond).[6] He played us recordings to allow us too to hear the difference, which *was* audible, even through a budget pair of headphones on the noisy trade show floor.

Without the data generated by that testing device's worldly interactions, arguably no one would know that there was something significant to measure that could not be measured already with less accurate (picosecond- or nanosecond-scale) equipment. Without a regime of experimental listening tests that produced repeatable results, there would have been no basis for Hill to correlate the measurements with significant auditory phenomena. Until Hill was able to produce a subsequent set of materials for prospective customers so they could learn to hear femtosecond-scale jitter too, few if any would have known that anything as fleeting as a femtosecond might matter to them. Many might not even have known how to hear jitter at all separate from the many subtle distortions—intermodulation distortion, quantization noise—that might become audible after converting between analog signals and digital data, and vice versa. Therefore, while femtosecond-scale jitter is an empirical thing in one sense, it is also a *constructed fact* that is dependent upon technological apparatuses. Unlike the imaginary phenomena one finds in consumer audiophile magazines such as transparency, soundstage, choking, and veil lifting, femtosecond-scale jitter is not pseudoscience, and Hill's converters are renowned for their clear and musical sound as well as their technical excellence. But the associated milieu around this test-produced phenomenon overlaps with other tests of converters, especially the hundreds of less rigorous converter shootouts on Gearspace's "Mastering" and "High End" subforums.

TESTING, TECHNOLOGICAL DESIGN, AND QUALITY ASSURANCE

Since gear is intended for musical applications—typically to help make recorded music sound better—we wondered how professional engineers conducted testing (from the prototype to design to quality assurance phases) and what role human listening and test equipment might play at different phases. As we found, regardless of how it is conducted, testing is not just an emplaced practice but relies upon years of cumulative experience. As Robin Porter of AMS Neve told us, "It's years and years and years of listening to stuff and making judgments . . . there's a good selection of building blocks and stuff like that which you know you can trust."[7] But who does the listening, what are they listening for, and how do listening practices relate to test measurements?

Equipment designers do not share a consistent professional and/or musical background, and therefore, who listens to prototypes can vary. Some engineers, such as Dave Hill and George Massenburg, have long undertaken recording studio–related work and trust their own "golden ears" to determine when a new design sounds good. Others, including Dave Derr (Empirical Labs), Geoff Daking, Ted Fletcher (Joe Meek / Ted Fletcher Designs), and Skipper Wise (Blue Microphones), were touring and recording musicians before transitioning into design and bring their own musical sensibility to bear on testing scenarios. Designers who began their engineering pathway through gear repair, including Marshall Terry (Terry Audio, Shadow Hills) often describe an arduous process of learning to listen to the nuances between working vintage gear and a piece they are repairing. However, designers such as Jakob Erland (Gyraf Audio), Peter Montessi (A Designs), and EveAnna Manley, at later prototype stages, involve a small group of trusted beta-testers who are established working recordists in the vicinity of their workshops (Denmark and Southern California, respectively) and revise prototypes, when necessary, based on this feedback. We also personally witnessed David Royer himself testing each microphone prior to it leaving the small factory floor (although maybe with a little assistance from the shop's dogs!). Even if there is no clear correlation between designers who personally possess musician-informed listening or recordist-informed listening and the creation of designs that are "more musical" or "better in the studio," for everyone we talked with, listening mattered at multiple stages of the process. In the words of Hutch Hutchison, "You have to breadboard. You have to try it. You have to listen, you have to tweak, and adjust by ear. Anybody who designs audio gear before listening is probably gonna end up with something mediocre."[8]

Most professional audio manufacturers, and even a few affluent amateurs, own precision test equipment, including audio analyzers, oscilloscopes, multimeters, bench power supplies, and signal generators. Such equipment can be essential for testing aspects of circuit prototypes, for measuring the performance of components, and for quality assurance testing of built products before they are shipped. Precision test equipment is often expensive: The lauded Audio Precision APx555B, a high bandwidth signal analyzer, costs well over $30,000. Beyond the basics of knowing how to use such equipment and interpret test results, we found a range of attitudes toward signal analyzers and related equipment, and the situations in which they provide useful information.

For example, signal analyzers are indispensable during quality assurance testing for determining if component variability caused a tube microphone to fall outside a company's internal specifications. However, for Rick Perotta, analyzers must be supplemented with human critical listening:

> Our tube microphones get burned in for a week before we even test them because that'll weed out any weak tubes, capacitors, et cetera. After they're burned in, they are placed in our semi-anechoic chamber with a calibrated reference microphone. We use a computerized system by Audio Precision that runs the performance curves on the microphone. Then, every microphone is performance and shock-tested before it goes to David Royer to be listened to himself.[9]

However, distortion analyzers will always show some kind of measurement, but it may not indicate if the distortion is appropriate for that application or if the distortion will be perceived by listeners as sounding good. For Jakob Erland,

> In the good old days, I invested a lot of money in distortion analyzers and frequency sweep analyzers. And to my horror I found out ten years ago that sound has nothing to do with frequency domain or even phase domain. It's somewhere else . . . the kind of, what you would describe as "magic." You can measure some distortion on the [Urei] 1176 and it'll be way higher and sounding good, then the same distortion figures in a power amplifier sounds horrible.[10]

This "magic" of "somewhere else" is not a rejection of scientific rationality on the part of Jakob, but rather an acknowledgment of the limited parameters—the constructed facts—that current test equipment can and cannot measure. It also acknowledges the subjectivity of assessing which professional audio equipment sounds good or not and the ways wider gear culture discourses attribute some sonic parameters to supernatural or otherwise unexplainable forces. The harmonic distortion of compressors and amplifiers can be repeatedly measured, but the threshold for suitability varies widely across applications. Moreover, despite the increasing precision of test equipment,

> Many of the AP tests are at the limit of the test equipment residuals. I'm convinced that there are dynamic characteristics that we simply can't yet measure. We can hear the problems, but the problems don't always correlate with static test results.[11]

As we noted before, testing produces the phenomenon it is testing. LaGrou's "dynamic characteristics" hint at the problem of analyzing musical sound that is always simultaneously the result of more than one characteristic through equipment that is designed to isolate individual ones. Even when using quality test equipment, human subjectivity enters when determining whether those measurements are significant.

Many designers and companies rely, to an extent, on external beta-testers and expert users to provide feedback on product prototypes. Samantha has undertaken such work in the past, and we were interested to discover, by chance, that this is another site where women have had a substantial yet publicly uncredited role. In our interview with Terri Winston, renowned San Francisco–based producer/engineer/recording artist, former professor of Recording Arts at City College of San Francisco, and founder of the successful educational nonprofit Women's Audio Mission (WAM), we had been chatting about perceptions she and WAM staff and students might have about vintage and contemporary gear.[12] We were familiar with their educational mission, which reaches nearly 4,000 women, girls, and gender-expansive (trans*, non-binary, intersex) people each year, but also knew that there might be facets to their work that were less known, including their role in beta-testing:

> **TW:** Women's Audio Mission has *become* the beta-tester, because all of the gear list is donated, and often a part of that is that they want us to try anything that is new. So we beta test *a lot* of different products.
>
> **SB:** That's a huge and diverse list of stuff!
>
> **TW:** It is and it's growing. Just today, two different microphone companies came: they'll give it to you if you try it. It's cool because then all of our students become basically a part of that process, they get to try all of the different gear. I was doing that for years earlier, so I think it kind of makes sense, so they're "now we're really interested" cos there's this wide audience [WAM students] that's going to use it and they'll get that feedback.
>
> **EB:** Has the feedback that you've provided been incorporated into the redesign?
>
> **TW:** It has! One in particular I remember: Countryman came with these really amazing omnidirectional tiny microphones . . . I said I wanted to put something permanently in the studio that we could just use that you

> couldn't really see. So they go "I have the perfect thing for you." And I said "ok, the only problem with it is I need right-angled XLRs on the head, cos otherwise it's sticking way out and you're bumping into this thing." And now you have an option, you can order it with right angle XLRs. So it's things like that, or "this didn't work and it's going to break."[13]

While a lot of professional testing equipment is used to test gear, we also see diverse testing methodologies. Taking microphones as an example, testing always includes frequency response and polar pattern measurements, but idiosyncrasies will depend on manufacturers' unique testing techniques developed over long periods and their perception of market competition, much of which is informed by a personalized iterative process informed by more than data collection. As Tim de Paravacini stated, "My listening over the years has been working with live music in my younger days but also designing my own system at home for listening and evaluating sound. I am never fully satisfied until I am sure it is better than the competition."[14]

TESTING SKEPTICISM

Amateur gear culture participants appear to be initially drawn to testing for several reasons. An amateur recordist might be unhappy with the sound of their recordings or mixes and become motivated to explore if one of their gear objects is the cause of the problem. A recordist might wonder which of the many Neve 1073 clones on the market sounds the most like an original vintage unit. Similarly, a tinkerer may claim that replacing certain components in an object will open up the sound, prompting other users to compare before-and-after recordings of unmodded and modded gear. An enthusiast might read about kinds of distortion that result from the analog to digital to analog conversion process and decide to become an amateur engineer-scientist. An experienced engineer may become upset about pseudoscientific claims made during a test of the above types and launch their own counter-test to "expose" the fallacy.[15]

While some test types make sense—a comparison of different mics recording a snare drum, for example—others are technobabble, pseudoscience performances. In the mid-2000s, a rumor spread on Gearslutz that mixes made using the same software and same settings would sound different on Windows XP than on Mac OSX. In another rumor from 2008, mixes sounded "wider" with more "definition and clarity" depending upon the model and interface connector of hard drive on which the data was stored. These rumors apparently died out by the 2020s, but myths surrounding the algorithms used in DAWs to perform the final mix (sometimes called "bouncing" a mix, "rendering" an output, or "summing" the tracks) persist. The myths are so pervasive that Lyn Fuston, proprietor of 3D Audio Inc. (not to be confused with the 3D Audio company mentioned earlier), sold a CD entitled the *Awesome DAW-SUM Comparison* that allowed users to hear the same mix rendered through dozens of different DAWs and launched his 3daudioinc.com message forum to discuss the seven such testing discs he released.[16] Do these aspects being tested impact the sound of recordings? Very few people would be able to perceive femtosecond-level differences in clock jitter, yet excessive importance may be placed on such microlevel parameters.

While our focus is not on DAW-based production and its tools, rumors and myths like these perpetuate the skepticism of digitization and of computer-based production that in turn constructs and reproduces "vintage, analog" gear as the logical other. They also define the formal conventions of testing within amateur spaces. The same testing tools (digital analyzer software) and methods (ABX testing, null tests) perform the informal tests of the (mostly analog) hardware gear that we do focus on here, where the very same skepticism looms: does the reissue microphone, preamplifier, EQ, or compressor sound like the original? These tests rarely prove anything, but they are *socially* essential. Notably, they allow gear enthusiasts to role-play. The cast of characters in the testing playbook includes the *golden ears* who can hear phenomena others cannot, the *perpetual skeptic* who will once-and-for-all debunk the flawed science behind digital audio, the *connoisseur* who has the discernment to know the difference between vintage gear and its modern clones and reissues, the *entrepreneur* who offers to sell mysterious technological fixes to solve any manner of audio malady, and the *amateur scientist* who argues with the myth and rumor perpetrators. An inordinate amount of time, money, energy, and material resources goes into developing, perpetuating, testing, and debunking audio myths. Despite the maturity of the science and the technology, little of the published research that could inform these questions is ever consulted. Yet again, we find a space of activity that, in this skeptical formation, is predicated on the fetishization of gear and the misattribution of agency.

SHOOTOUT ON THE EQ CORRAL

Shootout: A fight in which two people or two groups of people shoot at each other with guns. (*Cambridge Dictionary*)

Meanwhile, in Aotearoa, "THIS IS GEARSHOOT AND IT IS ABOUT STUFF TO DO WITH AUDIO GEAR" proclaims, in all-caps, the front page of Gearshoot, a website run by Waitakare-based Kog Mastering Studios. Like other staging grounds for gear culture shootouts, Gearshoot plays heavily on military and combat fetishization. "By using this site," they claim, "we aim to give you the crucial upper edge and Ninja like reflexes at picking which piece of gear is which blindfolded." Users should want to do this, since it provides "a way to break your myths and legends apart and dispense with the evil of all evils . . . 'Confirmation Bias.'"[17] We see the continuum here with our introductory example: A/B'ing gear with a blindfolded participant generates excitement in some gear culture members. "Blindfold testing" occurs in many kinds of research and is designed to elicit honest responses from participants when presented with choices, usually to pick something that, with visual deprivation, tastes, smells, feels, or *sounds* better.

Parsing the implications of this preamble, a ninja is a professional in Ninjutsu, a Japanese martial art developed in feudal times for espionage purposes, which hones skills in camouflage, stealth, and sabotage. The ninja is associated with mythological and supernatural powers (Kuji-kiri), weaponry including the katana, and concealment or disguise. While consumers do not need camouflage for gear acquisition, the implication is that acquisition begins with difficult choices that must be made—choices that require a high level of skill in fighting. By analogizing the ninja, rather than other types of fighters, one implication is that knowledge of gear and its attributes is hidden: only with a specialized, military-level training will those elements be revealed. Another implication is that gear knowledge is intrinsically linked to mental strength, physical capacity, and a somehow "higher level" of human endeavor. The Gearshoot user's newfound espionage skills will reveal audio secrets, all wrapped up in the spirit of war fetishization.

Gearshoot's enemies include a long history of botched shootouts and pseudoscientific methods for pitching one piece of gear against another. That said, Gearshoot itself is far from a work of science. They cheekily note:

> Gearshoot is aimed to be a semi scientific site. That is, it's as rigorous and anal retentive in its procedures as we can do without making it either a) really boring to collate, or b) impossibly time consuming to create. If you want peer reviewed scientific level shit we could recommend you go to a range of journals in the databases at, for example google scholar.

Gearshoot knows its methods are no match for the kinds of research undertaken in universities or appearing in the *Journal of the Audio Engineering Society*. "Scientific level shit," mocking and mildly derogatory, downplays the seriousness of such research. Gearshoot provides quite detailed, albeit anecdotal, background information on all pieces of gear available to be shot at and frequently cites Gearslutz as a source. Gearshoot knows its visitors want testing to be fun, accessible, relatable, and easily understood—and sees the large enthusiastic and communicative audience in Gearslutz threads. Forget about rationales, methods, and research questions. Which gear sounds better? That's all we want—that's all we *need*—to know.

What, exactly, is being shot out—or *at*? Compressors, microphone preamplifiers, EQs and effects processors: the main outboard equipment that constitutes gear. As they state,

> Here you can do shoot outs on a range of different audio toys and hear, in the comfort of your own home, studio, or wherever you have some monitoring system what they sound like in relation to each other. Shoot out your favourite plugin emulations versus the analog original and bust a few myths or preconceptions you have made.

By referring to gear as "audio toys," Gearshoot blurs the idea of gear as weaponry with ludicity. This conflation is not, however, all that unusual, as Peppino Ortoleva noted about the proliferation of paradoxical aspects of play without purpose in gaming. Ortoleva recognized how "controlling the theater of war is made easier through a clarity that would not be possible in reality,"[18] and while he refers specifically to combat gaming, such games evoke Gearshoot's GUI staging of gear combat. As Lakoff and Johnson demonstrated in war metaphors' deep embedding in society, Ortoleva notes, "The rise of the homo ludicus is happening in a fragmentary, complex and contradictory way."[19] Indeed, war and play have been inextricable in the gaming industry for decades, and the Gearshoot shootout GUI is playful, at least in some sense. The shootout interface is detailed and features extensive potential combinations of gear. Presets include Manley Massive Passive vs. UAD vs. the Pultec EQP-1A, pitting analog devices against digital emulations, where other presets pit legacy analog devices against present-day counterparts. Choosing a preset results in tracks appearing in a playlist window, which can then be A/B'd using a solo function. Additionally, the effect can be heard on a range of instruments by selecting an icon in the GUI: guitars, bass, drums, keyboard, vocals, and whole songs. Crucially, in the "Blind Shootout" option, the preset is blurred, and two options—thumbs-up and thumbs-down—encourage the user to select which gear sounds better. The colorful interface is easy to navigate but gives the user little to no idea about specific settings. Additionally, and in line with what Ortoleva recognized as the purposeless of such play, there appears to be no outcome to the exercise: no forum for users to discuss their gear combats, no points to score, no prizes to be won. This is shooting the gear out for its own sake and indulging in gear battles via a metaphorical play-war paradox.

CONCLUSION

Testing gear, whether blindfolded, with analyzers, or shooting out, is a primary means of gear culture socialization. From participants technobabbling among forum threads as they thrash out the pros and cons of a vintage vs. a reissue preamplifier to those voyeuristically indulging in a blindfolded microphone test at a trade show, gear testing has immense social potential in gear cultures. Academic research and the types of testing conducted at manufacturing stage by gear designers, and amateur ABX shootouts and forum-based comparisons, constitute two divergent testing milieux. The former are routinely dismissed by gear culture participants as "too scientific": perhaps the technically out-of-reach nature of this type leaves them less useful for gear culture participants in assessing the gas potential of technological objects. In online gear-testing discourse, rarely do participants refer to or share peer-reviewed research. Rather, testing functions as a simplistic means of performing an encultured kind of gear connoisseurship, in a social arena that prioritizes knowledge about and ownership of gear over its use value in recording workflow scenarios and practical work with real musicians. Contrast this with Larry Crane's description of testing as being an integral aspect of experience-based learning of audio engineering:

> If you have ten really great completely different microphone preamps and you did a shootout one day on one kind of mic and one kind of source, you get a certain feedback. Like, "Oh wow. I don't like that on vocals." But then if you do the same . . . a different mic on an electric guitar . . . and then you do the preamps on the overheads and different mics on overheads and different rooms and you do that on all these different sources for 20 years, you're probably going to have a very different opinion than if you do like one shoot out on one source at a Vintage King shop, which people do, and valid enough.[20]

For gear manufacturers, testing is the antithesis of the ludic kind. Involving rigorous and repetitive processes of "burning in," all manufacturers we interviewed mentioned their painstaking gear-testing processes. Often involving multistage procedures, on our visits to gear manufacturer sites, we witnessed many modes of measuring, listening, and response-testing applied to every piece of gear prior to leaving the factory. Testing means different things to different people in different corners of gear cultures—even in the far corner of the NAMM exhibition floor, whilst blindfolded.

11 ORGANIZATION

WILL YOU TAKE THIS GEAR TO BE YOUR WEDDED WIFE?

In late 2022, New York audio engineer David Silverstein posted to his hugely popular meme account—Audiohertz.[1] The meme began with a short video of two men, a groom and his best man, on a wedding day (figure 11.1). Struggling to contain his emotion, the groom turns away presumably from the approaching bride until, with a beaming smile full of pride, his best man lends him a friendly hand on his shoulder and slowly turns him around. The groom cannot contain himself, looking then pulling away in tears and grimacing with emotion as the object of focus becomes clear: draped in a white floral wreath is an image of two racks of gear. On the left-hand side, a Neve 1073 preamplifier sits atop a Pultec EQP-1A and two dbx compressors. On the right, a Furman power supply tops two Empirical Labs Distressors, a UREI 1176, and a lunchbox of API 500-series preamplifiers. If only men could marry gear, they would.

Internet memes have been described as "a piece of culture, typically a joke" that operate on multiple levels.[2] The "image macro" is the central image(ry), the base of the meme: in this instance, the wedding scene and its protagonists. Multiple layers fold into the meme: the overwhelming emotion of the objectophile groom, the pride of his best man, the floral arrangement, and, of course, the gear, which functions as the punch line. This is what Patrick Davison called the "meme ideal"—that is, the central concept or idea conveyed. In this instance, the meme ideal is around marrying gear; however, the meme operates on further levels and features "countless

Figure 11.1
David Silverstein's Audiohertz meme.

submemes."[3] Since the meme was posted to Audiohertz, the intended audience was Audiohertz's followers, presumably people involved in the audio industries—including many gear culture participants. Multiple gear culture values and ideas are portrayed in this single meme: gear is highly prized and sought after, specific types of gear are idealized, an image of gear can suffice as a replacement for real-life gear, owning gear affords significant emotional satisfaction, and there is great pride in witnessing gear or in witnessing others in possession of it. However, further gear culture tropes are embedded in this meme. First, as in most gear culture scenarios, gear replaces a missing woman (in this case, a wife)—objectophilia standing in for the institution of marriage and its commensurate ideologies of love, commitment, intimacy, cohabitation, and long-term interpersonal relationships. Second, the gear itself is not situated in this beautiful floral surrounding; rather, it is an awkwardly placed two-dimensional *photographic image* of gear. Third, the image does not feature just one piece of gear. Indeed, the crux of this meme is the image of two *full* racks of gear, each 16U (28") high and housed in a light pine floor-standing enclosure: not just any gear, but canonized pieces by Neve, API, dbx, and UREI. Though gear is fetishized in this meme, it is the aforementioned highly specific, idealized *organization* of the gear that makes the joke resonate so well with gear culture participants. Gear cannot be presented on its own. The meme presents a gear ideal: racks full of gear. Such organizational and presentational norms in gear cultures are reinforced in numerous ways, perhaps none more twenty-first-century than via social media memes.

PAIRING, CHAINING, RACKING, AND STACKING: HOW IS GEAR ORGANIZED?

In chapter 7, we examined a Sweetwater catalog advertisement that situated gear in a deliberately organized way. This advertisement is indicative of broader gear organization conventions in gear cultures. What types of gear qualify in a studio setup and how gear is displayed and organized are of critical importance. "Show us your racks!"—an infinite cry into the Gearslutz void within nineteen dedicated threads and in hundreds of additional ones—points to how gear cultures thrive on the reality and imagery of large quantities of gear stacked together, one on top of the other, in racks of many shapes, sizes, and designs. The ubiquitous rack-full-of-gear image, routinely shared on social media, reminds gear culture participants of organizational norms: one piece of gear is not enough, a rack filled with gear must be present in any gear situation, the rack unit may entail skilled carpentry and exotic woods like mahogany rather than the more commonplace steel, and/or the rack may act as a surface for accoutrements or ornamentation. From home, project and large-scale recording studios to higher-educational facilities, from retail demo rooms to trade show exhibition floors, the rack full of gear is second only to a mixing console in denoting gear culture status. However, such racks remain out of reach to many gear culture participants. One piece of gear is costly enough, but a full rack? When the rack full is not present in any of these locations, its imagery is.

Racks of gear primarily appear in professional recording studios. Indeed, the placement of the gear rack can be traced back to 1960s–1970s studios where racks featured Pultec EQP-1A equalizers, Teletronix LA-2A leveling amplifiers, and Fairchild 660 limiters.[4] Studios are often described as spaces for creativity, music, or production workflows, and this certainly provides the only plausible justification for them existing at all.[5] The ways studios warehouse and organize gear, the way gear is organized within them, and how gear is situated among other kinds of objects reveals much about the people who organized the studios. These two conceptualizations of studios, creative space or gear warehouse, can be profoundly at odds with each other. We have observed and worked in many studios where the quantity of gear and its prominent placement left little room or comfort for humans beyond one or two recordists.[6] When the needs of gear—ventilation for tube gear, heavy-duty racks to hold excess amounts, miles of cabling, or organizational schema that privilege visual effects over audible ones—take precedence over the needs of any musicians that hope to perform in the space, that studio is no longer primarily a space for collaborative music-making. Instead, these studios become *shrines to gear*. It is precisely these shrines that become replicated across gear cultures: in the pages of retail catalogs, in the promo ads for higher-education facilities, in online gassing, and in social media memes. Individual technological objects can be fetishized, but the fetish effect is most pronounced when gear is racked and when excess accumulations of gear produce a euphoria equivalent to the all-consuming emotion felt on a wedding day.

So far, we have mainly analyzed pieces of gear in isolation to understand why objects become fetishized, how they are designed, what their aesthetic regimes are, and how they form print, online, and event-specific gear cultures around them. After being purchased, however, they enter studios (professional, project, and home) where they are always in relation to other pieces of gear. Four of the most frequent verbs—pairing, chaining, racking, and stacking—exemplify these relations and come to describe how gear must be organized.

"Pairing" and "chaining" refer to two or more pieces of gear that their users configure to accomplish a singular task. Due to the modular nature of signal processers, and the peculiarity that microphones lack internal preamplification and thus need extra boxes to make their output usable, multiple pieces of gear have become the norm. This differs from many historical smaller studios where an analog console was the single object that accomplished preamplification, equalization, mixing, and monitoring duties. Today, gear forums host thousands of threads where prospective users wonder which preamplifier will pair with a particular mic in order to complement a baritone voice type, what signal chain (in other words, which combination of separate EQs, compressors, and other signal processing devices) will allow them to get the bass sound that a "legendary" engineer achieved on an album, or which monitoring controller will pair best with their studio monitor speakers. To be clear, using multiple pieces of gear to produce a result is not inherently fetishistic. What *becomes* fetishistic is when social interaction around questions of pairing and chaining becomes an opportunity for individuals to increase their social status by demonstrating connoisseurship and their "golden ears," to display their knowledge about who used specific technologies in the past, or to control visitors to the studio including musicians. These conversations excessively emphasize the causality of the paired and chained gear, rather than the musician, instrument, and performance.

What gear to use for what purpose? A user's overcoming of these social hurdles transitions into a gear organization problem when these objects are placed in a studio. Should that signal chain be racked together? Racking refers to one of two things. First is the placement of 19"-wide standard rack-mountable gear into custom equipment racks, which hold typically between 4U and 16U of gear, each stacked vertically; 1U is 1.75" high, and each piece of rackmount gear is typically 1U, 2U, or 3U high. Second is the insertion of 500-series "lunchbox" modules into a powered 500-series rack, which (depending on the rack) can hold between 3U and 11U of small modules stacked horizontally. Some 500-series racks fit within regular 19" racks, meaning that users may end up having racks within racks.

Users employ differing organizational schema for their racks of gear, sometimes organizing things around standard signal chains, sometimes organizing gear by type or brand or faceplate color, or sometimes even organizing gear in such a way that when all the room lights are turned off, the glow from the gear produces a pleasing visual effect. Stacks of racked gear, all turned on and fully glowing, emulate the awe-inspiring outdoor and indoor light spectacles dating back to the late nineteenth century.[7] It creates an impressive effect—for the owner/user, for the indirect users (musicians), for non-musician visitors to the studio, and, most of all, for members of online gear cultures. The effect will differ among these four groups, however. For musicians—indirect users who may have familiarity with some gear but who are not entrusted to use it in this studio—the racks may alienate and underscore the knowledge and power differential between the studio owner/recordist and the musicians. Due to the ad hoc organization, few visitors to the studio, even technically proficient ones, will understand how or why the gear was racked the way it was. Therefore, guild-derived modes of work organization continue to dominate.[8] The architectural and topological organization of gear enforces this alienation.

WOMBS, MAN CAVES, AND BUNKERS: GEAR IN SITU

Studios come in many forms and are designed to insulate people and audio technologies from different forces and phenomena. Historically, studios are concealed private spaces that harbor all kinds of technological and processual mythologies.[9] Aside from the best-known large multiroom studios and the teaching studios that support audio technology degree programs, the most common types are home or project studios—places where one recording musician, engineer, band, or producer works and plays. Such project studios are the most personalized and unique of all studio types, reflecting the aesthetic preferences of the owner/user, even in the cases of project studios with iterative acoustic designs (Westlake, LEDE, FTB) commissioned from professional studio designers.[10] Yet such spaces also conform to organizational norms in gear cultures.

While not formal types per se, common terms used to index three qualitatively different kinds of project studios are wombs, man caves, and bunkers.[11] Wombs suggest nurturing places, ones that appropriate the nurturing labor of mothers and assign that role to audio technologies and studio ornamentation. Once again, gear references women even while studios exclude female bodies. Man caves are a home subdivision (attic, basement, spare bedroom, or shed) where the man can be insulated from his family and family obligations, extending a long-standing battle over the organization, decoration, control, and use of home common spaces, especially in the Anglophone first world. Bunkers also insulate, providing a command center where one can be protected from battle and war—and from the encroachment of civil society and politics.

In visiting or working in examples of such places, and in comparing those places with pictures shared online of related studios, we are struck by how audio technologies—racked, stacked, hoarded, and placed in such a way that they fill much of the available space—are called upon to perform different kinds of emotional labor and to buffer against different kinds of human social problems. Wombs create an immersive yet escapist environment where the gear interfaces (especially the LED indicator lights and VU meters), alongside Persian rugs, lava lamps, wall tapestries, and fairy lights, complement a hopefully transcendent listening/audition experience—whether listening to something new being recorded or to classic albums. Most self-described man caves we have seen, even if they feature similar material culture to wombs, have a different gestalt: their clutter and material imposition are deliberately alienating to people who are not studio-interested (i.e., non-gear culture participants), and they are organized in such a way to be quite uncomfortable to most, even the man cave owner's studio-interested friends. The gear is called on less to nurture than to force all human participants into finding uncomfortable positions to inhabit and to ensure that the technologies must be the center of any human social interaction that transpires in the space. Therefore, the presence of gear and the man's prowess at successfully defeating "the wife and kids" is here staged through performing his ownership and command of technologies while the friends (or paying studio clients) uncomfortably look on as witnesses to a successful gear conquest. Bunkers may be command centers for imaginary spacecraft that thematize and fetishize the military origins or applications of much of the stuff of studio audio. Gear racks, here, serve as allegories for technological instrumentation found in "mission critical" environments where users face life-or-death situations: in planes, submarines, and rockets. Instead of retreating into the womb or staging social antics in a home, bunkers imply an ongoing external war, as well as command over that war requiring elite training in highly specialized technological operations. Some studio bunkers are literally bunkers, too. In one case, a pseudonymous Gearslutz user, during the first year of the COVID-19 pandemic, made a semi-advertisement on the site: a self-described prepper, he had built an underground bunker with a full recording studio to weather the upcoming "apocalypse" but had only one week to find a woman to live with him in the bunker/studio. We never learned if he was successful, nor what became of bunker man.

ORNAMENTATION

That gear must exist in large quantities, with those quantities organized in specific ways, is simply not enough. For the widespread dissemination of gear imagery to effectively reproduce gear's fetishization, gear must be depicted decorated with accoutrements or placed near other types of ornamentation. Just like pairing, chaining, racking, and stacking, gear ornamentation must conform to established gear culture norms; not any decoration will do. Veblen recognized how ornamentation related to specific types of consumption over and beyond that required for everyday life: gear decorations push those excesses beyond the gear itself.[12]

Since consoles are central to studio layouts, one coveted location for ornamentation is atop the console's meter bridge. The meter bridge—the studio's primary source of LED lights, with a top plinth usually at a 45-degree angle to the console—behaves as a mantelpiece in the gear setting. The mantelpiece has been a mainstay focal point in Western domestic spaces since the nineteenth century.[13] Objects positioned on mantelpieces represent "a material culture set which forms an integrated, organically related, or metonymic, unit."[14] Clocks, candles, framed pictures or photographs, objects made from precious metals including silver, crystal, and bone china are common mantelpiece ornamentation examples that lend matter to the broader relationship between one's environment and one's negotiation of social, economic, cultural, and ideological forces (figure 11.2).

Gear cultures' mantelpieces follow similar functions. Decorating gear draws attention to it, highlights its special importance, and creates a focal point within the studio. One of the most common meter bridge ornaments is actually a kind of gear. A pair of Yamaha NS-10 nearfield monitors sits atop

Figure 11.2
Meter bridge ornamentation relations (adapted from Pearce 1999).

numerous studio meter bridges, including A Sharp Studios and Studios 301 in Sydney, the ANU School of Music studios in Canberra, and Lush Studios in Brisbane. While typically considered to be low quality, their similarity to home hi-fi speakers makes them useful as "real-world reference" monitors. But why do we describe this gear *as ornaments*? In gear mythologies, numerous tales point to US recordist and mastering engineer Bob Clearmountain as the catalyst for the NS-10's widespread industry uptake. Since then, in online and trade show discourse, gear cultures participants learn a predictable script about the polarizing monitors—for example, "If your mix sounds great on NS-10s then it will sound great anywhere" or "the secret is to put toilet paper atop of the cones." In practice, the NS-10s are rarely used in many studios and serve as an ornament that marks gear culture connoisseurship.

Clearmountain's own Mix This! studio in California houses an excellent mantelpiece example. Although busy and cluttered, the meter bridge of Clearmountain's seventy-two-channel SSL 4000-series console features all kinds of gear accoutrements that reinforce social, economic, and material cultural aspects. Two lava lamps adorn each end of the bridge, with some patching extending out into even more gear. Numerous trophies, a dB meter, a smiley face stress ball, some books and model cars, and photographs of Clearmountain's wife are all positioned between three large Dynaudio monitors (but interestingly, no NS-10s) on the mantelpiece. Cartoon figurines of female musicians perched atop the monitors atop the bridge extend this gear ornamentation upward and outward. Validations of agency, tradition and heritage, creativity, and economic status are all carefully organized, balanced, and displayed at the center of Mix This!.

The lava lamp's significance extends further. Such lamps symbolize fluidity, creativity, and "glow," but also serve to validate the historical, cultural, and social norms within gear cultures. Since lava lamps are often found positioned on console meter bridges, they do fit in Pearce's functionalist analytical framework, though we often find lava lamps on the tops of racks full of gear or on desktops next to computers. The lava lamp is an iconic popular culture object of the 1960s that is synonymous with the hippie movement, psychedelia, peace and love, interior design, and, of course, music. An "affordable frivolity,"[15] the lava lamp is a fun icon, often chosen because it is "fascinating, decorative, and soothing."[16] As such, the historical affiliation of the lava lamp parallels the golden age of sound recording and its canonistic values. With its soothing fluidity and spherical, elliptical, shapeshifting gentleness, the lava lamp scaffolds gear in its predominantly masculine milieu as ornamentation depicting femininity and relaxation. Its gentle and gliding movements juxtapose the rigidity of fixed racks of gear and, as such, is analogous to sex. The lava lamp is analogous to the gear itself, too. Such lamps demonstrate "phenomena such as the transformation of energy, force and motion,"[17] precisely the operational role of the gear. The lava lamp, almost on its own, creates an essential atmosphere: the "vibe" of the gear milieu must be set off with these illuminated waxy globules bouncing around, preferably in a dark studio room.

Figure 11.3
Fairy lights and Timmy, the studio cat, at Black Mountain studios, Canberra. Photo by Samantha Bennett.

In earlier work, we recognized how illuminated gear, with its flickering LEDs and eye-catching glow-in-the-dark features correlated to sex work signage.[18] Strings of fairy lights are essential gear ornamentation and work on several levels to accessorize and fetishize gear. Fairy lights emulate, exaggerate, and multiply the illuminations of gear LEDs, thus amplifying the sexualized setting of the studio. In a darkened room, softer forms of illumination are normalized and, as such, fairy lights provide all-important "glow" where it is most needed. Fairy lights on long strings of cable can be draped over multiple gear racks, over a console meter bridge or computer, or over acoustic treatments and window frames (figure 11.3). These lights create beautiful bejeweled atmospheres and juxtapose the heavy, dark, and metallic materiality of gear (with its hard, rectangular lines) with prettiness, color, and a delicate ethereality. Here too, the fairy lights act as a substitute for missing women. Carolyn Marvin noted the numerous levels on which indoor illuminations have functioned since the late nineteenth century. Such light spectacles have been historically affiliated with supernatural, "fairyland," and "dreamlike" atmospheres. Illumination emblemizes intimacy, glamor, and jewelry and may conjure environmental spectacle including rain, rainbows, and flowers. Fairy lights also connote astronomical metaphor; gear culture words including "twinkling" and "blinking" describe both gear LEDs *and* fairy light ornaments.[19] Furthermore, fairy lights are perhaps most synonymous with Christmas—specifically, the Christmas tree. All these evocations—space, bejewellment, Christmas, sexualized settings (brothels and boudoirs)—are other-worldly in nature: they evoke imaginary worlds.

The cover of Wendy Carlos's trailblazing album *Switched on Bach* depicts a typical 1970s synthesizer-centric studio. While the Moog modular synthesizer is the central technology in this image, it represents broader gear-organizational norms in two significant ways. In her book *Switched on Bach*, Roshanak Kheshti analyzed this record cover and the symbolic meaning of its composition. Underneath a piano stool, pages of scores, and spiraling cable is a Persian rug, a mainstay floor covering in recording studios. As Kheshti points out, "When we consider the juxtaposition of the highly symbolic 'Persian rug'—with its capacity to not only bring another world into the studio but also carry those inside the studio to an elsewhere—we understand it as a portal to a different time and place from the otherwise closed world of the recording studio."[20] For Kheshti, the rug is an "orientational metaphor" that carries a "backward-looking, and Eastward pointing weightiness to the studio. The positioning of such a rug in proximity to gear is nostalgic, reinforces past values as present-day priorities, and symbolizes a move "away from digital toward a warm, fuzzy past".[21] To that end, the Persian rug is a kind of gear ornament that scaffolds colonial traditions with imagined "far away" lands. The Persian rug, as Kheshti states, also functions as a "magic carpet" that symbolizes the othered East to the centered West. It carries the potential of imagined transportation and journey, analogous to a creative process. The Persian rug-as-gear ornament is designed to be "felt"; as such, it comforts, softens, and absorbs the hardened edges of gear and studio furniture. Additionally, the rug offsets the prioritized gear in the workspace, transforming the recording workplace into a space for creativity, musicality, and production.

The second norm found on Carlos's record cover is the studio cat, or, in contemporary terminology, #studiocats. A quick scan of this hashtag across Twitter and Instagram reveals hundreds of cats in various poses, yet always in close or immediate proximity to gear (figure 11.3). Whether sitting at the producer's chair, draped over a gear rack, or spread out over console faders, the cat functions as gear ornament, bringing sentience into technological spaces. Against often inanimate gear, the cat also functions as a "helper" or "assistant" and may also be considered a "humble acknowledgement" of the "many forces at work" in a music production workflow. For example, Bob Clearmountain's cat is a mainstay of his many mix videos and social media posts. The sentience afforded to the gear site by such studio cats animates and enlivens it. These feline vital accoutrements are so popular because they can stand in for human agency and validate the gear as existing in and of itself—without needing to function or operate as part of a music-focused recording process. The widespread sharing of online gear imagery, or gear photography in catalogs, recording studios, and retail sites, rarely features gear owners or production personnel. Rarely does a gear culture participant include themselves in pictures of their gear; even "unboxing" videos almost always include only the unboxer's hands. Therefore, the studio cat's imagined agency alludes to—and illudes—the role of humans in gear settings.

MY PRECIOUS

Like Gollum in possession of his ring, some gear is particularly precious: it must be treated as if it were rare, highly prized, and in need of protection. Veblen recognized how "precious stones and the metals and some other materials used for adornment and decoration, owe their utility as items of conspicuous waste to an antecedent utility as objects of beauty." However, to an owner of such precious materials, the value is less in the "intrinsic beauty" and more about the "honor to which their possession and consumption confers" or the "obloquy" such ownership prevents.[22] Gear possession is certainly honorable in gear cultures, but some gear is more precious or honorable than others. As such, through several means, gear is presented as if it were ornamentation itself, or packaged like jewelry, crafted to emphasize its preciousness. Recall Manley's advertisement of "investment grade audio jewelry."

At the 2020 NAMM Convention, we visited numerous professional audio manufacturer booths that organized their gear with preciousness at the heart of their displays. Royer, for example, presented its dBooster preamplifier in a bell jar atop a shiny mahogany plinth, next to an award. Soyuz microphones displayed *their* microphone booster on dry ice, as if it would be ruined if kept at room temperature. UAD's entire booth experience was based on an organizational aesthetic somewhere between a museum and a jeweler, with each interface in a glass dome or plexiglass cabinet. In all these displays, lights were arranged to highlight and draw attention to the objects.

The most common preciousness in gear cultures pertains to microphones, particularly vintage microphones, their reissues, and other expensive contemporary microphones. Microphones are often presented as if they were jewelry. The Blue Bottle, a large diaphragm tube condenser mic, is operable with multiple interchangeable capsules. These are available to purchase individually but are also available as a flight case holding all eight capsules. Upon opening this case, each microphone "bottle cap" is snugly fit within a molded interior surrounded by crushed blue velvet. Against this luxurious, opulent material, historically associated with royalty and aristocracy, the capsules appear jewel-like and precious. These microphones are reserved for special occasions, not for daily use. Furthermore, the casing and fabric surrounding each capsule lends a visual appeal that invites technological voyeurism. There is no need to take these microphone capsules out of the box. Simply open the box; its matching blue ribbon will extend to hold up the lid, and the capsules may be gazed upon like precious stones. The excessive cushioning and coddling found in these microphone jewelry boxes is not unique to Blue. This presentation style was undoubtedly modeled on the cushioned blue satin and velvet interior of the 1960s Neumann/Gefell CMV563 boxes. The original AKG C-12 condenser microphone was originally presented in a molded box lined with gold satin or red velvet, and the Schoeps M221b small diaphragm condensers came in silver cases with red-velvet-lined molding and gold satin adorning the inside lid. Coles 4038 ribbon microphones are presented in molded plastic cases but inside are kept in place in small drawstring black velvet bags, as if collections of precious stones. German company Madooma specializes in reselling secondhand vintage microphones and retails bespoke luxury cases for microphones, each one lined in rich shades of dark purple or burgundy velvet.

These sorts of presentations are useful for gear cultures in delimiting precious microphones from cheaper models, many of which come with simple faux-leather zip bags or gray foam-molded plastic cases. Gear cultures have, therefore, successfully categorized their microphones into two tiers: those that are ubiquitous, cheap, for everyday use, and widely accessible, and those that are scarce, occasion-only microphones, expensive investments if not for practical use, inaccessible, and precious. Satin and velvet are important denominators for gear culture participants organizing their microphones since both materials are linked to exclusivity, and, critically, keep "bourgeois identity (materially) intact."[23]

CONCLUSION

From gear brides, to "show us your rack" forum threads to the nurturing or protective potential of gear in wombs, man caves, and bunkers, gear and its myriad ornaments are called upon to do a lot of work. While there may be disagreement about the relative sexiness of individual pieces of gear, images of studios chock-full of gear, with a fully lit meter bridge of large format analog console at the center of the frame, will always provoke the most affirmations of sexiness. The careful placement of Persian rugs, lava lamps, fairy lights, pinups, and (luck permitting) a studio cat completes the mood and organizes the gear—counterbalancing its sharp edges, metal fascia, and military technoaesthetics. Similarly, presenting microphones as studio jewelry within "royal" velvet-lined enclosures frames those mics as gear rather than as the studio's mundane "workhorse mics." Additionally, the fetish potential of the gear's locale is heightened if precious gear is known to be present.

We use the term "gear fetishization" as the overarching frame to theorize all these gear organization strategies, but the rest of the terminology is not our invention—it is common studio parlance in all Anglophone gear cultures. Gear fetishization attempts to materialize femininity, motherhood, and the cis-gendered, heterosexual female partner of marriage or relationships into the gear hoard of the studio. It carries the expectation that gear will provide the companionship and nurturing capacities of the missing woman and the feminine image to be gazed on to provoke sexual arousal. Fairy lights are ideal gear ornaments because they work on multiple levels in this materialization of femininity. This obviously seeps into the sexual metaphors of gear discourse. However, we immediately notice contradictions since the workplace's non-gear ornamentation is crucial for its womb, bunker, or man cave effect to work—both in producing those effects for the studio's owner and in structuring social relations if there are visitors to the studio. Despite several decades of experiments in aestheticizing gear interfaces and exteriors (e.g., Focusrite Red, the Blue Bottle mic), gear itself is neither sexy nor an effective surrogate for the missing woman. It only becomes so when placed within an ornamented studio, or if a gearslut *imagines* its placement in their own studio.

V GEAR ATTITUDES

All kinds of wondrous and fanciful ideas are baked into and projected onto gear. Gear materialities, gear stagings, and gear encounters are all grounded in fetishization. Gear attitudes that begin at the point of raw material extraction do not simply end in an online forum argument about who's got the sexiest rack. From war metaphors to tips and tricks, nostalgic reminiscing of the so-called "golden age" of sound recording to posthumous musical and gear canons, gear culture participation is framed around several ideologies, tropes, and metaphors. The gear attitudes that emanate from gear cultures feed back into the design and manufacturing of new gear, too. From designers to end users, all gear culture participants in some way recapitulate three specific and normalized gear culture attitudes—secrets, heritage, and obdurance—that also shape gear's social formations.

Since the turn of the millennium, access to compact digital sound-recording technologies has widely proliferated. Smartphones that record sound at ever-higher qualities, multitrack DAWs that ship as standard (free) computer apps, and free downloadable plugins have blown open the means of sound production to a growing user base. These curious discourses, then, stand in for the missing workflows and experience-borne expertise usually associated with gear.

Secrets are intrinsic to gear culture functioning, and the entire gear industry hinges on the notion that gear harbors a secret that can only be revealed upon ownership. Secrets conjure sexual ideations (in addition to those we've already examined) and function as particularly powerful psychological hooks that reel in gear users every single time. From secret locations to secret weapons, secrets-as-metaphors are easily commodified, substituting for the absent professional recording and production experience. One easy way to recognize gear as distinct from regular sound-recording and production technologies is to recognize the trace effects of heritagization. For new technologies to become gear, they must be in some way affiliated to old gear and feature heritagized aesthetics. Some gear is worthy of collections or displayed in plexiglass cabinets as if in a museum display, while some brand-new clones are marketed as if they were old gear. Gear cultures host gear competitions that bestow awards on gear and induct gear into halls of fame, and some very special gear, after going through a process we term "technological taxidermy," even finds its way into an actual museum. Although several extant critical theories pertain to the ongoing presence of the old in the present, in chapter 14 we theorize *obdurance* as the most apt frame for how gear retains its relevance in the twenty-first century. Concomitantly, valorization coupled with a vast accumulation of social values over time is intrinsic to gear obdurance. Gear histories routinely play out in retail sites, and gear is at the forefront of phenomena external to gear cultures, including heritage-inspired software technologies (plugins) and gear auctions.

12 SECRETS

THE SECRET INGREDIENT (BLOWING UP A GUITAR AMPLIFIER WITH POTATOES)

As is customary, the 147th Convention of the Audio Engineering Society featured several high-profile recordists giving presentations, doing product demos, and appearing on panels. One recordist we were excited to hear from was Sylvia Massy, renowned for her work with Tool, System of a Down, and Johnny Cash, and for authoring the widely acclaimed *Recording Unhinged.* In the convention lead-up, her talk had been titled "On the Run with Sylvia Massy" yet five days before the event was suddenly retitled to "Sylvia Massy—The Secret Ingredients."[1] Even though the presenter, timing, and location were identical, this seemingly innocuous title revision has several implications. A short synopsis elaborated, "What are the secret ingredients for a good session? [Sylvia] also reveals her secrets for making each session memorable."[2] From the synopsis, the reader deduces that this high-profile recordist possesses secrets; that those secrets specifically pertained to gear, songs, recording locations, and general recording session practices; and that during her presentation those secrets would be *revealed.* On the contrary, "On the Run with Sylvia Massy" implied something else entirely: she is running away to avoid being caught, she is constantly working and running between projects, or she is at a disadvantage in a competitive situation.[3] The original title presents Massy as an outlier, perhaps even an *outlaw.* In changing the title to include the word "secret," we argue that Massy is presented more powerfully to draw in a larger audience: the word immediately evokes curiosity, intrigue, and *anticipation* that something *unknown* will be revealed. During her presentation, Massy did indeed reveal some secrets. Although she asserted that microphones are *not* the secret, she did reveal a "drum mic'ing secret": to point all mics on a drum kit in the same direction when recording. She went on to discuss numerous recording and production techniques, some of which feature in her book, and she shared some highly entertaining and memorable anecdotes from specific recording sessions (placing a frozen trout in a kick drum and blowing up a Marshall amplifier with potatoes, for example). Ultimately, she concluded that the secret ingredient is "your participation in allowing the musician to be creative" and "it's about the song and the performance." In other words, the secret ingredients are neither gear nor recording locations.

Secrets, and their concomitant revelatory discourses, are intrinsic to the way gear is produced, situated, consumed, discussed, documented, valued, and iconified by gear culture participants. Secrets are everywhere and are just waiting to be revealed—often, of course, at a price. Particular gear, or even specific settings for that gear, are discussed as secret weapons. A recordist's accumulated experiential expertise is often promoted as their possession of "trade secrets." Perhaps most significantly, secrets represent a highly effective marketing strategy to promote gear as a ready-made or shortcut, a convenient yet implausible bypass of years of sound recording and listening experience. This chapter unpacks the notion of gear secrets, the production and manifestation of secrets in gear cultures, and how gear cultures thrive on revelatory discourses.[4] Secrets cut across many of the gear cultures we have addressed in this book: from circuit designs in gear manufacturing to the ways gear is discoursed online, from trade shows to gear heritigization. In fact, so ubiquitous is the notion of secrets and revelation in gear cultures that we even find it in academic studies of record production—without irony or critique. We learn much about how gear cultures function through interrogating how gear cultures hinge upon secrets.

WHAT'S THE SECRET?

"Kept from public knowledge, or from the knowledge of persons specified; not allowed to be known, or only by selected persons."[5] "Secret" is a word usually used in relation to information that would normally be kept private. Gear discourses mirror other broader presences and understandings of secrets in that there is the secret itself, the secret holder, and the revelation. In gear cultures, this is linked to commodification and consumption, and to the recipient, whether someone who does not yet know of the secret, someone who knows of and wants to know the secret, or the secret knower. Since secret recipients are rarely directly linked to secret holders, secrets become commodifiable to gear culture participants and create the simulacrum of this missing direct link.

For a discourse so ubiquitous in audio technology, surprisingly little research exists in terms of how the mechanisms of secrets operate. Studies in psychology orbit the ways secrets are bound up in trauma, embarrassment, disturbing personal experiences,[6] relationships, or other matters that an individual or group may not wish to disclose; individuals and groups may go to great lengths to ensure the secret is retained.

Secrets are routinely studied from the perspectives of secret holders and their concomitant burdens and, to varying degrees, the damaging nature of secret retention. Research suggests both advantages and disadvantages of disclosing secrets dependent on perception and attachment style.[7] Revealing secrets is an unburdening of psychological stress that can decrease rumination and increase self-esteem in the event that a listener's response was positive.[8] To that end, hearing revelatory information may reinforce self-esteem among gear consumers and users.

Secrets are, however, intrinsically linked to relationships. Most studies on the keeping and/or disclosure of secrets involve family relationships,[9] yet typologies of secrets emerging from psychology research show that secrets of a sexual nature are the most common and are often bound up in taboo; matters preventing revelation include fear of rejection, ostracization, and shame.[10] Some, but not all, secrets discourse in gear cultures revolves around the same kinds of psycho-sexual issues and metaphors. Secrets discourse in gear cultures is another, perhaps more subtle, means of sexualizing gear, and secrets discourse amplifies gear's fetish power. Furthermore, participation in secrets discourses helps participants be privy to what was once protected, tacit knowledge.[11]

Secrets are long ingrained in both the recording and technology manufacturing industries. In *When Old Technologies Were New*, Carolyn Marvin focuses on two key electrical developments of the late nineteenth century: lighting and the telephone. She details how electrical knowledge has always been guarded, and how the dissemination of electrical expertise via periodicals created a guild mentality whereby those with access to—and the ability to interpret and apply the information from within—these publications gained authority. "Aspiring electricians placed scientific textuality and certified interpreters of scientific texts at the center of their claim to public authority and attempted to persuade those less technically lettered of the validity of that strategy."[12] In Marvin's account, "technological literacy" separated experts and amateurs as insiders/outsiders in electrical culture, and in shocking ways, the holders of this electrical knowledge, almost always white middle-class men, used their expertise to exclude and demean Indigenous peoples, women, people of color, and those with disabilities. Gear and its cultures are part of a continuum, Marvin discussed, where "skills and techniques for performing particular literate practices are transferred from communities of adepts to less skilled communities."[13] Secrets discourse is highly effective for generating and perpetuating fascination around technological knowledge and its holders, and the desire for such knowledge to be transferred.

Since the secret is often bound up in a single piece of gear, or even a specific parameter on a piece of gear, gear culture secrets over-attribute agency to technologies. The secrets-secret holders-revelation-recipient continuum also falsely attributes technological agency to gear as opposed to writers, performers, recordists, and other technical personnel. Rather than discoursing gear as tools of the trade, secrets discourse promotes gear to center stage and asserts that gear ownership is directly linked to high-quality recordings. The success of secrets in gear cultures comes down to the retention and commodification of tacit knowledge—whether this knowledge is real or perceived—in a highly competitive technological landscape. Where the technological boundaries of professional/amateur status in the broader audio industry were clearly demarcated until the 1980s,[14] these lines are particularly opaque in the twenty-first century. As such, secrets become a powerful distillation of recording and production skill sets. In gear cultures, where excepting the cultures around extractivist and smelting operations, all participants have *some* access to recording technologies, the boundary lines have to be redrawn around experiential knowledge, thus maintaining power structures—even if these structures are based on perception and not reality.

Secrets do not, however, only manifest in gear culture discourse. Secrets surround the origins of gear materialities, the labor required to source and produce the raw materials, the labor required to build gear, and the differences between frontstage and backstage performances we have discussed previously. We find many correlations between the secretive nature of gear materiality and that of the materials used for manufacturing music media.[15] Who in gear cultures knows where the front panel aluminum comes from? From where are the plastics sourced to make all those knobs and switches adorning consoles? And once the gear is designed, who builds it, where is it built, and under what conditions? Keeping the origins of gear materials secret—as well as the associated labor involved in mining, smelting, and assembly—benefits gear cultures as it keeps the focus on the purported ingenuity of gear designers and recordists and obscures the problematic aspects (extractivism's environmental legacy, income inequality) of how gear is made.

THE SECRET HOLDERS

Scholars in record production studies have long understood how tacit knowledge associated with sound-recording technologies, processes, and practices is knowledge that was historically concealed. David Morton recognized how Edison's staff developed disc systems "*in secret*" in 1910.[16] In our coedited book, we recognized the "insatiable general interest appetite for 'behind the scenes' texts and documentary films that 'reveal' recording processes."[17] Susan Schmidt-Horning notes how bandleader Paul Whiteman was "careful not to reveal any secrets of the Victor engineers' techniques" in his 1948 book. In her interview with Frank Laico of Columbia Records, he revealed his "secret" of using early compressors: "It didn't sound like I was using it."[18] She goes on to describe 1960s Bell Sound Studios engineer Eddie Smith as in possession of a secret drum-recording technique. Other engineers, Schmidt-Horning claims, would "sneak into the studio, and try to learn the secret to getting the big drum sound," which is later revealed as "allowing the drums to leak into the string mikes."[19] But why should this be a secret? It could be argued that this is less of a secret and more indicative of 1960s limitations on multitrack recording and the inevitability of spill during live performance recording. More significantly, Schmidt-Horning described a 1968 advert for the B&K Model 123 Graphic Equaliser, which depicted a rock band, the technology, and their engineer. "That's between them, their audio engineer and the lamp post," declared the ad in relation to the sonic qualities of the band's sound, the implication being, as Schmidt-Horning noted, "the engineer and his technology were the secret to the band's . . . sound."[20] This frequent acknowledgment of secrets in record production as early as the 1940s is left uninterrogated. However, what becomes clear is that sound recordists possessed secrets, which were coveted by other engineers, pertained to both their techniques and audio technologies, and were rarely shared or revealed—perhaps only to musicians they worked with and trusted. In *The Poetics of Rock*, Albin Zak reinforces mythological discourse surrounding the work of recording engineers, suggesting they are "both craftsmen and shamans, harnessing the power of the machine and deftly manipulating its magical secrets to effect the transformation of the musical moment into a musical text."[21] The ways in which recordist processes and audio technologies are routinely analogized to supernatural and religious experiences has been critiqued yet may be summarized threefold:

> First, the historical role of the recordist–and their work–is one that is concealed. Secondly, the historical situation of sound recording technologies in the space of the elite recording studio confined operational understanding to a select few, highly skilled technicians. Thirdly . . . due to tensions surrounding the acquisition and preservation of technological knowledge.[22]

Since audio technology knowledge was, historically, "largely inaccessible to the general public,"[23] unsurprisingly numerous mythologies and ideologies prevail through its discourses. The guild mentality surrounding the protection of this knowledge exists to this day and is rhetorically embodied in the UK's Music Producer's Guild. Furthermore, while plenty of scholarly texts have recognized the discourse of the so-called "democratization of technology,"[24] there is little to no evidence that a concomitant democratization of skill sets has occurred in tandem. Thus, while the economic accessibility of audio technologies in some parts of the world may have improved since the 1980s, the same cannot be said for processes, techniques, and tacit knowledge associated with them, despite the rise of music technology education. Expertise, critical listening skills, and technological and processual knowledge are replaced with technological consumption, with gear being marketed as shortcuts to such knowledge, and the revelation of tacit knowledge "secrets" of recordists being commodified. This at least partly explains why the title of Sylvia Massy's 2019 AES talk changed.

REVELATION DOMAINS

In today's gear cultures, secret holders are professional recordists (tracking, mix, and mastering engineers), occasionally educational establishments,[25] and gear magazine contributors, some of whom have published books based around secrets discourse. Equipment manufacturers and other businesses have used numerous methods to exploit recordist knowledge and experiential expertise as the disclosure of secrets. As such, revelation of secrets cuts across multiple domains in gear cultures and is a highly lucrative business.

One of the most successful brands built on the commodification of recordist knowledge is Mix with the Masters (referred to from here on as MWTM), a France-based organization hosting online masterclasses and webinars for an annual subscription and week-long residential seminars with high-profile recordists at the Miloco-owned La Fabrique Studios in the south of France. These exclusive and expensive (E4000 / $4500US) seminars are "conducted by one of the world's top music mixers and producers, ready to share their professional secrets with a select group (typically fourteen to eighteen attendees) of carefully screened, professional-level participants who come from all over the world" (MWTM 2020).[26] The MTWM website currently lists more than seventy "masters"; fewer than 7 percent of those are women. Images of the week-long and one-day seminars depict an almost exclusively male arena, with participants gathered around the "master," invariably sitting at the Neve 88-R mixing console in the main control room within the mansion of La Fabrique.

Strangely, the seminars are aimed at semiprofessional or professional recordists who are screened prior to enrolment to ensure they themselves possess the necessary expertise. The incentive, then, is the promise that the master, in possession of multiple mixing credits on commercially successful recordings, will reveal secrets. This suggests that "mastery" can only be attained via commercial success in the mainstream Western popular music industry, commercial success is more valuable than length of time in the industry or mastery of noncommercial music genres, and exposure to "masters" as circumscribed before is highly valuable.

This is, of course, deeply problematic. MWTM isolates the *mix* engineer as auteur. MWTM leverages secrets and revelatory discourse to exaggerate the role of the mix engineer in a commercial recording process. At best, this downplays and, at worst, ignores the engineer's affiliations to major record labels, the role of the songwriter and/or arranger, the recording workplace and accessibility of technologies, the performers, musicians and standard of performances, and other factors that lead to success, including artist management, marketing, and publicity. But techniques and/or processes used by a single individual can only be associated with a specific set of circumstances that are impossible to replicate in their entirety; MWTM separates the mix engineer from the holistic music production process and attributes excessive agency to the role. Even learning multiple mixing secrets will unlikely equip a semiprofessional mix engineer with anywhere near the equivalent of years of experience in professional mixing. Yet the promise in such revelatory discourse perpetuates this myth. Another issue lies in the era in which these "masters" mixed records. Almost all are middle aged and possess recording credits dating back to the 1970s and 1980s. As noted by many scholars, the recording industry has shifted, particularly since the turn of the millennium, to drastically reduced recording budgets and a freelance gig economy that limits the number of personnel able to operate exclusively as professional recordists.[27] Indeed, the frequency at which some "masters" participate in these week-long residential seminars is testament to their current professional working conditions. MWTM also appears oblivious to recent campaigns to eliminate terminologies derived from slavery from the audio industry.[28]

For a discourse so prevalent, it is surprisingly absent from audio engineering and technology research. In the AES e-Library, a repository of thousands of journal articles, engineering briefs, and papers spanning decades, the word "secret" is mentioned in the title of just two: Barry Fox's "Nice Timing: Secrets of Successful Invention" (1980), and Anthony Tucker et al.'s "Perception of Reconstructed Sound-Fields: The Dirty Little Secret" (2013), neither of which purports to reveal hidden secrets. Academic research celebrates an ideology of ingenuity and its dissemination, particularly in conference and convention settings, and encourages sharing rather than concealing results.

In contrast to the culture of academic sharing, numerous trade books, some of which are authored by magazine contributors, exploit secrets and revelatory discourse: within these realms, efforts to distinguish "professionals" and "amateurs" correlate to those in possession of secrets and those who are not. Sadly, even though these books are aimed at those starting out in audio, they too often feature impatient, hostile, or aggressive tones, while scarce content inside the book reveals anything original. *Secrets of Recording: Professional Tips, Tools and Techniques* is one such example. With chapters on pitch and timing correction, replacing sounds, and emulating effects, the book traverses a DAW-based music production process. The introduction, "What Do I Mean by the Secrets of Recording?," asserts that no single engineer could know all the tips and tricks. The word "tricks" is used frequently, which is problematic in itself since not only does this replace one strange domain (concealed secrets) with another (tricks, implying magic or deception), but it reduces accumulated tacit knowledge into a series of "quick fix" or "bite-sized" techniques. The book's tone is negative and speculative throughout. For example, in discussing the Antares Auto-Tune, Bregitzer states, "Like it or not, (the 'Cher' effect) you had better be able to create the effect for them so you can talk them out of it, if possible," making no reference to how the device could be suitable depending on the artist, song, genre, style, or tradition of the music; used as a creative effect (as it has been within genres ranging from electropop to Egyptian *shaabi*); or specifically requested as an intentional feature. Framing the "problem" around an exceptionally popular and influential female performer adds a gender discriminatory aspect to the negative connotations of the effect. Another speculative statement concerns how effects processing in music production may be construed as "cheating": "These are really the 'grumpy old men' of audio chasing the kids off the lawn with an Ampex tape machine instead of a broom." This far-fetched, ageist analogy is unsubstantiated. Not only are secrets never revealed, but as a DAW-oriented "how-to," the book features little to no reference to recording, as promised in its title.[29]

Problematic work of this nature is widespread (table 12.1). Books by Mike Senior, a commentator for *Sound on Sound* for more than twenty years and author of titles including *Mixing Secrets for the Small Studio* and *Recording Secrets for the Small Studio* are further examples. Both are written with predominantly negative phrasing and feature particularly demeaning language toward those new to audio production. What are readers intended to gain from sentences beginning with "Probably the most reliable way to waste your time in a small studio," "Although I've done my best to make this book friendly to studio newbies," or "Until you get a proper grip on those issues, any discussion of mixing techniques is about as useful as a chocolate heatsink"?[30] His 2014 book, which continuously references "studio newbies," persists with this tone: "While I make every effort to fast-track the learning process within these pages, the art of recording is by nature tremendously complicated and full of nuanced decision-making," "Deafening yourself probably isn't the best way to begin a recording course," and "Even the vanilla setup presents you with an awful lot of variables to think about, which is perhaps why so many fledgling engineers come a bit unstuck with vocals."[31] But what secrets *are* revealed? "The secret to achieving a powerful rock vocal recording is learning to sing *as if* you're hammering it out, while actually delivering only a comparatively restrained acoustic level in the room so that the mic captures a well-balanced mixture of frequency content."[32] This revelation is flawed: it appears to be aimed at a singer who is learning to sing as opposed to a recordist or assumes that recordists are simultaneously rock vocalists, and it also appears to directly correlate "acoustic level" (amplitude) with frequency content. What "hammering it out" or "restrained" means in the context of a "powerful rock vocal" is vague, considering the breadth of rock subgenres and their stylistic nuances, the gender of the vocalist, and vocal range, not to mention the

Table 12.1
Revelation-oriented texts in the music and audio industries. This is not an exhaustive list but represents the books on the market today. Of the twenty-one authored titles, two are authored by women (9 percent).

Title	Author	Year
Inside Classic Rock Tracks: Songwriting and Recording Secrets of 100 Great Songs	Rikky Rooksby	2001
Mixing with Your Mind: Closely Guarded Secrets of Sound Balance Engineering	Michael Stavrou	2003
The Art of Digital Music—56 Visionary Artists and Insiders Reveal Their Creative Secrets	David Battino, Kelli Richards, Stewart Copeland	2005
The Golden Moment: Recording Secrets from the Pros	Keith Hatscheck	2005
The Secrets of House Music Production	Marc Adamo	2009
Secrets of Recording	Lorne Bregitzer	2009
Big Studio Secrets for Home Recording and Production	Joe Dochtermann	2010
Mixing Secrets for the Small Studio	Mike Senior	2011
Electronic Musician Presents the Recording Secrets behind 50 Great Albums	Kylee Swenson Gordon	2012
Think Analog—Work Digital: Tracking & Recording	Jonathon Evans	2014
Recording Secrets for the Small Studio	Mike Senior	2014
The Secret to Mixing Vocals [Exposed]	Dongatti Denero	2016
The Secrets of Dance Music Production	David Felton	2016
Audio Mastering Secrets: The Pros Don't Want You to Know!	John Rogers	2017
Song Mixing Secrets: How to Fix the Most Common Mistakes	John Rogers	2018
Secrets to Building a Home Recording Studio	Robson Green	2018
Recording Studio Secrets: How to Make Big Money from Home!	John Rogers	2019
Music Production Secrets	Calvin Carter	2020

artist's intentions for how the song should be performed. Is the secret of powerful rock vocal recording to sing quiet(er)? Senior's own website states his background as a professional recordist prior to his work at *Sound on Sound*. But to whom are his books aimed? Professional recording or mix engineers are not likely to consume revelatory discourse themselves, leaving only music technology and/or audio engineering students, enthusiasts, and prosumers as the intended audience.

These texts have numerous links to gear cultures; all were published in the twenty-first century. Coincident with the widespread take-up of DAWs was the emergence of revelatory texts about its processes. To some extent, this is a continuum of what Théberge noted about community-oriented discourses around home recording in the technical press of the 1980s.[33] However, while plenty of gear reviews and articles were printed during this time, they tended to be more aligned with the user experience. "Secrets" texts during the gear cultures period propose kinds of revelatory discourses associated more with gear fetishization than with skill or studio practice.

Skill sets and experience, however, empower professional engineers exploiting secrets discourse via programs including MWTM, gear endorsements, trade show talks, masterclasses, articles, and interviews. The possession of a secret derived from a prior successful professional recording is also a testament to that recordist having been socially involved with those legendary musicians. Therefore, more is at play than transmitting skill sets; secrets constitute an ideal construct targeted at loner, isolated individuals who can only dream of being in the studio with good musicians. As such, secret *recipients* are rewarded with the ability to perform reductive versions of that experience in various gear culture domains, particularly online. Secrets discourse, when exploited by professional recordists and commentators, therefore, gatekeeps professional/amateur boundaries.

RECEIVING SECRETS

In gear cultures, secrets are received in many ways. While there appears no shortage of recipients for secrets packaged as books or other fora including MWTM, secret recipients appear most engaged when secrets discourse is packaged into specific conceptual framings. Our opening example looked at **secret ingredients**, the idea that recording practice is parallel to a recipe of sorts, a bit like the seasoning in KFC. Numerous discussions concern **secret powers**, particularly gear as holding supernatural or telekinetic properties. Gear and its operational potential as imparting signal processing into a music production workflow, is routinely referred to as "magic" or "alchemy," or conveniently packaged into the "trick." The trick is prevalent throughout gear cultures and even spills over into academic studies of record production. Perhaps the most prevalent of these concepts is that of the **secret weapon**. Ubiquitous in gear cultures and far beyond, secret weapons function as an irresistible hook and are rife across web forums, social media, and the trade press. Recipients, particularly online, are only too happy to share their secret weapons, ask others to reveal theirs, and point to instances of secret weapon revelation.

> Weapon: Any object used in fighting or war, such as a gun, bomb, knife (Cambridge Dictionary)

In a 2020 tweet, recordist and *Tape Op* editor Larry Crane asked his Twitter followers,

> Why do people use the word "arsenal" when referring to their mic collection or audio equipment? I don't think of the recording studio as a war zone. Definition: "A collection of weapons and military equipment stored by a country, person or group."[34]

The responses were equal parts critical and varied. Across more than sixty likes, four retweets, and more than twenty responses, many agreed with Crane, some suggested alternatives (a "rainbow" or "quiver" of microphones), others felt the military analogy *was* appropriate for describing difficult recording sessions, and some pointed out the ways in which such language reinforces and normalizes the military and war in society. One respondent noted that Lakoff and Johnson's classic philosophical and linguistic text *Metaphors We Live By* explains how deeply military metaphor is embedded in society generally. While we are not primarily positioning our work in a linguistic framework, George Lakoff and Mark Johnson productively frame the discursive work that "secret weapons" do in gear cultures. Metaphor, they state, is essential to the way humans make sense of things and helps us understand and experience "one kind of thing in terms of another." Critically, the ways in which metaphors foreground some aspects of a concept while obscuring others demonstrate not only how we organize and prioritize particular attributes but also how those particular concepts align to our agendas and purpose(s). "The most fundamental values in a culture" are inherently bound up in structural metaphors.[35]

Secret weapons can be found among gear designers and builders, in the discourses of recordists and users, and in the marketing of all types of gear. Gear often is named after or like weapons, too, whether we consider the Burl Audio Bomber digital/analog converters, the IGS Panzer rack, the many mics by Telefunken that use literal weapon names (M80, AK-47, M16), the Soyuz Bomblet mic, Louder than Liftoff's Silver Bullet (a "tone amp" they claim "give[s] your mixes the analog MOJO"), the Shure Green Bullet harmonica microphone, or the entire product lines of Arsenal Audio (a prosumer division of API) and Airfield Audio. Software can be weaponized, too: Sugar Bytes Artillery, Bomb Factory plugins, Krotos Weaponiser, or Slate Digital's Bomber plugin. Prior to these naming conventions, location sound recordists talked about "run and gun" miking techniques, facilitated by mic clips with a "pistol" grip—but pre-1997, we find little evidence that such terminology was intended to evoke anything more than morphological similarities between objects.[36] Therefore, secret weapons discourse (not to mention the weaponization of gear through metaphor, analogy, and naming) ironically emerges in full force right at the moment when studio secrets and studio secret gear were first being revealed at a mass scale within nascent online gear discussions. This kind of secret, then, is a secret in plain sight—and available at the largest music instrument retailers.

Back to Larry Crane's tweet. The comments stemming from Crane's post exemplify differences in the ways values become bound up in structural metaphor. For Crane, audio technologies are not an "arsenal" of weapons and the recording process is not a battle or a war; for others, they are.

> A thing designed or used for inflicting bodily harm or physical damage.
> A means of gaining an advantage or defending oneself in a conflict or contest (*Oxford Dictionary*)

The latter meaning, especially, reveals a much broader aspect of gear-related sociability: the secret weapon is a means of advantage in the conflict or contest-like milieu and one-upmanship interactional norms that pervade gear cultures, and is only feasible via bourgeois notions of owning the means of production. Prior scholars have rigorously discussed competition and competitiveness among audio and music technology manufacturers. Yet the competitiveness between audio and music technology *consumers* is understudied. Earlier we covered the excessive competition surrounding the acquisition of, ownership of, and proximity to gear. However, where the focus is on a recordist or a process, "secret weapons" commonly frame the revelation of a secret *technique* involving use of a technology.

Table 12.2
As of September 2020, the frequency of the term "secret weapon" on gear discourse platforms and retailer websites

Platform	Number of "secret weapon" appearances
Tape Op	41
Sound on Sound	86
Vintage King	57
Sweetwater	728
Better Music	13
Gearslutz	~3,850

In *Sound on Sound*, the word pair "secret weapon" is found eighty-six times (table 12.2). The three highest-occurring subjects where "secret weapon" is found are: mixing/production (28), signal processors (12), and recording (12), although some of these are duplicated in two or more sections (table 12.3). Let's consider some of the uses of this term over a two-decade period in *Sound on Sound*:

"That's my secret weapon!"

—Hans Zimmer, film composer, on Creamware synthesizers (October 2002)

"The Trouper Series Limiter is my secret weapon."

—Tony Coleman, Drum & Bass producer (September 2003)

"The leakage suppression function is one of Trigger's secret weapons."

—Paul White, SOS editor, on the Steven Slate trigger drum replacement software (August 2010)

From the table and quotes, the trope has two clear dimensions. First, musicians, recordists, and reviewers attribute their tech-processual advantage(s) to a single piece of technology: a plugin, a microphone, a piece of hardware. This trope is particularly beneficial to the amateur/hobbyist worlds of gear cultures since it is much easier to win a gear battle if all that is required is a gear purchase. Second, reviewers suggest that a technology itself *features* a secret weapon—a particular setting, knob position, or switching sequence on a piece of gear. Once again, this benefits both gear consumers and manufacturers since it communicates the advantage a piece

Table 12.3
Twelve examples of the 86 "secret weapon" word pairings in *Sound on Sound.* Of these twelve, two (16 percent) are used by recordists and ten by journalists (84 percent).

Gear "secret weapon'	Agent	Year	Role
SPL Transient Designer	Mike Senior	2003	Journalist
Gem Dopamine plugin (Dolby A emulator)	Jason Elliot	2020	Recordist
Sonic Academy Kick 2 plugin	Paul Tingen	2018	Journalist
There was no "secret weapon" in the making of "Oh and O"	Markus Popp (Oval)	2010	Recordist
Nintendo Wii	Sam Inglis (on Matteo Scumaci and Robin Haller)	2009	Journalist
Yamaha SPX90 (reverse reverb)	Tom Doyle (on MBV Kevin Sheilds)	2018	Journalist
Sony C37A microphone	Mark Howard	2011	Recordist
White Gretch Falcon electric guitar	Mark Howard	2011	Recordist
Sansamp PSA1 plugin	Sam Inglis (on Arc Pilot)	2014	Journalist
Synth or guitar part	Ingmar Kiang (on pop backing vocals)	2013	Journalist
Analog tape	Jyoti Mishra (on Akira Kiteshi)	2012	Journalist
Trouper Series Limiter	Tony Colman	2003	Recordist

of technology will give to a user. It also narrows the secret weapon into something more nuanced: a particular function or parameter of the gear. This trope is clearly fueled by audio technology commentators whose roles are intrinsically linked to gear manufacturer advertising. The "secret weapon" is, therefore, a highly effective means of commodifying gear.

CONCLUSION

Secrets are gear culture glue. Despite the problems, they are a vital mechanism that functions on several levels to maintain masculine hegemony. Gear cultures benefit from the concealment of material and build labor; obscuring this removes any affiliation of gear to the realities of environmental and/or human exploitation. Relationship-wise, secrets demarcate knowledge boundaries among gear culture personnel, thus keeping the critical professional/amateur boundary line firmly drawn in the sand. Secret holders—magazine reviewers or editors, an author, a forum moderator, a recordist with commercial credits to their name—are mostly self-appointed yet are always in some position of power to exploit their accumulated expertise. It is, however, no secret that roles have changed dramatically since the 1980s and increasingly so since the turn of the millennium.[37] Simultaneous to the broadening accessibility of some technologies, in the commercial recording industry, it has become more difficult for recordists to make a living. Secrets, therefore, are a powerful and lucrative means to protect tacit knowledge, regardless of whether that knowledge will be useful to the recipient. Secrets also bind together past practices and techniques with gear for today's recording and production workflows. This scaffolds the gear canon and contributes to heritagization. As a commodity, secrets narrow the perceived proximity of the recipient to a canonized artist or recording. Additionally, gear featuring a secret parameter, such as the 1176 compressor example, can be profitably commodified. The promise and potential of gear is more seductively potent and has far more instantaneous, attention-grabbing clout when packaged up as a secret weapon, featuring a secret power or, as the AES and Sylvia Massy know only too well, a secret ingredient.

13 HERITAGE

SYLVIA MASSY'S MIND-BLOWING MICROPHONE MUSEUM

Surrounded by a vast array of vintage microphones and talking via Zoom, record producer Sylvia Massy leans into her microphone and quietly declares, "I have a gear problem." It is August 2020, and the Pacific Northwest branch of the AES is hosting an online seminar, "Sylvia Massy's Mind-Blowing Microphone Museum." Over the course of two hours, Sylvia presents examples and stories from her collection of more than one thousand microphones. She proudly displays a Columbia 47A condenser from the 1920s, an Altec lipstick mic, and an RCA 44 prototype. She demos a Western Electric 618A and shows off two humble Shure SM57's while telling not-so-humble stories about their previous home in The White House. She picks out an Astatic D104 crystal mic and meanders into a story about the origins of the Gefell microphone company during World War II. Shure's resident historian Michael Paterson is in attendance and drops anecdotes in the chat as audience members respond with joy and wonder, text boxes flying up the chat pane faster than this gear show-and-tell can keep up. As the seminar concludes, Massy acknowledges the origins of her collection: she bought the microphones from the late Bob Paquette, owner of the Bob Paquette Microphone Museum at Select Sound Service in Milwaukee, Wisconsin. Paying tribute to the former owner of the world's largest microphone collection, Sylvia notes "Bob was buried with a Shure 55."

Massy's seminar raises curious, potentially paradoxical questions around gear and its heritage. What types of gear are heritagized and why? How do some types of gear continue to attract a technological gaze beyond the respective use era? How does some gear become storytelling objects? In what ways is gear canonized, commemorated, awarded, and cloned? How does gear end up in museums? Who is doing the curating, visiting, and *gazing*? And why is some gear treasured so closely and personally that it ends up in the graves of its most dedicated collectors? Gear is heritagized in myriad ways, from its positioning as "star objects" in exhibitions through to seminars, trade show displays, manufacturer reissues, and clones.[1]

In this chapter, we also consider more nuanced aspects of gear heritagization: the presence of old gear in the new, the heritage discourse so prevalent in new gear marketing and promotion, and, of course, our own technological voyeurism of gear in the museum. We stroll through museums all over the world, and along the way, we meet a 1969 Datamix console at MoPOP in Seattle, a model replica of a Neve 8048 console at the Queen Studio Experience in Montreaux, and a Fairlight CMI at the Science and Media Museum in Bradford (UK).

Cultures, as sociomaterial formations, are held together through histories, myths, stories, and a shared sense of heritage. In gear cultures, the primary frame for these are certain memory-related practices, including the broader intertwined fields of heritage and technostalgia and the more specific practices of canonization and iconicization. These fields and practices are interconnected in that technological objects are called upon to be sites of or repositories for cultural memory. Moreover, this historiographical and memory work is central to the process of valuation. Valuation, here, is not just about individuals demonstrating connoisseurship (as we discussed earlier), but it is also necessary to lend coherence to a cultural formation—a gear culture—that otherwise would, more likely than not, fall apart.

GEAR HERITAGE AND MUSEUMS

Old gear is everywhere. It is not only the main draw in a twenty-first-century online AES event, but it also dominates the spaces reserved for so-called new gear. Retailers pepper their synopses and straplines with buzzwords including "classic," "vintage," "original," "new old stock," and "old school." Brand-new gear is marketed and promoted using its connections to the past. And in swathes of online commentary, gear culture participants pledge allegiance to the gear of old and loyally attest to its resilience and reliability; the ghost of gear past is always present. Long-dead gear justifies new products and inspires manufacturers to reanimate, clone, or reissue for the present-day market. At play here is a vast and complex process of gear heritagization. Materializations, discourses, and practices of heritage lie at the heart of gear cultures and are essential to the building, reinforcement, and maintenance of a gear canon. Heritagized gear is extremely useful to hardware gear manufacturers as it creates and maintains a space in the broader music technology industry, which otherwise is dominated by digital software technologies. While all technologies have their histories, software technologies cannot lay the same kinds of historical and mythological claims that hardware gear can. Through her stories of the history of the RCA-44 microphone to tales of Barack Obama making a speech into a Shure SM57, Sylvia Massy undertook this essential historicizing and mythologizing work as part of her AES talk. Beyond the discursive framing of gear history in the present, we are also concerned with the ways in which contemporary gear is propped up as historiographical tools and how heritagization can be a prerequisite for entry into the gear canon.

Heritage is not a phenomenon in and of itself; rather, it "is omnipresent, interwoven within the power dynamics of any society and intimately bound up with identity construction at both communal and personal levels."[2] Indeed, UNESCO's list of world heritage sites covers an international array of cultural (monuments, buildings) and natural (geological or physiographical) sites and works to ensure the "identification, protection, conservation, presentation and transmission to future generations."[3] More recently, UNESCO has moved to protect intangible cultural heritage, specifically oral traditions, performing arts, social practices, rituals, events, and other culturally specific knowledge, skills, and practices. Heritage is less an actualized entity than a process involving considerable labor and agreement surrounding "what we want to save."[4] Sylvia Massy's AES presentation staged this process: it questioned what gear is worth saving (old and valuable microphones), who says so (Sylvia herself, supported by high-profile microphone experts and manufacturers), and who willingly accepts this (participants). Widely recognized as "something that needs to be protected" for the public good,[5] there is always a degree of management, maintenance, and interpretation to heritage places, objects, and culture. Heritage, though, "benefits someone, and usually disadvantages someone else."[6] Historically and today, in most instances, heritagization processes serve colonial interests at the expense of Indigenous peoples and cultures. Heritage always involves some kind of conservation and/or preservation process; the identifying of what constitutes heritage in the first instance is politically charged and subject to critique.

Gear is not officially heritagized as such. There is no internationally sanctioned list of professional audio technologies with heritage status, and the intangible cultural heritage aspects of gear are not officially documented. That does not, however, prevent gear cultures from undertaking their own heritagization processes or stop participants themselves from attempting to canonize or iconize gear, leveraging its heritage to their advantage. For example, take Peter Rodriguez and his successful Spanish gear company, aptly named Heritage Audio, which manufactures professional outboard equipment that visually and sonically "clones" Neve, Pultec, and other historical gear. The Heritage Audio vision is to "create recording equipment that maintains the tone and character of the 1970s and adapt it to the needs of today's music."[7] From the company name to the signposting to the 1970s Heritage Audio demonstrates their commitment to preservation. This, along with an assertion of relevance in today's professional audio contexts, exemplifies how heritagization works to preserve and maintain gear culture discursive tropes, such as the so-called "golden age" of record production, and reinforce the standing of particular types of gear (in this case, mainly Neve) within them.

Heritagized objects, including gear, must be protected and displayed: that's where museums serve an essential purpose in gear cultures. Museums are "sites of politics and culture, where themes of power, citizenship, and democracy have played out in or been ignored by officially sanctioned spaces of representation."[8] Historically, museums have been sites of collections—of paintings, artworks, ceramics, and other objects—very often resulting from colonial theft. To counteract the "disadvantage" of heritage that Howard described, colonial artifact repatriation is today a significant aspect of museum praxis.[9] On the one hand, we might see gear in the museum as a way of foregrounding oft-concealed objects of sound (re)production. After all, gear does not hold much—if any—appeal beyond the boundaries of gear cultures; it must break free and become relevant to a broader audience if it is to end up in a viable museum. On the other hand, gear in the museum is an awkward fit. Museum objects must have obvious qualities and meanings to be selected for display, but these immediate qualities are inevitably bound up with other, related but not necessarily visible, qualities in what Webb Keane called "bundling."[10] To that end, gear must maintain its visual appearance but also signify embodied cultural, social, political, and musical values that bear relevance to audiences beyond the immediate gear cultures. Only the most valuable, meaningful, and representational gear tends to make it to museums. Usually, gear is displayed because it can be used to tell a much bigger story about an artist, technological trajectory, or corpus of recordings, and it can bridge the gap between the visitor and often posthumously canonized artists and/or recordings. One role of the museum is "to confer status and value to the contents of their exhibitions, a status and value based on the relationship between the museum, its contents and their visitors."[11] While museums can amplify the fetishization of technological objects, the museum cannot sexualize them. The lengthy process of acquiring, collecting, and displaying objects in and of itself imbues significant value into the gear regardless of its additional storytelling capacity.

Usually, gear is displayed because it can be used to tell a much bigger story about an artist, technological trajectory, or corpus of recordings. As Marion Leonard and Robert Knifton note, it may involve the "display of rare objects which hold the promise that visitors can experience a form of proximity to a star performer or an insight into the practices and performance of popular music which are normally only viewed from a distance."[12] To that end, we can consider gear in the museum as less about the gear than a convenient means of bridging the gap between the visitor and often posthumously canonized artists and/or recordings.

Annie Jamieson is in a rare job. As one of the world's only audio technology museum curators, she possesses remarkable expertise and a niche insight into how and why gear ends up in the museum. Responsible for acquiring, accessioning, and exhibiting sound technologies, she has worked exhaustively in this role managing the delicate balance of value perceptions, practicality, and provenance. The latter, she reflects, is the most important factor in considering how and why a sound technology might be acquired for the collection. Provenance is about both origin and documented records. While Jamieson does not subscribe to the social life of the technology being the be-all-and-end-all, she recognizes how, for the museum and its hierarchy of acquisition boards and governors, it often is. In the museum, the life story of gear means more than the gear itself: where did the gear come from and who used it? "Things are not collected because they are sound objects; they were collected because of their connections, because they came from the BBC studios."[13] This tension between the gear and its potential storytelling capacity is the main reason why particular types of gear end up in the museum. Additionally, the limitations of acquisition and collection policies mean that only specific types of gear may be acquired, and even then, there is a tension "between collecting 'star' items and representing everyday experiences."[14] Gear is anything but "everyday." It represents the tools of an elite profession and an extraordinary technical and artistic output, and it is largely exclusive—in locality, application, and cost.

We have written a lot about knobs, buttons, and switches and the flickering LED lights that animate gear in inviting sound settings. Gear that finds its way into the museum, however, stops breathing. Museums struggle with electronic objects because they rely on electrical current; for a television to be watched, a Walkman to be auditioned, or an electric guitar to be heard, they must be plugged in and switched on, or at least powered with trusty 9-volt or AA batteries. Almost always switched off, never again to glow or flicker in the dark, museum gear becomes inanimate and effectively unusable. Museums treat gear like rocks. The object arrives, it is formally acquired, cataloged, databased and shelved, and then—if it is lucky—it might get displayed. Nearly all museum objects are not on display and simply lie inanimate, archived, and hidden. Nonoperational gear in the museum loses its sex appeal, that being the sexual potential for it to stand in for missing women; how, then, does the gear fetish persist in its impotent and desexualized display in the museum? This absent sexual potential means that the economic aspect of the fetish must be excessive to compensate. The gear that does end up in museums is more often than not the very same fetishized gear that is lauded and revered in gear cultures: unique gear that was used on a canonized recording and/or built for a special purpose or place (for example, the EMI presence boxes displayed at the "80 Years of Abbey Road" exhibition) and/or objects that belonged to a canonized artist (for example, Jimi Hendrix's 1969 Datamix console on display in MoPOP). Even though gear in museums fails to work as a full fetish object in the same way that gear at trade shows, online, or in print media does, the museum still contributes to the power and historiography of gear: if it has made it into a museum, it *must* be important.

COMMEMORATING AND AWARDING GEAR: THE AUDIO TECHNOLOGY CANON

The gear canon, made up of a select few brands and objects/systems that act as benchmarks in gear cultures, is a discursive formation debated by online fora participants, lauded at trade shows and conventions, paraded by manufacturers, and displayed by museums. We neither advocate for a gear canon nor suggest that certain gear belongs in a canon. Instead, we consider the processes through which canons are formed in gear cultures. The kinds of gear that get promoted to the canon fall into recognizable patterns. With few exceptions, gear must be heritagized before being canonized, meaning it must be omnipresent, deemed valuable, and contain some heritage identity. In addition, the gear must be tied to a specific locality. This is most often Germany/Austria (e.g., Neumann, Studer, AKG), the UK (e.g., Neve, SSL, Trident), or the US (e.g., MCI, API, Ampex, Pultec). Mythology must also be attached to the gear in some way. This might be through particular "secret weapon" settings (e.g., "all buttons in" on the UREI 1176, see chapter 14), a particular artist (e.g., The Beatles and the Fairchild 660), or a particular recordist (e.g., the Josephson e22 microphone and Steve Albini). Stories need to accumulate around that piece of heritagized gear, and the gear must *repeatedly* appear in discourses that cut across gear cultures, for the gear to gain sufficient canonical traction. This might include appearing in print media gear lists, textbook references, online forum discussions, image circulation, clones, and interviews with professional recordists. Gear does not have to feature in *all* these realms to become canonized, but repetition across media is essential. Canonization may be accelerated by gear manufacturers who recapitulate gear heritagization tropes themselves (for example, through the release of anniversary and/or commemorative edition gear) and through industry awards (for example, the NAMM TEC Awards).

Canonization is implicitly linked to "power and authority"; its perpetual assertion reinforces which gear fits within the canon's bounds. Canonization is "a political procedure that enforces internal coherence and unity at the expense of diversity," and, as such, we see how canonized gear operates from its own island in the broader sea of gear cultures.[15] Canonization benefits *some* gear and its manufacturers since it constantly reinforces high-status positions, operates as implicitly exclusionary to genuinely new and/or innovative products, and ultimately "discourage[s] or even prohibit[s] us from creating deviant ideas and other identities."[16] The gear canon is largely made up of gear that held prominence through its affiliations with canonized workplaces and recordings and was inaccessible and financially unattainable in the 1960s–1980s. Such gear broadly fits within three categories. First are **large-format mixing consoles** from manufacturers including EMI, Neve, SSL, API, MCI, Cadac, Siemens, and Quad Eight, arguably at the pinnacle of the gear canon. Second are **outboard processors** made by Fairchild, Pultec, and UREI (including the 1176), the Teletronix LA-2A leveling amplifier, and other effects units made in the 1960s–1980s by AMS, dbx, and Eventide. Third are the **microphones** of Neumann and its East German offshoot Microtech Gefell, Schoeps, RCA, and AKG/Telefunken, among a few others; these extend the start date of the canon back to 1932 (with the RCA 44). Ancillary gear including analog tape machines, particularly those made by Studer and Ampex, are also firmly positioned in the gear canon, as are some microphone preamplifiers (although most canonized preamplifiers began their lives as channel strips within consoles, not as standalone units). Occasionally recent gear becomes canonical, too, including Manley gear, such as the Massive Passive EQ, and microphones made by David Josephson including the e22 and c700s. All post-1980s canonized gear is prized for its build quality, locality, circuitry, and materiality—and is distinguished by its high cost, too.

The gear canon recapitulates itself via two key mechanisms: commemoration and awards. Gear cultures love celebrating anniversaries and the occasion of (re)releasing "new" commemorative gear since they remind gear culture participants what matters and provide an(other) opportunity for the participant to engage in gassing and discourse. The "anniversary edition" object serves multiple purposes: it reaffirms a technology's heritage, it uses that heritage to reassert the object's place in the canon, and it demonstrates to skeptical consumers that long-standing companies are still capable of making gear just as good as they used to. Prior to 1997, the recorded music industry was seemingly focused mainly on commemorating recordists and workplaces, landmark recordings of popular and rock music, or long-running instrument series (e.g., the Gibson ES-350 T 1992 Sunburst Limited Edition electric guitar, released to celebrate the guitar's one-hundredth anniversary).[17] The anniversary edition gear phenomenon appears to be wholly contained within the gear cultures period we analyze, having begun in 1997 with Neumann's limited-edition re-release of a gold-plated U87 microphone and Beyerdynamic's limited-edition "Classic" reissues of their popular M88 and M500 microphones. By 2005, anniversary edition gear was a regular trope in *Sound on Sound* and Gearslutz.

Let's consider Microtech Gefell's UM75 large diaphragm tube condenser microphone. Released in 2003 and featured at the 2003 AES Convention, it is dubbed a "75th Anniversary" edition, *not* of the original UM57, which was released by Neumann in 1957, but of the date, 1928, when Georg Neumann founded the company. The press release is less a description of the microphone and more a history of the company and its "founding father," "inventor," and "scientist."[18] We are taken through a series of significant historical events, from the development of the Neumann "bottle" with its revered and oft-imitated M7 capsule to how such microphones were the primary broadcast instrument at the 1938 Olympics. From World War II and the destruction of the original Neumann factory to Gefell operating through the Cold War and repositioning itself following the fall of the Berlin wall, this microphone release embodies Gefell's legacy via Georg Neumann's history. The release is peppered with heritage buzzwords: "special," "famous," "legendary," "vintage," "milestone," "favorite," "tradition," and "antique." Built-in scarcity—only seventy-five of these individually serialized microphones were made—ignites online forum discussion and gassing about the transformers, polar patterns, and price. This press release shows how gear canon status must be reaffirmed to gear culture participants (in other words, the company continues to make high-quality microphones like they used to), recapitulated with "special" product releases, and reinforced within an accompanying historical, heritage, and mythological context (figure 13.1).

Another significant contributor to the gear canon is the TEC Awards. Founded in 1985 by *Mix* magazine and originally held at AES conventions, the TEC Awards proudly proclaim themselves to be "widely recognized as the highest honor dedicated to the pro audio and sound recording industry."[19] Some award categories honor individuals for their outstanding technical or creative achievement in professional audio. Others bestow technological innovations with awards, each year recognizing multiple finalists and an ultimate winner in categories including hardware signal processors, microphones, and preamplifiers. Related to this is the TECnology Hall of Fame. Founded in 2004, around five technologies are inducted each year into a hall of fame reserved for that which has "made a significant contribution to the advancement of audio technology."[20] In recent years, gear including the Neumann M50 (1950) and Neumann U67 (1960) microphones, the 3-M 16-track multitrack tape recorder (1968), the Eventide 1745 Digital Delay (1971), the API 500-series Lunchbox format (1985), and the Rupert Neve Designs Portico 5015 preamplifier (2008) have all been inducted alongside the original Edison cylinder (1877), synthesizers, guitar pedals, and PA gear.

Heritagization processes, especially commemoration and awards, comprise a key facet of gear canon maintenance. These processes are especially useful for museums since most of the labor involving the affiliated agents, workplaces, history, and mythology has already been done. As such, the gear canon narrows the pool of gear through which stories may be told, it

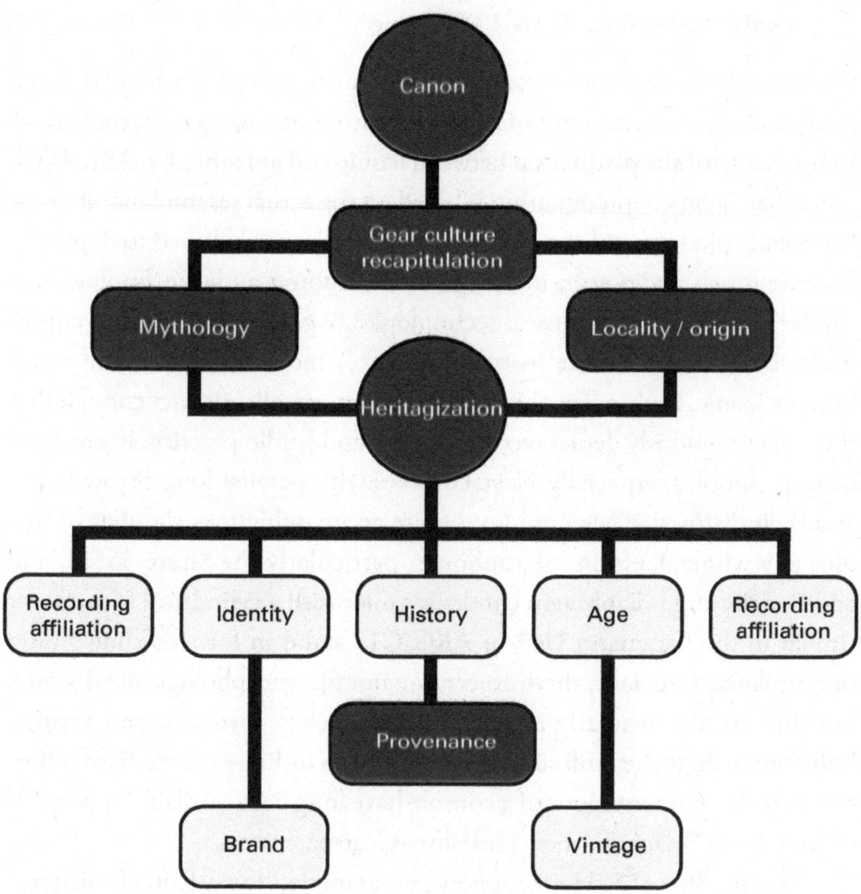

Figure 13.1
The relationship between heritagization and canon in gear cultures.

provides the curator with a small and easily navigable range of potential go-to "star" objects, and it allows objects to come fully loaded with embodied social, cultural, and political meanings. The gear canon is at the heart of gear cultures, and its participants are expected to contribute to its maintenance. It behaves as a compass with which gear cultures may be navigated; gear discourse often begins and ends from the point of canonized gear no matter how far participants stray. There is, however, a demarcation between gear in the canon and gear considered "iconic."

TECHNOLOGICAL ICONICITY

For gear to become iconic, it requires two features: formal attributes that can withstand abstract representation and enduring mythological associations. In his theory of the distinction between iconic and aniconic art, Alfred Gell notes that "iconic representation is based on the actual resemblance in form between depictions and the entities they depict or are believed to depict."[21] Because much audio gear, including all rackmount audio technologies, is similar *in form* to non-musical technologies (e.g., test and medical equipment, industrial machine instrumentation), most audio gear can never become iconic. Only a few objects are morphologically distinct enough that they unambiguously depict recorded music and studio practice. Recording/mixing consoles, especially Neve consoles with parallel long-throw faders and their distinctive Marconi knobs, are an unambiguous signifier of studios as a whole. Certain microphones, particularly the Shure 55SH, and additionally large-diaphragm condenser mics with a cylindrical form factor similar to the Neumann U47 or AKG C12 stand in for recording-studio performance. Crucially, these objects are not just morphologically distinct, but they are also regularly represented in sketches, cartoons, and abstract reductions where they still successfully depict recording practice. Two different cartoon representations of a console have long been available as possible default avatar images for new Gearslutz/Gearspace users.

Take the Shure 55SH microphone, for example. "Every Icon Has a Story" is emblazoned across Shure's website landing page, which leads us to "meet an icon." By no means an expensive or inaccessible model comparative to some microphones, its iconicity is bound up in its history and identity. Likened to "skyscrapers, jukeboxes and cars with fins" and inspired by a 1937 Oldsmobile grill, Shure deems its 55SH a "style icon." Beyond its many visual associations, Shure describes its 55SH as a "living legend," the microphone into which Martin Luther King declared "I have a dream," Frank Sinatra crooned, and "The King himself" Elvis Presley clutched on stage as he sang and swaggered. The microphone's formal distinctiveness and mythological connotations provide the necessary preconditions for becoming iconic, but the cycle was completed only when representations of the Shure 55SH circulated in other domains beyond gear cultures. A brief sweep of microphone tattoos, for example, reveals pages of Pinterest and Instagram posts with people around the world showing off their mic tattoos, where nearly all depict the Shure 55SH. No wonder then, Sylvia Massy ended her AES talk, full of extraordinary stories, with the anecdote that Bob Paquette took his beloved 55SH to the grave.

Iconic gear entails canonical gear that has taken on an extra degree of mythologization due to its accumulation of a large degree of social and cultural value. Having accrued stories over time, iconic gear exudes historical significance that manufacturers often leverage to promote new gear, and museums will play off to attract and engage visitors. In gear cultures, both real and projected notions of iconicity play out.

Rarely does a piece of gear accrue enough significance that it is unambiguously recognizable in broader society. More common, we find manufacturers marketing and promoting gear *as if* it were iconic, thus attempting to imbue gear with iconic status. In 2020, just prior to the 2020 NAMM convention, AMS Neve re-released their RMX16 digital delay unit by leveraging almost four decades of accumulated cultural value; the device, in their words, "characterized an enormous number of seminal recordings." SSL, aware that mixing consoles have iconic potential, often refer to theirs as "iconic studio classics."[22] Retailer Vintage King, whose long-term business model has included selling the most expensive used vintage gear, routinely describes various Neumann microphones as "iconic." Even AVID, best known for their ProTools DAW, named its live control surface ICON—the same name as one of Earthworks's current microphone ranges. This preempting of iconicity is an attempt to position new product lines among not just any gear, but *iconic* gear.

Reverence and idolizing are part and parcel of gear cultures and interactions between its participants, to the point where the display of knowledge of iconic (or not) gear holds social, political, and cultural value in itself. As Clinton Lanier and Scott Radar noted in their work on the synthesizer, all icons "ultimately have their foundation in some form of iconoclastic behavior."[23] Iconic gear is no different. It undermines the ubiquity, convenience, accessibility, and portability of digital music technology. Iconic gear routinely defies its own lifespan, original purchase price, and workplace: it is gear that has moved from the professional recording workplace (studio) into mainstream consciousness. While the visual and formal aspects of the object are, as noted before, the first conditions for iconicity, when iconic gear becomes the focal point of gear cultures, the gear's material, haptic, and aural attributes matter as well. Touching the material iconic gear and listening to its signature sound couple with the accumulated mythology to create a transcendent state—whether that situates the gear user in the golden age of recorded music or inspires the user to create their own new music.

CLONING GEAR

Heritage Audio is one example of cloned gear, but there are many more. Most of the gear we find on the market today consists of devices inspired by canonized gear, with (in some cases) dozens of straight-up clones all claiming to sound the same as the originals. Obviously, there is a pragmatic aspect involved: if one can reverse engineer the schematic and find contemporary components that perform similarly to vintage ones, a company can bring a product to market with practically no research and development. But twenty-first-century clones of heritage gear exhibit different approaches, ranging from devices that are intended to look and sound nearly identical to historical objects, to ones that look nothing like the originals but feature similar circuitry and components, to skeuomorphic software emulations, to plugin presets that model the circuit but within the context of a plugin that can emulate many different related objects. Clones run the gamut from free (certain plugins) to relatively inexpensive (DIY kits and clones manufactured outside of the US or Europe) to very expensive (boutique clones of the rare gear of yesteryear). Even the heritage gear companies themselves are getting in on the action: it is arguable that some of the many outboard products sold by AMS Neve, who now markets dozens of unique pieces of gear using the external look and model-numbering conventions of certain pieces of their most famous gear, are clones since their internal components and manufacturing processes differ significantly from those used when Rupert Neve was heading the company in the 1970s.

As of the year 2000, it has been relatively easy to create a new company with only one or two permanent employees, where the design aspects of cloning are outsourced to one of many techs who have repaired the objects in question and the manufacturing is outsourced to a contract assembly house in China or elsewhere (Chile, Spain, and Latvia have become preferred contract assembly sites for some jobs that require more stringent quality control, and some work is subcontracted within the United States or across the border to post-maquiladora factories in northern Mexico). "Made in Europe" can command higher retail prices than "Made in China," sometimes more because of ethnonationalist ideology in the US and UK than any demonstrable quality issues.

Since little has been written about the Chinese audio manufacturers, a bit of background is in order. When we talk of Chinese-made clones, this refers to objects made in part or whole by the companies Alctron, Beijing 797, Shanghai Feilo, ShuaiYin, and Soundking. Beijing 797, the first of these, was initially owned by the Chinese government and, in the 1950s, acquired training from the East German company Gefell. According to David Josephson, 797 produced their own original mic designs in the 1960s–1970s but from the 1980s onwards were "pressured" *by Western companies* to produce knockoffs of Neumann and AKG microphones. When China's "opening-up policy" came into effect, several 797 employees left and brought their knowledge and experience to a Shanghai-based factory called Feilo. In 1990, Feilo began making the CR-3A mic for US-based company VTL (later known as Manley Labs); this mic was later rebranded as the Langevin CR-3A, reviving a heritage gear brand of the 1960s, and was one of the first mass-produced <$500 mics in the style of the Neumann U87 (which at the time retailed for over $2,000). By the year 2000, 797 was the original equipment manufacturer for some of the mics released under the brands ADK, Marshall/MXL, Studio Projects, BPM, and Rode; in 2000, Shanghai Feilo produced some of the MXL, Pacific Pro Audio, Carvin, Apex, Yorkville, and ADK brand mics.

The contemporary retail landscape, therefore, is characterized by dozens of brands of clones of the same original products, with most brands being unknown to many consumers. Only customers at Thomann, the largest studio gear retailer in Europe, would be familiar with t.bone-branded microphones, and only consumers at US-based Guitar Center and their online subsidiary Musician's Friend would likely know the Sterling-branded microphones. Since they are manufactured with significant economies of scale, compared to boutique makers at least, their retail price—typically one-fourth to one-twentieth the price of boutique products of the same type—has turned them into attractive targets for modding. Companies such as Microphone Parts, Revive Audio, and in the 2000s–2010s Michael Joly all offer either DIY kits to improve the performance of select cloned products or a service to mod the mics. The modded mics and outboard gear pieces are promoted well enough that they occasionally get reviewed in *Tape Op* or other recording magazines. Notably, modded products still look, on the exterior, the same as cloned gear, which in turn is closely inspired by the external look of heritage gear.

TECHNOLOGICAL TAXIDERMY: GEAR GOES TO THE MUSEUM

The Queen Studio Experience in Montreaux, Switzerland, is tucked away in a dark corner of a casino. In part a re-creation of the original Mountain Studios, glass cabinets line its entrance, each featuring re-creations of recording-studio sites. Brian May's Red Special guitar is propped up against a Vox AC-30, augmented with a Styrofoam coffee cup and a garish yellow Ricard ashtray complete with butts. Lyric sheets and an abandoned game of Scrabble adorn the flight-cases-cum-coffee-tables, and plenty of nostalgic descriptions of Queen's time at the studios adorn the museum cards. For such a museum to work, it must be what Charles Fairchild called "emotionally convincing"—that is, convincing enough to create "a credible space around the missing bodies of absent musicians for spectators to inhabit."[24] The studio itself is, of course, remarkably still and quiet, devoid of the creative energy, sprawling cables, exhausted but determined musicians, and workflow glitches synonymous with such an arena. A 24-track Studer A-80 is recessed into the wall, backlit but inanimate. A rack of cutting-edge gear sits motionless, no signal flowing, LEDs flickering, or meters alight. Intriguingly, this rack of gear leans to creative effects processors as opposed to nuts-and-bolts dynamics and EQ. An Eventide Instant Flanger, two AMS DMX 15–80 delay units, a Francenstein Stereo Enhancer, and a rare Russian

Dragon (rushin' draggin') time identifier are stacked in the defunct rack: the latter unit features a sellotaped photo of drummer Roger Taylor on channel two, and the former units are an obvious nod to the layers of choir-like harmonies so intrinsic to the sound of Queen records. In the center of the studio sits a replica of the 48-channel Neve 8048. Built in 1974, the console was updated in the early 1980s to include Necam automation. Yet in this studio experience, the console is not "the console" at all, but a photographic cardboard overlay of the console's *appearance*. Smooth and flat, the board bizarrely erases the tactility and haptic desire of all the knobs underneath, rendering its functionality to mere imagery and imagination. All that is exposed are ten red faders. A video screen featuring narration by drummer Roger Taylor invites visitors to "mix" tracks like "Made in Heaven." The mix is a reductive multitrack version, split into major stems of drums, percussion, guitar, bass, vocals, and effects. The mix process is transformed into ludic, interactive curiosity as visitors sit at the desk and play with the console. But what is going on here is a kind of technological ventriloquism;[25] the sounds do not emanate from the mixing console as displayed, but from a concealed playback system underneath the console.

Ironically, this "studio experience" is about as far removed from the realities of an "irl" scenario one could get. Much of that distance is created by the absence of personnel, but the space is rendered unrealistic by the gear's positioning. Disemboweled, knobs and buttons flattened, its circuitry drained of power, this gear is lifeless. Devoid of the signal required to operate it, or even a powered-up standby status denoting its readiness for such use, the "Queen Studio Experience" somehow lives on in the context of a (posthumously) canonized workplace and artist.

> I've kind of joked sometimes that I feel . . . like David Attenborough in the nineteen-fifties, going out into the rivers or whatever and finding the beautiful birds of paradise and killing them and putting them in a glass case. Because I feel that ninety-nine percent of the stuff that comes into the collections will never be turned on again.[26]

Jamieson's analogization of the acquisition process to the preservation of an exquisite and endangered creature is commensurate with the life and death of gear that is central to exhibitions such as the Queen Studio Experience. In our interview with Annie, we were reminded of another moment of gear display, this time at "Pink Floyd: Their Mortal Remains," exhibited in 2017 at the Victoria and Albert Museum. Here, we found another eviscerated-console-as-interactive-delight where visitors could mix *The Dark Side of the Moon*'s song "Money" to their heart's content. But that was not the point of contention for Annie. A Midas XL-3 live mixing console "was displayed in a case, upright against the wall: no mixing desk is ever like that. And it was actually on! It was working! It had flashing lights and everything, but you couldn't hear anything. And I was thinking that's the most bizarre thing."[27] This display of animate gear is, however, an exceptional case.

This pattern of gear display we term **technological taxidermy**. From the Greek meaning "order and arrangement" (taxis) and "skin" (derma), the peculiar practice of displaying electrical audio technologies without the required electricity has much in common with the ways dead creatures are stuffed and displayed. These are practices full of paradox, bound up in eccentricity yet signifying preservation, and can be "characterized by proximity and intimacy."[28] Common in the nineteenth century, as Rachel Poliquin notes, taxidermy was intrinsically linked to colonialism and the display of an imperial archive. A largely female pursuit concerned with aesthetics and natural beauty, taxidermy has served as an "educational tool for presenting visions of nature" yet is seen as perverse, something that induces squeamishness. Poliquin does, however, critically reappraise taxidermy as being about the practice of collection and mounting, and the embodied cultural, political, and ideological forces at play.[29] Taxidermized gear exudes excessive heritage. This recapitulation of technological canonization affirms that only certain kinds of technologies are in fact *gear* and worthy of gutting and indeterminate display.

During the 2016 EMP Conference, Samantha took a walk around what is now called MoPOP, Seattle, one of the world's biggest popular culture museums. A gigantic tower built from acoustic and electric guitars dominates the museum floor, as do oversized displays of Nirvana and Sub Pop record label memorabilia. "The MoPOP Permanent Collection now contains over 100,000 artifacts," boasts its website, "but it all started with this":

the object being a 24-channel 1969 Datamix mixing console, custom-made for Jimi Hendrix, originally installed in his New York Electric Lady Studios before being acquired by MoPOP in 1991. The console is displayed behind glass, inanimate, and with some of the original studio acoustic paneling as a backdrop. It is ornamented with various images and personal belongings of Hendrix yet is the star object in the collection—and the entire museum. Photographs of the console in its original location depict Hendrix with recordist Eddie Kramer and/or cofounder Gary Kellgren happily perched at the desk, cigarette ash precariously hovering over the faders and the console alight and alive with the guitarist's music. Patch cables spill out from the large panel of sockets, hands and fingers adorn the pots and switches, and the backlit VU meters cut through the black-and-white imagery with a warm glow. Yet here the desk sits in MoPOP still and inanimate, dull and contextless, despite the museum cards that proudly proclaim its provenance. Without electricity, the console is functionless; without patching, the console is disconnected; and without the personnel, the console is inoperable. What, then, is the point of this star taxidermy? As Poliquin stated, it is equally about cultural, political, and ideological forces, and precisely those values are conveyed through the console. Taxidermized objects may be conceptualized "as specters of a dissociated past and as evidence of fractured relationships with nature, spirituality, and labor."[30] This same analogized religiosity is deeply embedded in studio discourse, as the romanticized labor of the recording-studio personnel is embodied in the console.[31] MoPOP invites its visitors to travel back in time, to "travel alongside the legendary Jimi Hendrix at the height of his fame," to that dissociated past Youdelman describes as emblematic of the taxidermized object.

Taxidermy is also symbolically aligned with "hunting and trophies,"[32] and technological taxidermy extends the gassing and gear fetishization so prevalent in gear cultures. These objects are unique, prized, inextricable from the rock canon, and so excessive in their cultural, political, and ideological values that they transcend gear cultures and become sensible in mainstream Western tourist sites. As such, taxidermized gear represents stage two of figure 2.3 and indexes Veblen's conceptualization of goods as status: taxidermized gear symbolizes *prowess, skill, and courage* in its acquisition (even if that acquisition is by a museum curator); *conquest* in that the acquired object resides in a singular unique locality as opposed to somewhere else; *value* in terms of its provenance and embodied social, cultural, political, and ideological ideas and connections; and *status* in that the dead gear is a trophy for exhibition, the star object of the museum. Furthermore, taxidermized gear is so part and parcel of heritagization that it feeds back into marketing and promotion campaigns for new gear. After all, despite its elaborate promotional World War II paratext, we never saw an operational Shadow Hills mastering compressor; it was depicted as belonging to a dissociated past, detached from any workplace or workflow. We witnessed the same staging of brand-new yet stagnant Universal Audio interfaces in the plexiglass cabinets at NAMM. Companies now foreshadow gear heritagization, where technological taxidermy becomes a premonition, or even a goal, in the product launch of new gear. It is as though the museum is the ultimate destination for gear, the only possible way gear can live on: heritage *as* timelessness.

The awkwardness, embarrassment, and discomfort of taxidermy makes it applicable to gear heritagization. Dead yet proudly displayed gear can be synonymous with high-profile progressive British rock bands of the 1970s, but it also symbolizes waste, obsolescence, and disposability. For every piece of gear that ends up in a museum, thousands of pieces of audio technology end up in landfill. Taxidermized gear reminds us of our inability to detach ourselves from technological objects when they reach the end of their lives, especially those objects we associate with famous musicians or recordists. It also acts as a means for music fans to be proximal to and intimate with posthumously canonized artists. Technological objects that once carried the signals of long dead voices, despite normally being concealed from public view, retain significant cultural, political, and ideological value. Audience proximity to taxidermized gear can only be achieved through the comparative accessibility of the museum, where intimacy, if present, is achieved through ludic and interactive displays (even if they are mere ventriloquism) and close-up imagery. Unique by definition, the star objects of gear museums live on separate from (but parallel to) the gear canon: as the ultimate conquest, they are now unobtainable by even the most hardened gasser.

CONCLUSION

Successful gear is heritagized gear. That is, to maintain relevance, appeal to gassers, and ultimately sell, gear must have some link to its past and must preserve and conserve the values of "golden age" large-format studio recording. Heritagization processes are laborious and time-consuming yet essential to the functioning of gear cultures. But participants may not realize they are undertaking heritage work since most of the process is inextricable from gear fetishization and gassing. In the absence of official heritage lists, gear culture participants pick up the slack in affirming gear's heritage status. Whether conducting an ABX shootout between an original and a clone or indulging the show-and-tell of a famous record producer with an even-more-famous microphone collection, gear culture participants have all the necessary tools to heritagize gear. No gear can be truly canonized prior to having heritage status, and newer canonical gear significantly resembles the gear of old. Since iconic gear is drawn from a small pool of canonized gear, heritagization is a prerequisite for that elite strata, too (despite some companies' attempts to market a false iconicity).

Museums play an important external validation role for gear cultures since the process of acquisition and display reinforces and reaffirms gear culture discourses and grassroots heritagization work. But taxidermized gear in the museum, silent and inanimate, detached from recording work or the clutches of a gasser, lacks the sexual fetish potential of gear in other sites. Therefore, to make it into the museum, gear's embodied sociocultural, political, musical, and economic values have to exceed that of even the most canonized and iconic gear. Without heritagization or canonization, gear cultures would likely not endure. If gear cultures were more centralized and not strewn across online fora, the trade press, trade shows, and museum exhibitions, perhaps the gear would be subject to a more formalized heritagization process. Because if gear cultures could award gear with heritage listings and blue plaques, they most certainly would.

14 OBDURANCE

A LARGE KNOB

Following the conclusion of the fortnight-long intensive sound archiving course at the ANU School of Music in July 2021, former AIR and current 301 Studios maintenance engineer Stephen Crane knocked on Samantha's door. "I have a present for you! For getting through all the tape challenges!" he gleefully announced. He handed Samantha a plastic object that looked like a giant hollowed-out switch. "It's a 3-D print of a TG series console knob!" he beamed, delighted at its likeness. The printout was light in its frame, yet around ten times oversized.

In the months prior, Samantha had been running around attempting to get four Revox tape machines operational to teach the niche biennial course. It was a long process of begging, borrowing, and stealing: parts were ordered from Europe, a couple of machines were sent to Peter Mony at Nagravox in Sydney, and Stephen Crane arrived at the school every other day with an update as to the capacitors on machine one, the belt on machine two, the motor on machine three, and the pinch roller on machine four. It was a race against time to contact various secondhand suppliers, eBay sellers, and parts specialists to ensure the machines were operational.

Tape itself is not going away, and it can be bought from numerous retailers. As Dave Harries notes, "People still use tape, they still use the valve microphones—the same microphones that we were using at Abbey Road then are still preferred today."[1] Tape as a medium is glorified and valorized; students are already asking if they can do various projects other

Figure 14.1
TG series knob. Photo by Samantha Bennett.

than archiving—tape composition, stereo recording, re-creating Beatles recordings—as part of their course. There are, however, significant issues in maintaining the machines. Stephen is integral to the process of maintaining gear so it may be used in contemporary recording and production contexts, and in this case, three of the four machines were operational in time for the course, which ran for the fortnight without incident.

The 3D printed gift (figure 14.1), however, symbolized something deeper and resonated with Samantha as an indicator of what this laborious process of maintaining gear represented. As replica knobs go, this one was loaded with meaning, being symbolic of the original TG Series console's part that had been scanned for the printout. This knob also instantiated the valorization of EMI audio technologies long into the 2000s. Since the original console was built in-house by EMI engineers, the knob represents engineer agency and labor, including the enduring relevance of Stephen's labor as a technical engineer—formerly of an EMI legacy studio and currently of one of the last remaining large-scale studios in Australia, and, at the time, a visiting technical engineer at the ANU School of Music. Of course, the knob also represents the detached interfacial elements of a mixing console. To 3D-print the knob (as opposed to any other element of the TG Series console) is to render it iconic of gear design and gear interfaces; as such, the knob indexes the haptic ideation so intrinsic to the console's operation and the phallic connotations so sexualized and fetishized in gear cultures. Gifting the knob, a materialization of gear obdurance, materializes knowledge transfer, representing a continuum of technology and process through generations, and recognition of a shared understanding of gear values, labor, and representation. Who would have thought a knob could carry such significance?

THE PAST IN THE PRESENT

Gear persists in the 2020s despite professional processes being predominantly dependent upon software technologies, digitization processes, and computer-centric workflows. For all but a few specialized recording scenarios, high-quality recordings can be made with nothing more than a DAW, a well-specified computer interface with built-in preamps and the requisite number of suitable microphones. Every kind of outboard gear, including recording consoles, exists as relatively accessible software emulations—whose quality has improved year upon year. Gear, in contrast, is cumbersome, expensive, heavy, spatially consuming, and energy inefficient. Gear persists less because of its aural enhancement capabilities than its extra-audible potential. We have already charted the gear-specific properties that contribute to its persistence, including its materiality, look, haptic feel, and playful potential. But this alone is not enough to account for gear's continuing appeal.

Gear is enmeshed in a multifaceted schema of representations, some of which (secrets, heritage, canon, iconicity, war) we have already discussed. Gear culture participants themselves theorize those very concepts, and, as researchers, we strove to consolidate the theory-building work that gear participants undertake every day to make sense of their relations to gear. Here we will take a different approach and delve deeper into a novel theoretical lens that we see at the heart of all relations between people and gear—the concept of **obdurance**. Obdurance, a word that connotes stubbornness, resistance to influence, and unyielding, implies holding onto the past and onto a tradition or custom or habit. Where heritage implies the death of gear and a retrospective gaze toward historicized gear pasts, obdurance accounts for the present and the future: why gear is still here, and why it will continue to be here.

Obdurance is a distinctly human disposition; no piece of outboard gear possesses sentience that would allow it to come up with its own representations. However, as we have shown throughout the book, due to its fetish nature, people routinely map human properties onto gear including attributing it agency. But throughout, obdurance more accurately describes a central quality of the relations between people and gear, especially relations to the past that are kept alive in the present.

The concept of obdurance relates to, but differs from, several recent theories about how the past lives on in the present. Indeed, studies spanning history, media, STS, sound recording, music technology, critical theory, and philosophy provide various ideas that might frame old gear in terms of its present-day relevance. Take, for example, scholarly and journalistic takes on analog media's presence in the 2000s, which have been described around dozens of terms including "revival," "nostalgia," and "technostalgia."[2] In such work, analog media is regarded as having receded from view, later to (re) emerge. Vinyl records' ongoing appeal is the main "analog" object thrashed out in this discussion, which makes sense given its fluctuating sales trajectories since the 1970s. But, as we have asked throughout, why analog *audio* technologies?

We recognize several existing frameworks that to some extent overlap with our obdurance concept. "Nostalgia (from nostos–return home, and algia–longing)," begins Svetlana Boym in her seminal book, "is longing for a home that no longer exists or has never existed. Nostalgia is a sentiment of loss and displacement, but it is also a romance with one's own fantasy."[3] Boym frames the concept of nostalgia as one of "returning": to real and imagined places and times that may be historicized and/or fantasized. She suggests nostalgia is beyond conventional understandings of time and space and, as such, is atemporal and a form of "rebellion against the modern idea of time."[4] She does, however, suggest that where nostalgia mourns distances and detachments from time and places, technology is an enabling medium that promises "speed, ease and oblivion of everything except the technological products themselves [and] offers solutions and builds bridges."[5] This idea is useful when it comes to thinking about gear and its omnipresence.

We can think of gear as being bridges to times, places, and people that are no longer there: the so-called golden age of record production, large-format recording studios, and a buoyant Western record industry. When the signals that flow through gear are synonymous with the aura of (often posthumously) canonized musicians, this buoys the prominence of high-profile auctions of recording equipment belonging to leading recordists or known to have been used on significant recording sessions. In his work on analog nostalgia, Dominik Schrey focuses on an object's potential to convey imperfection in "the continuing process of digitisation." He suggests such nostalgia is "directed towards the noise, not the signal. In the broadest sense, it operates as a strategy of re-enchanting an object through aesthetic de-familiarisation as it is characterised by deliberate imperfection."[6] We can certainly equate the fetish for "vintage," "classic," and "old" gear with the pre-digital era when, due to equipment limitations, recordists managed signal chains to optimize noise reduction. However, the present-day gear fetish stems more from its accumulation of embodied social values and romanticized labor (which may or may not be nostalgic). Additionally, much new gear prides itself on having reduced or eliminated noise.

Sound quality, while aesthetically present in gear cultures, does not feature prominently in gear fetishization discourse. We concur with Pat O'Grady that "the nostalgia for analogue and the resurgence in analogue products can be understood as an expression of class-based capacities to purchase boutique products."[7] A more nuanced take on nostalgia is found in Timothy Taylor's theorization of *technostalgia*, where technological anxiety manifested in "past visions of the future, a future that never arrived."[8] Taylor's work is, however, pinned on synthesis and a millennium-era fascination with 1950s–1960s sci-fi. (Tech)nostalgia does not entirely account for the current and continuing presence of gear; romanticism and sentimentality do play some role in gear cultures, but more complex aspects of fetish, status, canon, and heritage are at play.

"The time is out of joint," a lesser-known phrase from Shakespeare's Hamlet, appears in Jacques Derrida's *Spectres of Marx*.[9] As Mark Fisher noted, it is the origin of *hauntology*: the ways in which specters of the past exist in the present. Fisher suggested hauntology as being of two strands: identifiable patterns and repeats, "that which is (in actuality is) no longer, but which is still effective as a virtuality" and "that which (in actuality) has not yet happened, but which is already effective in the virtual (an attractor, an anticipation shaping current behavior)."[10] In gear cultures, hauntological ideas play out in the hardware cloning and digital domains, where long-dead gear is reimagined and brought to life in skeuomorphic guises. Additionally, the intensity of gear imagery display in online forums and in retail catalogs acts as an attractor to gear consumers. Gear exists but is not necessarily in the hands of those who aspire to own it; precisely this mechanism underpins how gear fetishism drives GAS. Jamie Sexton recognized hauntology as a broader musical movement in electronic music where artists "foreground technologies because of their ability to resurrect the past and distribute historical traces across networked platforms."[11] Sexton also suggested that hauntological ideas are bound up in memory and intertextuality, yet, when we consider gear cultures, many participants would lack firsthand memories of professional recording processes or of a process involving a particular piece of gear in its original form. To that end, gear cultures project valorization through imagery, instilling a sepia-toned memory that is largely detached from consumer realities: accessible in the virtual but aspirational in the physical. Hauntology does help with explicating digital skeuomorphs but fails to frame the ongoing presence of hardware gear, and how and why it is necessary in the 2020s.

As the obsession with new media, particularly computers and software, grew toward the millennium, Charles Acland noted a tendency to "confront yet another triumphal proclamation of our age of freshly minted media trinkets." Alongside technological utopianism's ongoing persistence, Acland recognizes a parallel desire "to remember through the objects we elect to save and treasure long past their initial use-value has been expended," which he framed as *residual media*.[12] The concept of residual media is not about users persisting with old technologies in the same way that some writers might reject Microsoft Word for a typewriter. There are politics at play, but these are around cultural and economic value, particularly "retrieval and revalorization," as opposed to a conscious rejection of what new digital technologies stand for. Some gear, particularly that bought and sold on the secondhand market, is indeed residual. It belongs to a past era of recording, and it might be redundant, but in the present, it represents a novelty. But what of new gear as a historical continuum of old gear? Acland alludes to a sense of denial in new media studies that fails to recognize new technologies as "taking account of continuity, fixity, and dialectical relations with existing practices, systems and artifacts."[13] This is imperative to us in our study of gear and its persistence; it is precisely its links to the old that justify and maintain its continuing existence.

Might the presence of gear be related to what some scholars have called "slow media"? Jennifer Rauch wrote that "principles associated with Slowness include mindfulness, humanism, localism, serenity, sensuality, simplicity and self-reliance."[14] Some of these facets exist in gear cultures, too, particularly gear's sensuality—either in literal or ideated states. Slow media is "alternative media" that must "(1) be produced by small organisations rather than big companies, (2) use traditional technologies as well as new ones, (3) not be motivated by profit alone, and (4) advocate for different societal values."[15] Gear since the 1990s has been similar to slow media with regard to "accompanying themes of 'hand made,' 'high cost,' attention to detail and slow manufacturing process" in marketing and promotion materials.[16] However, gear's slowness is bound up in valorization: the labor involved in gear's design and manufacture. In use, gear is anything but slow. Interactions with hardware are often momentary—a quick switch, button press, or potentiometer turn. Digital tools often require more labor in terms of installation, registration, updates, or linking to security accounts like iLok. In fact, Kelli Coyne of San Francisco's Women's Audio Mission championed the ease of use and fast labor aspects of gear: "I also like the workflow better, working with outboard gear, cause it's so easy to plug in things. You actually take a step back and you're like 'I really need a compressor on this.' I feel like the workflow is a little different. I love using the older gear."[17]

The rejection of certain new audio technologies, and the widespread reference toward historical studio practices is reminiscent of luddism. Steven Jones described *neo-luddism* as a "back-formation based on the received idea of a historical labour movement."[18] In gear cultures, the closest equivalent to a labor movement is bound up in the ideology of handmade and boutique hardware, and the workshops or studios of known designers and/or canonical recordists, typically located in the US, in the UK, and in Germany. However, more than luddism is at play. While technological valuations and romanticizing of guild-era studio economies are operative, so too is the seduction of the manosphere. Labor histories and laborers are not "sexy"; only gear is. Additionally, while "vintage" and "classic" gear is valorized, so too is brand-spanking-new gear.

OBDURANCE: PAST GEAR IN THE FUTURE

The late Tim de Paravacini, professional audio technology designer, manufacturer, and maintenance engineer, was keen to speak to us about the lasting presence of some technologies. "Some or many classic products from the past have stood the test of time," he said. "Everyone still uses classic microphones or variations on classic microphones."[19] In other words, gear did not disappear to later return; gear persisted and endured in an increasingly digital and digitized technological landscape. Indeed, while the company names may have changed, plenty of gear manufacturers have *not* ceased manufacturing, and several continue to rely on old machinery. New boutique manufacturers have entered the market. What accounts for this lasting presence?

The Merriam-Webster dictionary gives two definitions of "obdurance":

1. unmoved by persuasion, pity, or tender feelings; stubborn; unyielding.
2. stubbornly resistant to moral influence; persistently impenitent: an obdurate sinner.

While obdurance is most often used to describe people and behavior, we suggest it can also be applied to objects, especially considering two phenomena: the widespread ascription of agency to objects and gear's *resistance* to the influence and seduction of digital technologies, particularly computer-based software. Despite the convenience, comparatively low cost, and space- and time-saving benefits, gear resists this influence and prevails. Obdurance is, however, closely tied to a broader philosophical discourse around endurance and perdurance. Sparked by David Lewis, something *perdures* "if it persists by having different temporal parts, or stages, at different times, though no one part of it is wholly present at more than one time; whereas it *endures* if it persists by being wholly present at more than one time."[20] However, "endurantism is the theory that objects are three-dimensional, and objects persist by being 'wholly present' at every point in the time of their existence."[21] Some gear endures *or* perdures, but neither concept is universally applicable. For example, the Fairchild 660 exists in its original guise (albeit in ever-decreasing numbers), in physical clone objects released since the 1970s, and in skeuomorphic plugin form. In the Fairchild's atemporal existence, in different spaces, in different times, and in different form(ation)s, it could be regarded as perdurant. Other gear shifts over time; not being "wholly" present at each stage of its existence, it is not especially endurant. For example, gear may be animate or not, sometimes it ends up taxidermized, or later in its life it may accrue canonistic values that were never there from the start and therefore its meaning and presence takes differing forms over time. Hence, we land on obdurance as the most useful lens through which to view gear in the present day. Why and how is gear obdurant?

First, gear is **unyeilding.** It takes up its own space in music and audio technology cultures and remains steadfast in an evolving digital audio landscape. It yields neither to marketing pressures nor to trends in digital technology. Rather, it is insistent on its continuing presence, appearing side by side with computer-based technologies in retail catalogues, on websites, and at trade shows. Second, gear is **(omni)present and visible**: in design, in manufacturing, in the press, at trade shows and exhibitions, online, in events, and in and around heritage sites. Third, **gear fetishes do not diminish** over time. Gear's economic and ludic potential is as valuable today as it ever was. If anything, gear has accrued new social values, including sexual associations, and is becoming fetishized in more ways in the gear cultures period than previously. Fourth, **gear is not seduced by computer-based software technologies.** We see the opposite going on; software companies and digital audio workstations continually position canonized gear as both aural and visual benchmarks in their design, manufacture, and marketing of new technologies. Fifth, **new gear relies upon a connection to its past.** This is different to heritagization, which is about the celebration and display of the past in the present. Obdurance is, therefore, about the ways in which new gear cannot exist without its past, whether in terms of components, design, interfacial elements, marketing and promotion, events, or discourse—or a combination therein. Sixth, **gear is manufactured, consumed, and fetishized today.** Despite heritage and canon practices, or (tech)nostalgic takes on how gear was used in sound-recording practice in the past, gear is wholly present in the now. Seventh, since brand-new gear is still being made, **gear is not anachronistic.** For example, it sits side by side with a ProTools DAW in a recent Sweetwater ad, it is the main attraction at AES, it is aspirational, and it keeps numerous businesses afloat. New gear cannot only be something from the past existing in the present; new gear by AMS Neve and Manley are not hauntological ghosts or specters.

Obdurant gear has four key characteristics (figure 14.2). First, gear obdurance is about its symbolism and accumulated social values over time. These include relationship(s) to recordists, musicians, studios, eras, and economic and sexual fetishistic values. Social worlds are created and maintained around gear that exceed the boundaries of professional audio workplaces and the professional audio industry. For some, gear is collected, invested in, hoarded, and exhibited. The accrual and reinforcement of social values maintain and elevate gear within gear cultures.

Second, gear represents an embodied, romanticized labor. As Marx recognized in what he called the valorization process, goods accrue added value, particularly via value-forming labor: "In order to translate money into capital it is transformed into commodities which constitute *factors of the labour process*" (emphasis original).[22] The valorization process means that excess values are created; when it comes to gear, this is often bound up in

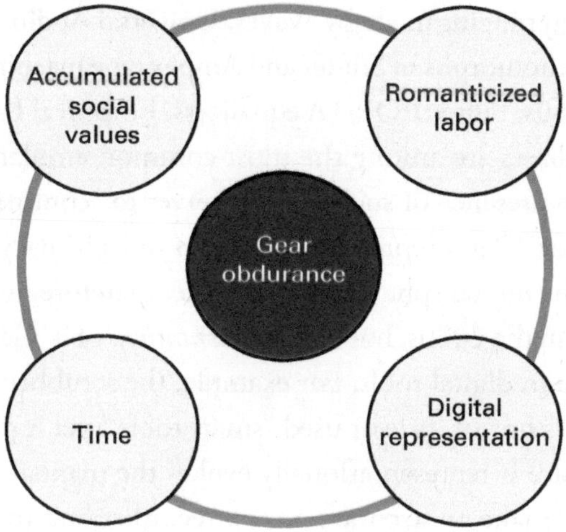

Figure 14.2
Features of obdurant gear.

the "slow labour" marketing techniques and imagery used to promote gear to consumers.[23] Additionally, when gear designers become the face of the company, their labor is embedded in the product, meaning that buying into the gear is to buy into the designer and/or manufacturer who built it. This form of gear valorization is separate from the labor of recordists and other studio personnel. Interviews with gear designers and their participation at trade shows and events reinforce the romanticization of *some* labor within gear cultures; this in turn reinforces obdurance as a property of design and manufacturing.

Third, computer-based software technologies recognize and capitalize on gear obdurance through skeuomorphic representations of gear. Digital representations of a compressor do not *need* to feature digital image representations of the controls and interfacial elements of analog counterparts; as Adam Bell et al. recognized, "skeuomorphism in software design is unnecessary."[24] Pat O'Grady noted that skeuomorphism can lead to a worse user interface, where "aspects of user interface functionality are traded for the visual aspects of the hardware."[25] In terms of functionality and sound, a digital likeness can be superfluous, which reinforces our thesis that the *excesses* of audio technologies are what constitute the objects as gear. We see skeuomorphic representations in plugins made by Waves, Universal Audio, Steinberg, and many others. Skeuomorphs of Studer and Ampex tape machines, 1176s, LA-2As, Fairchild 660s, Pultec EQP-1A equalizers, Empirical Labs Distressors, and GML equalizers are among the most common emulations. O'Grady suggests that the presence of such plugins serves to "continue to shape the sound of pop music" and represents "a politics of legitimacy" that serves to shape value in the digital sphere.[26] Obdurance, therefore, is not simply the presence of gear in the 2020s, but also *representations* of it. Gear valorization, too, is bound up in digital tools. For example, the scrubber tool in AVID's ProTools DAW is one of its least used "smart tools," yet it prevails upgrade after upgrade since it representationally evokes the manual labor of analog tape editing. To make an accurate edit, a recordist had to move the tape slowly over the playback head and mark the tape with a chinagraph pencil at an exact edit point before taking a blade and physically cutting it.

Fourth, for gear to be obdurant, it must exist over time and in different spaces or contexts, not necessarily in whole (endurant) or atemporally (perdurant). Time is essential to gear obdurance since technological objects cannot quickly accumulate the necessary social values and meanings necessary to become gear. Rather, they must evolve over time and connect to a historicized past. This is not necessarily nostalgic or hauntological, but there must be a continuum of some sort present in the gear trajectory.

OBDURANCE INTERNAL TO GEAR CULTURES

"We're not just vintage gear," proclaims the "About" page on US retailer Vintage King's website. "Little did we know back in 1993 that our little vintage gear refurbishing business would grow into what it has become. Today, we carry everything from vintage classics to cutting-edge digital gear." Since the late 1990s, Vintage King has grown into a foremost gear retailer with showrooms in Los Angeles and Nashville and a high-profile online business. Originally established as a gear repair business which later grew into a secondhand vintage gear retailer, Vintage King has recently expanded to include software, musical instruments, and high-end playback equipment. As a paragon of obdurance, the relationship between old and new gear in Vintage King is symbiotic; it is now a given that vintage and new gear exists alongside digital equipment in modern-day recording. Retail is an important part of gear cultures, and Vintage King plays an integral role; many of our previous observations are present in its operations. Its recently launched print magazine *Playback* features interviews with leading audio industry practitioners, recording studio profiles, and gear reviews. "In this issue, we take a look at some iconic pieces of gear that are bursting with soul, like the legendary Fairchild 670 compressor, the beloved Helios 'Type 69' channel strip and the Q2 Compex F760X," beams company president Ryan McGuire, packing in all the right gear names and terminologies likely to resonate with gear lovers. Vintage King's website is-jam packed with gear news, features, and content pinned to its retail operations. A Pro Audio Hall of Fame features the usual suspects, and even the store itself is split into new gear, marketplace, and "vintage and used latest arrivals," the latter representing the retailer's original core business. Vintage King also serves gear consumers as a site of aspiration and is an important cog in the GAS mechanism. Vintage King's operations consistently romanticize recordist and studio labor, blur the historicizing of technologies, recognize how gear exists within the digital domain, and load up new products with social values drawn from broader gear cultures. As such, their business model is reliant upon and perpetuates gear obdurance.

Certain gear accumulates such an excess of value that it becomes an investment opportunity. Here, valorization is hard-wired into chosen pieces of gear that embody a highly prized coalescence of gear culture values. Such gear is out of reach to most gear culture participants yet can be gazed on from a distance. Auction houses even adopt gear cultures' technological exhibitionism tropes to appropriately stage this very special gear, as noted by Robin Porter: "If you've got a recording console that did Dark Side of the Moon, you know, it's going to be worth a lot more money than a model that didn't. Old cars are exactly the same. If you have a Ferrari that won Le Mans, as opposed to the same model which didn't, you've got a car that's worth twice as much."[27] Porter was referring to Bonham's March 2017 auction of the EMI TG12345 MKIV console. "The console was used extensively between 1971 and 1983 in studio two [Abbey Road] and was used to its full potential by rock royalty Pink Floyd to record their landmark album, *The Dark Side of the Moon*."[28] Valorization is baked into this auction listing: the dates the console was used, the facility where it was housed, valuation terms "extensively" and "full potential," a nod to a canonized artist and album, and, most of all, the term "rock royalty." This discourse, gear culture discourse, mirrors those we saw at Vintage King and reverb.com. Initially, the console was listed with a guide price of $700,000 USD, yet it sold for more than $1.8 million. During the gear cultures period, there has been an upswing in auctions of the types of gear to which Porter refers. Gear auctions perform obduracy, in this instance, the unyielding staying power and valorization of prized gear that, over time, accrues excessive value(s)—due to its heritage and *provenance*.

Gear auctions are gathering pace to such an extent that a specialist company now exists to manage them. Dubbing itself "The world's finest selection of rare, one-of-a-kind studio gear and memorabilia," ANALOGr was launched in 2021 "to provide the equitable and responsible transfer of prized music industry items of historic and functional value." Their website states, "ANALOGr is the destination for highly curated, music-related assets of pop culture's most important moments. ANALOGr partners with the world's leading artists, collectors, and estates to tell the inside stories of their unobtainium. We travel the globe in search of these one-of-a-kind artifacts to present them to the world. Our platform connects artists, collectors, and fans with a mutual love for the iconic assets of music history."[29]

Numerous aspects of gear cultures are mixed here. Fetishization is intrinsic to gear auction success and sensible through themes of secrecy and mythology (inside stories), scarcity framed as materiality ("unobtanium," "one-of-a-kind"), iconicity, and economic value. Gear auctions are the apex of *technological exhibitionism*. If the deepest, most unconditional gear love is expressed through performance, what better milieu to perform than in a competitive, global public showdown of who loves this gear the most? *Technological voyeurism* on special gear, particularly online, generates social engagement both within and outside of gear cultures. This is enhanced when large quantities of gear are on display. Loving gear till it hurts is central to this auction mechanism; here, the rarest, most precious piece of gear exudes value, history, secrets, and provenance. And, most likely, you can't afford it.

"Al Schmitt's 'The Magic Behind the Music' Historical Gear Collection" is one recent example of how multiple gear culture values coalesce in obdurant gear that transcended gear cultures. Held online in mid-2022, this collection listed more than 150 pieces of gear—microphones, outboard hardware, monitor speakers, and other professional audio technologies—all of which belonged to the late US recordist Al Schmitt. Let's take the listing entitled "Holy Grail Pair of Neumann U67 Microphones from Al Schmitt." The desert-island-style Al Schmitt quote on the auction listing states, "If I could only use one microphone—just one mic—it would be a Neumann U67." Other supporting quotes frame Schmitt as "a living legend" (stated by Abbey Road Studios apparently with no irony), Sir Paul McCartney refers to the "mystery" of Schmitt's "magical process," and *The New York Times* suggests Schmitt was a "maestro." Here, the auction listing for an iconic object is reinforced with connotations of secrecy, valorization, and romanticized labor. An included discography cites multiple recordings by Sam Cooke and Jefferson Airplane, Frank Sinatra's *Duets*, Toto's *IV*, Steely Dan's *Aja*, and Bob Dylan's *Shadows in the Night* among a host of others as having Schmitt at the helm of the recording process.

Accompanying the listing is a 50-second video of recordist Steve Genewick discussing the "holy grail" pair of microphones, set to footage of Schmitt operating the faders of a large-format SSL console and turning a knob on a piece of outboard gear marked "Capitol Studios," and close-up photographs of the serial numbers and the "Made in West Germany" insignia. Now, social values, including locality, heritage, and affiliation to canonized workplaces, have been added to the already oversignified mix. It goes on. Twenty-six additional professional photographs further solidify this *unobtanium* gear. Two highlight a sticker placed on the back of one of the microphones: "Custom Modified by Klaus Heyne [503]" along with a phone number and an additional serial number (this one is 0645). Heyne is widely recognized as the foremost microphone modder in the world, having refurbished and reconditioned microphones for numerous high-profile recordists in the professional audio and film industries. As a performance of technological exhibitionism and instantiation of gear obdurance, this auction goes to excesses to stage the gear collection, its provenance, and the romanticized labor (both Schmitt's and Heyne's). Vintage Neumann U67s without provenance are not cheap items; their price on the secondhand market raised from a 2019 average price of $13,000 USD to $25,000 in 2022. But Schmitt's pair of Neumann U67's eventually sold for more than $60,000 USD, above market value and beyond the reach of most online gassers.

OBDURANCE EXTERNAL TO GEAR CULTURES

Gear obdurance begins within and around gear cultures but produces effects in outside milieux—for example, in the design of contemporary software plugins. A plugin itself is not gear since it is a computer-hosted, massless technology. However, many plugins exemplify, signal to, borrow from, and capitalize on gear culture aesthetics. Indeed, such is the embodiment of gear obdurance in skeuomorphic plugins: they become some of the most sought-after technologies in the computer-based domain.

Let's take the example of the UREI 1176 compressor, a gear obdurance exemplar. Designed and built in 1967 by Bill Putnam (d. 1989), this then state-of-the-art FET compressor featured none of the Bakelite knobs and tubes so intrinsic to its predecessors. Putnam's 1176 was a sleek, then-futuristic, silver-on-black brushed steel design with a blue stripe highlighting the golden backlit VU meter. With no threshold control and a fast attack time, the 1176 was considered revolutionary and quickly became a staple of studios. Since the late 1960s, the 1176 has been cloned, historicized, (re)-imagined, heritagized, mythologized, and valorized. The 1176 is perhaps the most sought-after, copied, and gassed compressor of all and retains its value and status as a prized piece of gear.[30] Its obdurance, however, lies in its transcendence of gear cultures and into the world of software plugins where it became the single most replicated piece of gear.

Bill Putnam Jr., founder of Universal Audio (better known as UAD), focused on reproducing the 1176 as the first UAD product because "it was the product my dad was most proud of and it was also still a very sought-after piece of gear."[31] It has been cloned by numerous companies including Warm Audio (WA76), Purple Audio (MC77), Klark Teknik (76-KT), and Black Lion Audio (Bluey). The UAD 1176 won the "Outstanding Technical

Achievement" award at the 2004 NAMM TEC awards and was inducted into the NAMM TEC Hall of fame in 2008. The 1176 appears in the title of more than 450 distinct Gearspace threads and is present in thousands of posts related to everything from history to components, clones to shootouts, and plugins to recordists. Through its history and remaking, the 1176 has amassed a significant amount of social capital.

The UAD Apollo audio interface (2020) ships with a sepia-toned card featuring two men in white shirts and black ties, one wearing black-framed glasses. Both are looking at an open 1176 on a bench surrounded by test equipment. Including this card with the Apollo interface (a new studio technology object for interfacing a computer DAW with external inputs and outputs) explicitly connects UAD's past with its present. The original UREI 1176 is nothing like the Apollo in design, application, purpose, look, or shape, yet is bound to it by romanticized labor. But a different mode of romanticized labor underpins academic papers and cultural commentaries of the 1176 as it was used by canonized recordists including Chris Lord-Alge and Jim Scott.[32] The 1176 features four ratio settings—20:1, 12:1, 8:1, and 4:1—to the immediate left of the VU meter. One technique involving pressing all four ratio buttons at once, "all buttons in," results in an overdriven and aggressively distorted compression with faster attack and release times.[33] The technique has also been termed "nuke mode" and "British mode," embodying war and terroir fetishes, respectively. "All buttons in" is commonly referred to as a processual "trick," a "secret" of tacit knowledge that once constituted concealed labor.

In the world of software plugins, the 1176 is perhaps the single most copied piece of gear. Companies including Waves, UAD, Arturia, and Pulsar all make software skeuomorphic emulations of the original 1176 compressor. Softube's "ground-up remake" extends the skeuomorph through new knobs and switches. Lindell Audio and DMG Audio's compressor plugins eschew the skeuomorphism of others but still reference the 1176 hardware object and its sound. The plugin market is flooded with this highly sought-after effect. One of the more interesting 1176 copies is designed by Rıdvan Küçük, who operates his Analog Obsession brand exclusively via Patreon. His aptly entitled "Fetish" bundle plays on the technology of the FET while embracing its social values. The Slate Digital Monster Extreme Dynamic Processor (2022), in contrast, was designed to emulate only the "classic 'all-buttons-in' extreme compression sound of the famous vintage FET limiter hardware"—and none of its other features.[34] In addition to the numerous historicizing tropes, this turns the romanticized labor of a once-concealed studio "trick," a haptic and interfacial encounter, into a new object. Such digital representations, skeuomorphic or otherwise, not only model the technological facets of gear, but they reflect the values, discourses, and fetishes inherent to gear cultures.

WHAT GEAR REPRESENTS—AND WHAT IT DOESN'T

Gear obdurance is a combination of gear culture values that coalesce over time. While gear obduracy is dependent upon fetishization, it lacks some of the sexualization connotations and instead depends upon the economic aspect of the fetish. "Sexy" discourse mainly circulates within online gear cultures and is absent in plugin representations or auction listings. Select labor (who designed the gear, who modified it, which recordists owned and used it) and workplaces (where the gear was made and which studios used it) all contribute to obdurant values. Labor and workplaces are foregrounded through interfacial insignias, marketing materials, retail brochures and websites, and product synopses and blurbs, and in the technobabble of gear users, particularly online.

Gear design does not, however, need to explicitly acknowledge labor for gear to be obdurant: this happens in social framing of the gear, the valorizing discourses within gear cultures, and in derivative technologies. Paradoxically, gear maintains its visibility through the ubiquity of digital technologies. Even though plugins are not gear, they serve a critical function in keeping gear relevant, reinforcing histories and mythologies and perpetuating gear culture participant gassing and aspirations. Through increased accessibility (a plugin 1176 is much cheaper than any hardware 1176 version), plugins unnecessarily perpetuate hardware's technoaesthetics and sonic characteristics. Product descriptions add additional contextual information, as can be seen in this description of the Abbey Road plate reverb plugins, made by Waves:

> State-of-the-art modeling of the four legendary reverb plates housed at Abbey Road Studios and used on recordings by the Beatles, Pink Floyd and Radiohead.
>
> Introduced in the 1950s, plate reverbs have been a fixture of recorded music ever since. Used most prominently in the '60s and '70s by pioneering bands, including the Beatles and Pink Floyd, Abbey Road Studios' original reverb plates were first installed in 1957 to complement the fixed reverberation times of the studios' echo chambers. These beautiful-sounding plates, with a variable reverb time of up to six seconds, were then tweaked to perfection by Abbey Road's technical engineers. To keep noise to a minimum, EMI's Central Research Laboratories designed unique hybrid solid-state drive amps for Plates A, B and C. Plate D was fully valve-powered on both drive and output stages, allowing a versatile array of sonic characteristics, from warm and dark to lush and smooth.[35]

Such a description ticks all the right gear culture boxes without the product actually *being* gear: "legendary," "Abbey Road," "The Beatles," "Pink Floyd," "1950s," "pioneering," "60's and '70s," "original," "Abbey Road's technical engineers," "EMI," "valve," and "warm." The full representational litany is present in this modern-day technological synopsis. While the haptics are removed (a mouse click is the same action regardless of whether someone is operating a plugin or a spreadsheet, and not at all like interacting with a several-hundred-pound plate reverb unit), the interfacial elements are skeuomorphically rendered to keep the object familiar. Moreover, the Waves plugin does not lessen consumer desire to one day own a "real" plate reverb.

Even though gear is omnipresent in new professional audio technologies, gear and its derivatives remain conservative in their symbolism. They stand for the retention of *past* materials—European and US design and recordist labor, materiality and its sourcing, the "golden age" or audio—in today's digital music technology sphere. Gear is not about innovation or futurism, utilitarianism, or utopianism. Gear works through a mythological demarcation of the boundaries between amateur and professional recordists widespread during a specific period in sound-recording history—hence the value of the Al Schmitt plugin collection. Gear that once belonged to a recordist and/or that was used on sessions with musicians who are no longer living, especially if part of the Anglophile record industry's posthumous canon, are inscribed with excessive value(s). But bear in mind the irony that obdurant gear, understood vis-à-vis the contemporary landscape of available alternative technologies, is missing key attributes. Such gear is not cheap, computer-based, or necessarily efficient, and it is from the past, not the future.

While we have documented gear's wide variety of *kinds* of representational valences, we have also shown that these tend to reduce to the promotion of a small subset of recorded music—the "golden age" of Anglophone pop/rock—at a time when transnational awareness of recorded music's heterogeneity has never been greater. Gear *could* represent the music, recording practices, studio sites, cultural memories, and values of the Global South, Black and Brown and Indigenous recordists, non-Anglophile nationalities, and LGBTQIA+ people, but rarely does so. In saying that, at the time of going to press, we note Manley's savvy product launch for their Massive Pride equalizer, which sold out immediately, and we certainly see an increasing presence of women, Indigenous, Black, non-binary, and trans gear designers in the broader music technology worlds of guitar pedals, synthesizers, and software design. What is unclear is if gear cultures will be able to adapt and fundamentally shift their values, or if the political, economic, and environmental instability of the 2020s, and the changing demographics of music technology users, will lead to new social formations around audio technology that are unlike the gear cultures of the last twenty-eight years.

CONCLUSION

Obdurant gear represents a labor of love, and a love of labor. Its symbolic value is one of high status, of affiliation to and knowledge of a "golden age" of Anglophone record production, of gear's design and recording labor, and of a mythology that does not simply endure but *obdures*. It is precisely these social and economic values that are embodied in Stephen Crane's gifting of the large-scale 3D-printed TG Series console knob to Samantha. The knob itself represents the labor of love and the love of labor: in this case, the labor both Samantha and Stephen undertook to maintain old tape machines and synchronize them with the new, the intergenerational transfer of knowledge and skills, and the all-important *continuum* of labor without which gear would not survive. The fact that gear persists is not about a "return," and it is not *only* about heritage and provenance either, although that plays a part. Obdurance makes sensible how gear got here and why. It also explains how gear will still be here. Long after we've gone.

15 CONCLUSION

CONFORMING TO GEAR CULTURE IDEOLOGIES

What might a gear culture participant do with more than $250,000 to spend? In 2016, prior to the start of work on this book, Samantha and ANU School of Music technical manager Matt Barnes found themselves in this enviable position. As part of a large-scale project *Wired for Sound*,[1] from which various pedagogical and research outcomes would emerge, Samantha was tasked to refurbish the School's recording studio facilities. Where would we begin? With a gear wish list, of course! Combined, Samantha and Matt had more than thirty years of experience working in recording studios, recording facilities at educational establishments, teaching sound recording and music technology, and working in project and home studios. Both Samantha and Matt are long-time gear culture participants—technophiles, even—with strong ideas about the gear that was required.

First stop, The Studio Shop in Highgate, London, for a meeting with James Townend of AMS Neve. The revered 88-R console is out of our budget, but the Genesys—a stunning in-line analogue console with on-board DAW control capability, a compact channel strip with centralized and customizable equalization, and VCAs—may fit the bill. After donning Samantha's head with a Neve baseball cap, James runs a Bob Marley and the Wailers multitrack through the console. It sounds as good as anything ever *could*, and unmistakably Neve: beautifully warm and rich around the lower midrange, with crystal clear top end. We, and our budding sound-recording students, could not ask for more. Samantha seals the deal on a 48-channel version, confirms the layout, orders dozens of add-on 88-R equalizers and loads up the first eight channels with 1073 equalizers and VCAs. Meanwhile, Matt places an order for two Mathmos lava lamps for the meter bridge. Priorities.

Back in Canberra, there is more gear to source. Matt and Samantha already decided to repurpose some of the old studio's gear: an Ursa Major Space Station signal processor, a Lexicon PCM90 Digital Reverberator, two Avalon VT-737 SP preamplifiers (in case 48 channels of Neve 1073 was not enough), two dbx 165 over/easy compressors, a Joe Meek SC-2 stereo compressor, and monitors—DynAudio M3s, a DynAudio AIR Series Surround Sound system, and, of course, a pair of Yamaha NS-10s that would join the lava lamps' globule glow on the meter bridge. The School's EMT 140 Quadraphonic plate reverberator and an AKG BX-20 spring reverb were anchored in a corridor outside the studio. Of course, this was nowhere near enough gear. Both Samantha and Matt had numerous browser tabs open and refreshed the pages of secondhand sites daily. "No Pultecs today" said Matt, just about every day for a few months, until an original EQP-1A popped up on US retailer Vintage King's site and we hit the proverbial button. Meanwhile, Samantha negotiated to buy a 1" Ampex ATR 2-track analog tape machine that was originally housed in Rhinoceros Studios, Sydney, before being bought by songwriter Andrew Farriss; this ATR had significant Australian rock music provenance, having been used to master multiple records by Farriss's multiplatinum selling band, INXS. After agonizing over large diaphragm condenser microphones, we eventually decided on the Blue Bottle and caps. After a few more gratuitous compressor purchases including a Chandler TG1, a Manley Vari-Mu, and UAD 1176 and LA2A reissues, our once-in-a-lifetime gear hoard was just about complete.

Our first gear culture clash occurred when Samantha asked Matt to research a rack-mountable device that could switch between preamplifiers and compressors. The ability to route signals through multiple devices and to hear the differences is important in teaching situations. "I've found something but it's got a weird name," said Matt, as he unveiled Third Floor Lab's Switch Maid. With a bright white façade and four large square buttons illuminating green when switched in, the device might seem innocuous but for the implications of "maid": originating with the middle-English "maiden," "maid" has three broadly accepted meanings: a female servant, a girl or young unmarried woman, or a virgin. Samantha decided to keep this problematic device and use it for A/B'ing gear, and as a touchstone for regular discussions of sound recording, its industry, and culture, and how apparently subtle technological inscriptions like the Switch Maid are representative of a much broader gender diversity and masculine design problem.

As a woman and therefore a minority participant in gear cultures, Samantha still managed to replicate just about every gear culture cliché in the book—in *this* book. From excess material imposition to buying British heritage, from ornamenting the Neve meter bridge to focusing almost exclusively on an obdurant gear wish list, the ANU studio redesign conforms to well-known, rigid, and conservative gear culture values. That said, as Samantha and Matt can attest, it is possible to have this gear *and* to reject the more problematic aspects of what it stands for. Centering the sonic identity of gear, its breadth of sound, and application to pedagogical, research, and commercial recording workflows is one way to undermine gear cultures' obsession with gear looks. De-fetishizing, myth-busting, and simply being *present in the space* are also important to Samantha—with the latter most visibly and physically undermining what gear cultures try (and in this case, fail) to do, which is to erase women's work, technological expertise, and bodies from professional gear-centric spaces.

Our conclusion reflects on gear cultures, the gear within them, and the ways that technologies reproduce gendered norms while being gendered themselves. We recognize just how problematic some gear cultures are—and we want to change them.

GEAR CULTURES

Gear cultures are not just a heuristic. Across online fora, international trade shows, and the readerships of several print and advertising media, we found full-fledged durable cultures engaging in their own historiography around mythologies and stories of legendary gear. Hundreds of thousands of people are connected by shared beliefs, practices, rites of passage, language including argot and humor, and codified social roles—gear pimp, gear slut, lolcow, "real professional," newbie, modder, perennial skeptic—all contingent upon a defined material culture with its own distinctive materialities, aesthetics, and canons. While recording-studio equipment gear cultures are not unique in being a cultural formation held together by shared interests, they are distinct from "taste cultures" such as fandoms in their ascription of agency to technological objects, and their insistence on gear—a special class of fetishized objects—being the primary social mediators.

Gear cultures produce palpable and tangible effects in the world. This starts with technology's considerable extractivist legacy, where opaque supply chains for ore and "raw" materials mined in dozens of countries feed a transnational manufacturing assemblage of component manufacturers and gear assemblers. On account of gear, labor power and natural resources are consumed in the periphery so that gear can be enjoyed and indulged in the first world. As we extensively documented, part of this is the dependence of many industries—mining, component manufacturing, and electronics assembly, but also trade shows, print media, studios, and educational programs—on women's labor. Gear cultures also appropriate many practices, such as nurturing and care associated with women's work, in the design of studios. That said, gear cultures, as hegemonic masculine formations, depend upon the erasure of this "hidden" labor. Gear cultures influence gear manufacturing and design and how gear looks, feels, and sounds; in doing this, they also shape the bodies and subjectivities of gear culture participants, their sensory modalities, and their social values. In all, gear cultures constitute a fundamental reorganization of social life related to music and music technology.

When we began dedicated collaborative work on the Gear Cultures project in 2017, we quickly came up with several questions that stuck with us, in some form, throughout the research phase. Why are technologies fetishized in the way that they are? Why do audio technologies take the form and have the aesthetics that they do today? What accounts for the perseverance of analog technologies during a time when digital simulations and emulations—cheaper, recallable, easier in some cases to use, and less space-demanding—were widely available? What are the range of cultural formations in which audio technologies are key actors? And what roles do technologies play in each? We approached the project as two people with substantial prior experience in recording studios, ranging from working at amateur and professional studios in several countries to teaching audio technology in university programs to performing as recording artists. However, we knew that our experiences with these technologies were not necessarily representative of the range of cultural formations and did not wish for the book to reflect *our* attitudes toward technologies—as recordists, educators, scholars, and music fans. We also wanted to provide a useful alternative to the "producer narratives" that had become a veritable cottage industry by that time and that we had critiqued in our previous edited book.

The beginning of the period we analyze coincides with the end point of Paul Théberge's book, *Any Sound You Can Imagine*, a work of unusual clarity that was a springboard for our analysis. The year 1995 is also when gear "went online," and is just one year before the software cloning of hardware gear began (with the Focusrite D2 plugin). Therefore, an ancillary interest was in assessing how the situation had changed. Whereas in the 1980s, 4-track tape recorders and other "prosumer" music technologies were integral to musical practices of emergent genres, in the twenty-first-century gear cultures period, much "gear" is unnecessary to perform music production workflows. The fact that it still exists, obdurant values and all, is a curiosity, but one that we have relished investigating.

Prior to 2017, we had both been researching gear cultures without realizing it, having taken photos and fieldnotes on trips to numerous sites (Samantha and Eliot) and having accumulated a trove of online gear culture discussions (Eliot). Once we started to take stock of the mass of data we already had, we immediately realized the scope of the theoretical and methodological problems at hand. Neither the normative representations of technologies within music and sound studies nor some of the most widely adopted approaches to technology within science and technology studies adequately accounted for the ranges of practices and attitudes we saw play out at trade shows, online, and in print publications—or even some standout aspects of recording studios, for that matter. Realizing the extent of fetishistic practice, agential delusion, and masculine performances that structured some of the gear cultures, we knew we needed to expand our questions to include ones that had not previously been asked. We also developed and tested unconventional methods for analyzing our data. We took to heart David Graeber's frequent complaint that siloed academic fields and institutional structures often impeded the imagination and radical potential of researchers and research.

Gear cultures are not only found in the audio technology industry. Indeed, gear may refer to objects central to all manner of pursuits, hobbies, and pastimes. For each gear culture, some manner of fetishized commodity becomes the fulcrum of social interaction, and fetishization is staged through a combination of print media, online discussion fora, multimedia platforms, and trade shows. While fetishization does not always take on the sexualized aspects as audio technologies do (although both digital and film photography equipment appears to be sexualized even more than audio, both in online fora and in review videos), it will always constitute more than just an economic fetishization. Gassing for gear, even if by a different name, is a precursor for social involvement, and prevalent "secrets" discourses help maintain interest in the space. In the broader music industries, several distinct gear cultures have formed around electric guitars and effects pedals.[2] Such cultures cluster in similar ways; collectors scour the pages of Reverb.com and eBay for coveted "vintage" instruments, they share stories and images of acquisitions on dedicated message forums, they congregate at trade show events, and they produce virtual reams of discourse, including web 2.0 multimedia content, which is distinct from that surrounding the musical uses of gear objects. Indeed, the biggest forum for guitar, strings, and pedal enthusiasts is called The Gear Page.

Similarly, alongside the same twenty-eight-year history of audio gear, synthesizers have been the focal point of different gear cultures.[3] Starting with newsgroups, mailing lists, and neighborhood meetups, synthesizer engagement expanded into trade shows, festivals, reblogging sites (with comment sections) and message forums, and, by 2020, the formation of over a dozen regional modular synthesizer societies. There is some crossover with audio gear too, since the NAMM show and Musikmesse attract a similar number of guitar- and synth-interested participants as does professional audio. These gear cultures would contain the closest parallels to the heritage, canon, and obdurance discourses and practices of audio gear, albeit developed on an accelerated timescale.

Several other electronics devices attract related attention to audio gear and musical instruments, sharing a cohesion brought by fetishized gear but differing in key aspects. Let's consider the vibrant computer-related communities including lanparty and overclocking enthusiasts, mechanical keyboard groups, or "Hackintosh" (DIY assembled computers that emulate Apple's consumer offerings) forums. Since these computer-related gear cultures are fixated on "upgradeitis" (rapid obsolescence cycles requiring yearly upgrades to one's gear), they have little concern for the past, heritage, or industry awards. But even for gear cultures without a strong historical bent, they all share extensive metaphoric, analogic, and mythological framings, and those who wish to participate in the gear culture in question are expected to know this lexicon. In Anglophone countries, camping equipment is only ever referred to as "gear" and countrywide outdoor pursuits cultures orbit around the sale and acquisition of camping gear, both new and on the secondhand markets. Indeed, backpackinglight.com has an entire forum dedicated only to gear: "gear lists," "make your own gear," "multiple use gear," and "gear (general)" are four such forums, the latter featuring almost half a million posts. Similarly, the "snow sport gear" offshoot takes in everything from skis and snowboards to snow harvester machinery. The big difference here is that while equipment might be gassed over (not to be confused with outdoor cooking), we find no evidence of sexual fetishizing, and little emphasis on heritage.

GEAR

What is gear about and what does gear *do*? We originally thought gear entailed the tools by which professional recordings can be made, but we were enamored, surprised, and also *alarmed* at what gear actually *is* relative to what we originally thought. After researching gear materialities, stagings, encounters, and attitudes, we have witnessed gear operating on whole new levels. Gear is much more than professional audio equipment and, even though we both find ourselves at once participants in gear cultures and agents for undermining them, the enduring power of gear overwhelms even the most determined gear culture contrarian.

Gear is, by and large, about three key themes: war (including military technologies and shootouts), imagination (of mythologized gear histories and an imagined present), and replacement (including themes of sex and sexualization). First, war. Whether considering the direct historical connections to military technologies (and their materials and components), the incessant metaphorical apparatus required to keep up with the arsenal of secret weapons that gear culture participants hoard, or the naming-and-framing of gear in World War II contexts (we're looking at you, Shadow Hills), to engage with gear is to *literally* go to war. Secrets and espionage permeate gear cultures so deeply that gear economies rely upon them. Competitiveness and the pitching of one piece of gear against another in shootout scenarios plays out in all gear milieux. Teutonic fetishization, and even gear with origins in Nazi Germany, have a particular fetish appeal in gear cultures. While some areas of gear cultures work to erase or tiptoe around this history, others embrace it. Conquest, here, is bound up in the acquisition of conspicuous goods, and male gear culture participants set up all kinds of fanciful narratives of how they ultimately won the gear after overcoming the unreasonable girlfriend/wife/missus who holds a firm grasp on the household purse strings. These hard-fought gear battles hinge on role-play with characters internal and external to gear cultures, which forms the basis of parables that structure social relations around gear.

The war concept provides an instance of the broader domain of imagination. Our first chapter recognized the Delorean-level capabilities of gear to allow its users to time travel. All gear is heritagized in some way and is representative of some kind of tradition, historical continuum, terroir (e.g., UK, USA, Germany), or, most often, the "golden age" of sound recording. In the brief period between roughly 1955 and 1980 when dedicated large-scale recording facilities existed as part of the broader operations of US- and UK-based record labels, and recording was almost wholly reliant on analog equipment, there were likely only a couple hundred working recording professionals, few of whom continue to work in the industry today. Affiliations to this tiny elite of workplaces, recordists, and associated recordings, however vague or tenuous, are essential in the design, promotion, and uptake of new gear, and in the discourses surrounding it. Whether considering cloned gear named Heritage Audio or Vintage Audio, plugins claiming to replicate the sound of plate reverbs once housed in Abbey Road studios, or a clone of a Fairchild 660 cutting limiter once used on a Beatles recording session, gear must somehow be linked back to a small pool of people, places, or records from the record industry's golden age—hence the finding that gear cultures hinge upon imagination. For gear culture participants, gear is a conduit, a portal to a time that no longer exists. As the missing link to a whole range of (mostly dead) agents, gear works to continually reinforce conservative canonical and ideological ideas. Without the near-continual mythologization of workplaces, recordist processes, and recordings, participants would not remain actively engaged with or find themselves gassing for gear. Of course, most of the ideas around gear-as-sex and gear-as-war are also imagined, too. Occasionally, a gear culture participant will divert from general discourse to pipe up that "gear is not sexy" or "you can't have sex with gear." Since this disrupts the imagination of most participants, typically this diversion is shut down. Furthermore, most of the sexualized gear settings are not places where actual "irl" sex takes place; they are constructed from the imaginations of gear culture participants to further imagine scenarios where sexual activities *might* occur. Similarly, we found no evidence that gear is used as an "irl" weapon in a fight. The gear-as-war imaginary is metaphorical in nature and discoursed around imaginary battles, for the most part fought—and won.

Gear is about replacement—mostly of professional recordists, posthumously canonized agents including musicians, and women. Gear replaces professional recordists, their skill sets, and their practices. When gear is so stripped of its use value, its intended purpose for professional audio-recording workflows becomes almost irrelevant. Gear that uses displaced agency claims to package the skills and expertise of that recordist. Ownership of the gear, therefore, becomes evidence of ownership of the associated designer or recordist's skill set by proxy. Interestingly, in gear cultures, we find little motivation to understand or adopt the skill sets of professional recordists when interacting with gear. Instead, we find an artificial inflation of the importance of other types of knowledge—including technobabble and connoisseurship. For example, knowledge of gear (down to individual components and wiring) and its history is a ubiquitous type of connoisseurship. Performance of that knowledge—who used the gear, on what famous records, and where—is more important to many gear culture participants than for what the gear is intended.

Gear also replaces the missing women, and the gear sexuality discourses comprise one performative trope where we see this happen. Our drawing so heavily on masculinity and sexuality studies throughout this book reflects the depth to which cisgendered heterosexuality is normative in gear cultures, and how gear replaces sexual activity. Much sexualization of gear is found in forum threads that routinely analogize and objectify gear as if it were a female body. We find personified technologies, sexy racks, sleek looks, and even booth babes all over the place—and let us not forget that the "number one" pro audio forum, Gearslutz, did not respond to mounting pressure and petitioning and change its name until 2021. Gear is sexy: looks that can kill. At the beginning of our research, we thought that gear looks were perhaps on par with their aural capacity. But in gear cultures, in contrast to in music-focused studio settings, gear looks are a nexus of social interaction, and the look of the gear replaces its aural potential. Not only gear looks are fetishized, since the studio stagings, arrangements, and ornamentations of gear create settings with the potential for sexual ideation to take place. LEDs that glow, flicker, and blink, set off with strings of fairy lights and gently floating globules of neon lava, decorate gear so as to conform with organizational norms; such norms are, of course, structured around cisgendered and heterosexual ideas of sex and sexual encounters.

The multifaceted fetishization of gear perpetuates the erasures—of land, labor, and liberty—that the combined forces of capitalism, colonialism, and masculine gear cultures have enforced since the invention of electrical recording. Returning to the question of agency, in chapter 9 we focused on agency at the individual level and concluded that the fetishization of gear meant that the primary site of agency was bourgeois consumer spending—to act or to not act on GAS. Let's expand our definition in light of our findings concerning extractivist and anthropocenic problems, the typical erasure of hidden labor, and the promulgation of hegemonic masculinities in milieux ranging from print to online communities to events, and around practices ranging from testing to organization to heritage. Considered from this broader perspective, agency, understood as the ability to make a change, to direct the narrative, to influence technoaesthetics, or to shape the social structures of gear cultures, is possessed by a tiny minority of people. Miners, uncredited assembly workers, women and LGBTQIA+ individuals, and everyone else outside of the hegemonic masculine formations are unable to flip the script on gear or gear cultures. Rather than a democratization of technology, the evidence overwhelmingly points to a technocracy governing the social formations around music/technology cultures.

GENDER, SEXUALITY, AND TECHNOLOGY

One of gear's primary roles is as a replacement for women, who are largely absent in masculine, heteronormative gear cultures. Unwelcome in gear fora, in minority numbers at trade shows (except for Women's Audio Mission and the wives of gear designers), and grossly underrepresented in the pages of the trade press, gear cultures are worlds constructed for male socialization. When women *are* present in gear cultures, then they are posited as protagonists or gear enemies: negative in character, playing an obstructive role to gear acquisition, and needing to be told or persuaded that gear is indeed good for the betterment of men. Women are almost always cast as having full control over household finances and domestic spending, which is rarely true. As such, women become the easy stooge on which men can blame their lack of financial power to buy—and own—the gear.

Men exclude women from gear cultures, yet, through gear, they reconstruct an idealized and sexualized female presence everywhere. "Ideal" here is feminized gear that does not communicate except for the possibility of sexual enticement. Most object-specific online gear threads will at some point veer toward objectification, where gear is often referred to as a "she."

But in the broader worlds of gear, women are everywhere. They are responsible for a significant amount of the invisible or peripheral labor without which gear would not exist: in extractivism, in manufacturing, at trade shows, online, in magazines, and in studio manager roles. The male/masculine face of audio manufacturing conceals the labor necessary to make these gadgets, most of which is undertaken by women with their "nimble fingers." This has been the case at least since RCA's radio manufacturing in the 1930s. But why are women hidden from view? In most cases, concealment is necessary to preserve the myth that the male figurehead of the company really deserved all the credit for "inventing" the gadget, when, in reality, certain women were the only workers with the competence to do highly specialized and essential tasks. On a broader level, to admit that women play a role in the manufacturing of gear would be to severely disrupt the masculine hegemony of gear cultures and would bust the illusion that only men understand technology. This is no different to Marvin's findings around women's work in telephone exchanges: men simply cannot admit to the fact that women possess technological expertise. In some cases, women were hidden from view to protect them from sexual harassment at trade shows—information that was volunteered to us by several exhibitors at AES and NAMM even though it was not one of our interview questions. Sometimes it appears to be a strategy to conceal the low wages and poor working conditions on the shop floor, as is widely reported in the case of some East/Southeast Asian and Central American factories. When speaking with gear designers, many credited their wives with important aspects of their work, whether that be building gear, "doing the difficult tasks," financing, or organizing other aspects around the gear business, yet rarely did we see any documented evidence of the contribution of "the wife." One exception was BeesNeez, where designer Ben Sneesby detailed how his wife, Veronica Sneesby, takes on the more meticulous and high-risk aspects of the microphone build work such as setting the diaphragms. Veronica is also visible on the BeesNeez website where her work *is* accurately credited.

Our interest in understanding gender in gear cultures goes beyond accounting for the missing women. Material culture has a formative but underrecognized role in male homosocial formations, and we have shown how gear cultures are defined by media- or site-specific hegemonic masculinities. Utterances are one entry point for recognizing or understanding each hegemonic masculinity, and here the use of tropes in engineering education, like the shockingly violent and misogynistic pneumonic for the resistor color codes "Bad Boys Rape Our Young Girls but Violet Gives Willingly" (included in a Udemy-hosted class), does a lot of extra work—toward defining who is and who is not supposed to participate in such spaces. But these multiple masculinities are not just discourse formations or discourse networks. They are dependent upon massive amounts of material culture and dependent upon multisensory modes of interacting and intra-acting with material technologies. It is largely through technological intra-action that the subjects and objects of gear cultures are formed.

Due to the attributed, delusional, and displaced agency problems in these social formations, and the relentless reliance on object fetishism, gear is typically positioned as the only subject that can mediate all social activity. Gear does a poor job at that, and gear cultures ultimately are not just bad for women, LGBTQIA+, and Black and Brown people; they can be toxic places for men, too. Due to its toxicity, the most popular of gear forums, Gearslutz/Gearspace, has experienced a downturn of posting frequency: the considerable majority of the 425,000 members are inactive, almost none of the "pros" who used to occasion the site are active posters anymore, and many gear-interested people we met during our research "fled" the site.

So what is the outcome of this peculiar gear/discourse formation? We have studios without musicians, control rooms without music, and a social milieu where, devoid of female participation, men have left themselves with no option other than having a massive circle-jerk on the Neve console no one remembers how to use anymore. Audio engineering, in this formation, is the apex of unproductive labor; piles and piles of expended and unused gear, to be ogled over until the shiny luster of the extruded aluminum knobs has worn off and even the most ardent gearslut can't get a boner anymore.

GEAR CULTURE ALTERNATIVES

The problem with gear cultures is that they exist—at all. Gear cultures are unlike technology-interested maker communities where individuals are interested in understanding technology and its workings so they can make it themselves; gear cultures depend upon widespread delusional beliefs about technological agency, they construct and are co-constructed by violent online communities of pseudonymous participants nonprofessionals, they maintain patriarchal and heteronormative professional formations, they replace experience-based recordist knowledge with performed connoisseurship, pseudoscience, and technobabble, and they entail massive amounts of material resource and energy waste in the name of GAS. Not to mention the many ways they fail to contribute to music-making.

But even for those at the center of these formations, and who might appear to be the least harmed by participating in them—white affluent men of the twenty-first-century version of the leisure class—they are counterproductive and prone to toxicity. The gatekeeping that depends upon belittling those who lack the culture-specific (pseudo)technological lexicon vocabularies or the correct connoisseurship (which objects are gassable and/or sexy, and which are not)—or in other words, who fail to believe in the correct episteme—belies the strong hostility toward those with or who work in formalized education: "Those who can't do, teach." Hegemonic masculinity, here, reproduces a culture of mediocrity rather than one of healthy competition, technical competence, and musical aptitude. Men can—and need to—do better, but will not, within a community where a self-appointed group of men work tirelessly to drive out women, LGBTQIA+ individuals, East Asians (especially Chinese), at times Black and Brown people, educators, and working professionals in electrical engineering, recording, and music.

Are we living in an age of unprecedented technological change? The obdurance virtue of gear cultures presents a counterexample. But despite the concentration of capital in high-tech industries, consider that the science fiction fantasies of the 1950s–1980s never came true. We haven't colonized the moon, we neither commute in flying cars nor have fifteen-hour work weeks, and life expectancy is decreasing in some G20 countries. Instead of the promise of technological acceleration, we have achieved new domains for technological simulation, creative stagnation, and the continual reproduction of late capitalism and modernity.[4] Jaron Lanier pointed out something similar: if the greatest achievements of the Internet until 2010 were a behemoth shopping mall in the cloud and an error-filled encyclopedia where none of the authors were paid for their labor, then neither the corporate domination of Internet governance nor open-source initiatives are leading to new futures for arts and sciences, radical alterities in society, or politics, or achieving planetary goals like reduced inequality or climate justice.[5] As David Graeber shows, while in the "first world" the landscape might seem different since many large-scale mines and manufacturing plants have been offshored and placed out of sight, smell, and hearing, at a planetary scale all that has really happened is the displacement of industrialization onto "the periphery," and the reallocation of human labor to postcolonial subjects.[6]

What about audio? Over a century of marketing has promised that the next technological development in audio recording will bring us closer to feeling like we are in the room with the actual musicians, but most consumers today experience music through *lower*-quality media (mp3s, and YouTube's Opus codec streams) than CDs. We listen through seventy-year-old transducer designs to recordings made with seventy-year-old signal processor designs: we seem further from tech-utopian audio promises than ever before. Within recordist and engineering communities, the nostalgic gloss of 1960s–1970s-era recordings as representing *the* golden era in audio effectively transforms what were, at the time, often countercultural musical projects into a "classical" canon that is to be preserved in service to a conservative and corporatist political agenda. This agenda—of heritage, canon, and obdurance—buoys gear cultures, underpins gear design and manufacture, and provides the subject matter for socialization. This does, of course, implicate the considerable popular music studies publishing industry, which features no shortage of books and articles on the very canonized period we critique. Indeed, Samantha has even contributed to it.

Professional audio gear cultures are regressive in nature. If the olden gear is the golden gear, then where does that leave innovation? In our interviews, we met some of the visionary innovators of new audio technologies: people like Hutch Hutchison who came up with new analog circuit topologies; like EveAnna Manley who has driven valve gear into the twenty-first century; like David Josephson who has pioneered surround sound mics and digital microphony specifications; and like Dave Hill who advanced the state of the art in digital-analog conversion. But there has been comparatively little uptake of their designs, since apparently the market for newly built audio technologies is more interested in cheaper clones of vintage gear. The R&D of the larger manufacturers—Music Tribe/Behringer, Yamaha/Steinberg, and all of the Harman International subsidiaries now collectively owned by Samsung—predominantly goes into production efficiency and "clever new ways of combining existing technologies" or straight up copying of existing technologies.[7] This became a new business model for Chinese manufacturers such as Alctron, 797 Audio, and Feilo, whose products are sold under dozens of brands. "The market" is, in this case, really a shorthand for an observation about how cultural norms—heritage, obdurance, technostalgia, fetishization, and gear cultures—lead to economic trends and the manufacturing of hoards of stuff that is too often *not* used for producing the music of the future.

So far, we have arrived at a somewhat depressing outlook for professional audio technology and its cultures. But we also see signs of change that give us some hope and recognize areas where existing gear cultures could be greatly improved. How could things be different? We have a number of ideas and recommendations: changes to trade show and professional society structures and values, inclusive gear, myth-busting, collaborative work, and software.

DEI (diversity, equity, and inclusion) initiatives at AES, while a welcome development, have largely been restricted to increasing the visibility of women and Black recordists and mixing/mastering engineers, and conducting research into harassment within the society specifically and the professions more broadly.[8] Organizations such as AES and NAMM need to fully integrate DEI initiatives at all levels: through standards committees, conference and convention representation, event design, and governance—and be held accountable for this. While we see signs that things are changing, more needs to be done. "Booth babes" and other sexist tropes at trade shows need elimination. This could be done by vetoing personnel, display, and marketing materials in advance. We suggest that both AES and NAMM publicly commit to improving diversity and inclusion in their organizations, as well as eliminating damaging gear ideologies from their spheres (see further recommendations below). They could further accelerate this urgent issue by formulating codes of conduct, member expectations, and other instruments that set industry-wide *behavior* standards—not just *technical* standards. DEI initiatives need broadening to encompass manufacturing industries, too. In our fieldwork, we witnessed several diverse and potentially inclusive organizational structures in gear manufacturing. However, the realities of these diverse workplaces, and in particular the reality of women's labor, are not yet visible outside of the shop floor. We believe that if more were done at organizational levels to include women and other underrepresented groups, we would see industry-wide improvements on recent broader music industry workforce statistics.[9]

In our analysis of the extractivist legacy of all gear, we attempted to cover the range of issues—environmental, Anthropocenic, economic, political, social—that surface when attending to gear's reciprocal landscapes. Whereas manufacturing trade associations in other industries have been implementing limited supply chain tracking, conflict-free certifications, sustainability educational resources, carbon neutrality, and the like, AES and NAMM have done nothing on this front. Even if manufacturers wanted to ensure that there was no "blood in their microphone," or hoped to move toward "carbon neutrality," they wouldn't know where to begin. AES and NAMM need to step up and expand their standards and publications efforts to provide resources on supply chain sourcing and corporate sustainability, create standards and certifications that are enforceable, and devote some of their public events to rigorous discussion of the environmental and social consequences of gear manufacturing.

Despite the exclusivity of most corners of gear cultures (and noting that the cost of gear is exclusionary by its nature), we have met people and worked with gear that, in its design and manufacturing aesthetic, undermines broader gear culture ideologies. From BeesNeez microphones making their whole family the face of their brand, to EveAnna Manley's Massive Pride EQ, to Marshall Terry's white and primary color facade for his Terry EQ, we are hopeful that enough progressive gear manufacturers are present in its cultures to set the stage for inclusive gear design. We recommend gear designers look at these three examples as to how gear can embed values beyond the militaristic and technoaesthetic. Technological and interfacial design does not have to conform to extant gear culture ideologies: it can be different and, as such, may assist in producing a wider variety of musical-sonic-aesthetic outcomes. If we agree that more than 3 percent of professionals working in audio engineering *should* be women,[10] then let's use engineering (circuit) and design (industrial) as part of a solution to the problem rather than being limited/constrained by it.

Another means of destabilizing and dismantling gear cultures is for music communities to absorb technologies into their own cultures and communities. Of course, many music genres are inextricable from the technological means to create them—hip-hop, dance music, hyperpop, and many more. However, commercial popular and rock music is still tied to structures that separate musical performance and writing from technology and production.[11] A lot of new gear in the 1980s was oriented toward musicians. MIDI, 4-track recorders, synthesizers, and emergent software sequencers were aimed at creative musicians and became absorbed into musical communities. To reduce gear fetishization, the gap needs to close between *all* music creators and the technological means by which music is made.

Broader music technology cultures have a role to play in dismantling problematic gear cultures. Regardless of the regressive nature of gear culture ideologies, quite likely digital technologies will ultimately prevail; after all, in the grand scheme of things, nearly all music and audio production in the 2020s is computer-based. Digital modeling of analog gear continues to improve, and even though a large proportion of plugins still are based on hardware gear, they do not generate the same kinds of fetish problems we see in broader gear cultures: they are socialized more as audio technologies than as gear. Plugin hoarding is present in broader music technology cultures but does not carry the same kinds of sexualized or combative implications of its hardware equivalents. Perhaps, as gear becomes more expensive and more exclusive, "in-the-box" production, with no outboard gear at all (save for microphones and monitors), will become the unapologetic, defacto standard. Computer-based production undermines gear culture ideologies: no one ever said Pro Tools was sexy.

Understanding gear through an agential realist lens is also crucial to defetishization. Clearly, gear culture participants know little about the historical origins of gear, its labor and material chains, and its environmental impact. Exposing these facets is something we have tried to do in this book, but more could be done. Gear is unnecessary for a whole variety of reasons, but if the true costs of gear were more broadly recognized and understood, this could serve to defetishize it. To challenge gear cultures' reliance on secrets and revelation, we advocate for reorienting this around the revelation of (particularly) women's labor as fundamental to gear manufacturing. Hence, we recommend further studies and exposés on gear's materiality and labor formations.

One area that needs a culture overhaul is Gearslutz. Despite the name change to Gearspace in 2021, we did not observe a commensurate culture change. There are positive aspects to Gearslutz in that it brings a global community of enthusiasts together. However, the perpetuation of fetish discourses and gassing undermine potential positive community values. For Gearspace to be more inclusive, it needs to clean up: taking a zero-tolerance approach to sexualized gear discourse, devising a rigorous code of conduct, and more consistently applying sanctions for breaches of this. Additionally, Gearspace could set the tone(s) of its threads by shifting the narratives away from the gear itself and toward gear in context. We believe Gearspace (not Gearslutz) has an opportunity to demonstrate leadership in the professional audio technology industry by embracing its own DEI initiatives and raising their expectations of participant behavior.

We have the wrong kinds of online gear communities. One partial model for how we might flip the script of gear cultures can be seen in the electronic music gear forum, lllllll.co, typically known as Lines. Using Discourse software, with an interface that encourages users to reward each other's positive contributions, the forum disincentivizes interpersonal conflict. They prohibit sexualized gear discourse, or any homophobic, transphobic, misogynistic, and racist rhetoric even if couched within an artistic endeavor, and users behave much better. Moreover, since much of the forum is dedicated to users creating free open-source scripts for open music computing platforms, social capital is gained not only from gear fetishization rhetoric but from making things that other users can use for creative purposes.[12]

Gear cultures presently revolve around individual ownership of gear and the enactment of combat with other gear owners—gassing and conquest. Quite simply, "the measure of men" should relate to men's ability to proactively contribute to music creation instead of amassing gear hoards and competing with gear enthusiasts. Where people are the focus of music and audio production, less agency is attributed to gear. Again, we know this *can* be achieved since print media such as *Tape Op* have centered such approaches. Audio technologies must be present to make recordings, but *gear* does not need to take center stage. If collaborative and inclusive production processes were championed in all milieux—online, at trade shows, in magazines—people would be rewarded for group work, as opposed to the large number of individual "producer" awards that flatten a production team to the effort of one individual.

At the end of this phase of our six-year project into gear cultures, we issue a call to arms for like-minded technophiles, audio professionals, and gear culture participants: conduct anti-fetishization work, disengage from gassing on Gearslutz, and champion women's labor. Above all: myth-bust. Where gear is present, it can be explained and demystified without resorting to canonistic tropes. Focusing as much as possible on music and sound production processes is one means of doing that, and ensuring that the *people* involved in professional performances, recording, and mixing are central to workflows is critical. This might appear contradictory—we're at the end of a book about *gear*—but deemphasizing gear and diverting attention toward the talents, skills, and techniques of human agents is key to dismantling fetishization and all the problems arising from it.

Glossary

ABX test—A method of comparing two anonymized audio signals, A and B, to a third source, X (identical to either A or B), to determine differences

AD/DA converters—Devices that convert analog signals into digital data, and vice versa

agency—The ability to act, intervene, or produce a relational effect

audio interface—A device used for connecting audio inputs (microphones, instruments) to computers (DAWs)

Bakelite—A synthetic shellac-like material used for gear controls invented by Leo Baekeland as an ivory alternative

black box—A system understood by its external features as opposed to its inner workings

breadboard—A construction base used for building prototype circuits

burning in—The process of loading and testing devices / components prior to sale to eliminate faults

canonization—Continual reinforcement of omnipresent, valuable, heritagized, and mythologized gear through media and discourse

capacitance—The capability of a capacitor to store an electrical charge

capacitor—A passive two-terminal electrical component that stores energy

clipping—Distortion caused by an analog or digital audio signal exceeding the level of a system

clone—A replica of gear design, circuitry, and build

compressor—An audio signal processor for reducing a signal's dynamic range

condenser microphone—A professional studio microphone featuring a small, medium, or large diaphragm and requiring an external power source to operate

continuous encoder—An interfacial element (usually a knob) that rotates 360 degrees and converts the shaft's position to an analog or digital signal

DAW—Digital audio workstation, an aggregate of computer software and hardware used for recording, editing, mixing, and storing audio and MIDI data

diode—An electrical component allowing flow of current in a single direction

displaced agency—Excessive (mis)attribution of agency to gear exceeding the dyadic relation between user and technology

dissipation factor—The measurement of power lost when a signal passes through a capacitor or insulating material

ESR—Equivalent series resistance

equalizer—An audio signal processer for adjusting frequency bands, usually via boost and cut controls

fader—A linear potentiometer, most often used on mixers to adjust audio signal levels

farad—Unit of electrical capacitance

FET limiter—Field effect transistor-based dynamics signal processor

GAS—Gear acquisition syndrome, characterized by fetishization and obsessive fixation on acquiring large quantities of gear

gear—Professional hardware studio recording technologies including microphones, mixing consoles, and signal processors that have absorbed extra-audible meanings and aesthetics

gear cultures—Predominantly masculine social formations around gear

gear fetishization—Combinations of commodity fetishism and sexual fetishes embodied in gear

gerber—A standard vector format for printing PCB designs

graphic user interface (GUI)—Facilitates interaction with a range of visually represented icons

harmonic distortion—Nonlinearities in an audio signal that occur at overtone harmonic ratios

hegemonic masculinity—A form of masculinity that dominates any specific social space, even though only a minority of participants may personally believe in it

heritagization—A conservation or preservation process for transforming gear into heritage

interface (colloquial)—Controls protruding from gear's front panel fascia

interface (ontological)—Point of contact between two independent entities

jitter—In digital audio, deviation from timing intervals relative to a reference clock signal

knob—The covering, usually made of metal, Bakelite, and/or plastic, for a rotary potentiometer or switch

loss tangent—Ratio of energy lost to energy stored in capacitance

lunchbox—A rack for housing multiple 500-series format modular signal processors

materiality—The constituent raw materials used to make gear

meter bridge—The upper part of a large-format mixing console housing the console's visualization of signal levels

microphone capsule—The top, sometimes interchangeable component of a microphone housing the diaphragm and encased by a grille

microphone diaphragm—In a condenser microphone, a thin membrane set parallel to a metal plate that vibrates relative to sound pressure

mod/modding—Modification of gear components or interfaces

NOS—New old stock, relating to discontinued but unused stock, usually of components including vacuum tubes

obdurance—Persistent presence defined by accumulated social values, romanticized labor, time, and digital representation

opamp (operational amplifier)—A differential input, single-ended output amplifier used for boosting weak signals and for performing mathematical operations

outboard—Signal processors housed outside a mixing console

PCB—Printed circuit board used to connect electrical components

personified technologies—Gear featuring the name and/or persona of a musician, recordist, or designer and characterized by displaced agency

plugin—Software code with a graphical user interface that performs signal processing within a DAW

potentiometer—A manually adjustable three-terminal resistor

preamplifier/preamp—A device that converts weak electronic signals into stronger output signals while converting the signal's impedance

quality factor (Q factor)—The ratio of energy stored versus energy dissipated in a capacitor

recapping—The process of replacing capacitors in an electronic device

resistor—A passive two-terminal electrical component for limiting electrical current

ribbon microphone—A microphone that converts sound pressure via a thin corrugated metal ribbon suspended within a magnetic field

rotary switch—An interfacial element that, upon turning, switches between a number of fixed states

shootout—A mode of testing gear that compares two or more functionally similar objects

signal processor—A system designed and used to manipulate frequency, dynamic, or time-based attributes of an audio signal

skeuomorph—An imitation of an object that retains its original design features, often a digital representation of an analog device (see plugin)

slew rate—The change of voltage or current over time

spectral repair—The process of eliminating unwanted noise or sounds from audio

technobabble—A socially constitutive discourse characterized by the misuse of technical terminology and the propagation of pseudo-technical terms

technophilia—Enthusiastic attraction to technology often combined with futuristic or utopian values

technological exhibitionism—Public display and staging of excessive quantities of gear, often in stylized settings that (re)create historical contexts

technological iconicity—Canonical gear with enduring mythological associations and formal attributes that can withstand abstract representation

technological taxidermy—Preservation and presentation of unpowered and/or inanimate electrical technologies in a museum or other location

technological voyeurism—Ways in which people look at audio and music technologies, often for extended periods of time and with imagined sensuous and/or haptic experiences

technoporn—Visual and textual representations of technologies in media that draw upon the representational frames of pornography

through-hole components—Electrical components with conductive metal leads that are inserted into holes in printed circuit boards before being soldered to conductive metal pads

transformer—A passive electrical component transferring energy between circuits, used to change voltage levels and galvanically isolate circuits

transient sculpting—Manipulation of the attack and sustain elements of a sound's envelope

transistor—An electrical component for amplifying or switching signals in a circuit

U—Measurement for mounting gear in racks where 1U is 1.75" high. Gear is usually 1U–4U and racks are usually 4U–16U high

vacuum tube—A glass bulb that, when heated by a coil-based cathode, results in a controlled flow of current

valorization—Enhancing the value and/or price of a commodity by including value-forming labor in the production process

vari-mu—Variable gain (μ), usually in relation to compression

vintage—Old, particularly in reference to pre-1980s analog audio technologies

Notes

CHAPTER 1

1. Interview, Stephen Crane, 2019.
2. Downing (2008); Barad (2007).
3. Pinch and Trocco (2002); Sterne (2012).
4. On audile technologies, see Sterne (2003). On technoporn, see Bennett (2012b).
5. On the actor-network concept of durability, see Law (2009).
6. On technostalgia, see Taylor (2001); Pinch and Reinecke (2009); and Bennett (2012a).
7. On technoaesthetics, see Masco (2006) and Simondon (2012).
8. On subcultures, see McRobbie (1980). On scenes, see Straw (1991).
9. Marvin (1988); Gaston-Bird (2019).
10. Mitchell (2012: 69). See also Simondon (2017 [1958]).
11. On produsers, see Bruns and Schmidt (2011). On producers-as-consumers, see Théberge (1997).
12. Meintjes (2003: 88).
13. Simondon (2014: 321), translation by Eliot Bates.
14. Akrich (1992).
15. Barad (2007: 140).
16. Barad (2007: 139).
17. Bates (2012a).
18. Grajeda (2015).
19. Haring (2007: 7).
20. Marvin (1988); Cowie (1999); Schmidt-Horning (2013); Gaston-Bird (2019).

21. Connell and Messerschmidt (2005); DeCook (2019); Clark-Flory (2020).

22. Holland et al. (1998); Richardson (2010); Seidman (2001).

23. Law (2004).

24. These include The Science and Media Museum, Bradford; Royer Microphones, Burbank; Studios 301, Sydney; The Queen Studio Experience, Montreux; Funky Junk, London; Guitar Centers and other gear retailers; and the Met's 2019 rock'n'roll exhibit, NYC.

25. On sensuous scholarship, see Pink (2015) and Plotnick (2018). On visual anthropology, see Ruby (2000).

26. For example, our archive includes corporate reports, manufacturer websites, electronics assembly manuals, component datasheets, journals and newsletters of professional and trade associations, product advertising of several kinds, and patent applications. We also draw upon our own qualitative tagging of gear magazines, online forum posts, and electronic mailing lists.

27. On studying up, see Nader (1972).

28. Marvin (1988); Théberge (1997); Schmidt-Horning (2013); Bennett (2018).

29. We do, however, recognize how this hegemonic, cultural formation is largely led and informed by US, UK, German, and Japanese histories and ideologies of sound technologies, recording, and production.

30. See, for example, Massy and Johnson (2016).

31. Pinch and Trocco (2002); Sterne (2012); Harkins (2019).

32. Schmidt-Horning (2013).

33. *Sound City*, for example, was titled after the Los Angeles studio of the same name and was loosely based around the Neve 8028 mixing console housed in its control room. However, the film focused on musicians and musical agency and charted the trajectory of the studio via canonized rock recordings.

34. Book-length studio ethnographies include Wallach (2008); Moehn (2012); and Bates (2016). On tech-processual analysis, see Bennett (2018).

35. In our works to date, we have conducted related ethnographic and analytical work and, in some instances, blended these methodologies. Eliot has spent twenty years writing about music materialities. Samantha has spent more than twelve years analyzing the sonically discernible presence of audio technologies and production processes in recorded popular music. Both of us have published on recording studios (Bates 2012b; Bennett 2016), recordists (Bates 2016; Bennett 2015, 2017), and sound technologies (Bates 2020c; Bennett 2012a, 2018).

36. At the time of writing, Eliot is winning.

37. See Pickering (1995).

38. For field-specific analyses, see Dion et al. (2018) and Earhart et al. (2020).

39. Both of us hold permanent positions in research-intensive universities. As part of our contracts, we are allocated research time, which was used in part to produce this book. As such, we acknowledge we are in positions of significant privilege.

40. Brooks et al (2021).

41. Devine (2019).

42. Wolfe (2019); Gaston-Bird (2019).

CHAPTER 2

1. "Victory Is Our Business!," Shadow Hills Industries, accessed December 2018, http://www.shadowhillsindustries.com/.

2. On economic theories of fetishes, see Veblen (1899), Baudrillard (1998 [1970]), Hornborg (2014), and Graeber (2005). On critical theory, see Lakoff and Johnson (1980) and Feenberg (1999).

3. Pietz (1987: 23, 30). See Matory (2018) for a contrasting contemporary perspective on fetishes, and material objects more generally, among Yoruban religious practitioners in a broader Afro-Atlantic geography.

4. Pietz (1985: 7).

5. Pietz (1987: 40).

6. Gell (1998: 61).

7. Pietz (1987: 43).

8. Marx (1996: 83).

9. Schmidt (2000).

10. Hornborg (2018: 6).

11. Hornborg (2014: 122).

12. Pels (2010: 629).

13. Feenberg (1999: 213).

14. While some of the phenomena we consider have been discussed by other scholars through the concept of mediation, in this book, mediation is not a primary keyword. As we hope will be apparent, more important than the intermediate positionality of a person or an object between two other entities are the qualities of the relations between entities, the sensory modalities through which human–object–human and object–human–object interactions and intra-actions unfold, and the underlying power structures, historiographical work, and symbolic invention that get articulated through fetishistic encounters.

15. Veblen (1899: 52, 21).

16. Connell and Messerschmidt (2005: 837).

17. Holland et al. (1998: 10).

18. Connell and Messerschmidt (2005: 831–832).
19. Richardson (2010).
20. Ebitz (1988: 207).
21. Elliott (2006: 235).
22. Sweeney (2014).
23. Schrock and Schwalbe (2009).
24. Mumby (1998: 180).
25. Marvin (1988); Cowie (1999).
26. Bennett (2012b); Annetts (2015).
27. Clare (2001).
28. Loe (2002).
29. On masculinity in crisis, see Bronfen (2002). On Korean soft masculinities, see Jung (2010).
30. Pascoe (2014); Moreton-Robinson (1998).
31. Solnit (2008).
32. Baudrillard (1998: 129).
33. Théberge (1997: 153).
34. Cheng (2019: 2).
35. Osiceanu (2015).
36. Richards (1993: 190).
37. Bennett (2018).
38. Marsh (2010). See also Motschenbacher (2014).
39. Motschenbacher (2014: 54).
40. Simpson (1994).
41. Miller (2005).
42. Mogan et al. (2012: 306).
43. Wheaton (2016: 45).
44. Herring (2014: 53).
45. Newell (2014: 306).
46. Nordsletten and Mataix-Cols (2012).
47. Marx (1996: 144).
48. "Gas Haiku Thread," accessed December 2023, https://llllllll.co/t/gas-haiku-thread/5192/470.

49. Herbst and Menze (2021: 13). Walter Becker's original article for *Guitar Player* magazine was long available on the Steely Dan website: https://web.archive.org/web/20170321042549/https://www.steelydan.com/gas.html.

50. Brounley (2022).

51. Rob Power and Matt Parker, "The 7 Key Stages of Gear Acquisition Syndrome," Music Radar, August 18, 2022, https://www.musicradar.com/news/7-stages-of-gear-acquisition-syndrome.

52. To get a sense of how this actually manifests in conversation, see the Gearspace thread here: https://gearspace.com/board/high-end/1328294-high-end-gear-over-rated.html. Note that we did not reverse engineer this one thread for either of the analyses; instead, we surveyed across numerous product-specific threads.

53. Lakoff and Johnson (1980).

CHAPTER 3

1. Davis (2016b).
2. Davis (2016b).
3. Davis (2018b).
4. Macho B (d. 2009, euthanized), El Jefe (last seen in 2015), Yo'oko (d. 2018, killed).
5. Davis (2021).
6. Ascarza (2020).
7. Loibl and Tercero Espinoza (2021).
8. Davis (2018a).
9. Those supporting Hudbay's claim have included the Arizona Game and Fish Department, the US Forest Service and the Army Corps of Engineers, while opponents have included the US Environmental Protection Agency, Pima County, the US Fish and Wildlife Service, and the Arizona State Historic Preservation Office. NGOs include the Tucson-based Center for Biological Diversity, Save the Scenic Santa Ritas, and four other environmental groups. Lawsuits intended to stop the mine were filed by the Tohono O'odham nation and Pascua Yaqui and Hopi tribes in 2018.
10. Davis (2016a).
11. Levick et al. (2008).
12. Smith (2021).
13. Forelle (2022).
14. Bates (2020c).
15. Devine (2019).

16. Nash (1979); Godoy (1985).
17. For a survey of issues within the anthropology of mining literature (1985–2018), see Jacka (2018).
18. Asmarhansyah et al. (2017).
19. Schoneveld et al. (2018: 30).
20. Jacka (2018).
21. On fossil capital, see Malm (2018). On the Anthropocene, see Malhi (2017). On e-waste, see Gabrys (2013).
22. Devine (2019); Roy (2020); Argabrite et al. (2022).
23. Smith and Mantz (2006: 78).
24. Smith (2021).
25. Auty (2007); Gilberthorpe and Rajak (2017).
26. Hornborg (2014).
27. Hutton (2020).
28. Bates (2020c).
29. "TIERRA Green," accessed June 2022, https://web.archive.org/web/20210615100929/https://tierra.audio/tierra-green/.
30. Sheller (2014: 5).
31. Sheller (2014: 85, 251–252).
32. Sheller (2014: 195).
33. Passive electronic components, which include resistors, capacitors, transformers, and inductors, are differentiated from active components (e.g., transistors, vacuum tubes, and operational amplifiers) in that they do not require an external power source to function. Passive components, therefore, cannot amplify a signal in a circuit.
34. *Panasonic Technical Guide: Aluminum Electronic Capacitor,* version 2019.4 (Panasonic Device Solutions, 2019); "General Descriptions of Aluminum Electrolytic Capacitors," Technical Notes CAT.8101E-1 (Nichicon, n.d.). An example of their patent filings is: Komatsu, Akihiko, Tetsushi Ogawara, and Shigeru Uzawa. Electrolytic solution for driving electrolytic capacitor and electrolytic capacitor. US7539006B2, 2006.
35. Nichicon, "General Descriptions of Aluminum Electrolytic Capacitors."
36. Knierzinger (2014).

CHAPTER 4

1. Hutton (2020).
2. Théberge (1997).

3. NAMM (2020: 19).

4. Reese (1988); Vágnerová (2017).

5. On the history of and differences between Neve input modules, see Geoff Tanner's post in the thread: "1064/1066/1073/1081 Differences?" GearSpace, accessed December 2020, https://www.gearspace.com/board/so-much-gear-so-little-time/54617-1064-1066-1073-1081-differences.html.

6. "V-251 Tube Mic Platinum Edition," Microphone Parts, accessed December 2020, https://microphone-parts.com/products/v-251-tube-mic-platinum-edition.

7. For an introduction to Wilson Capacitors, who provide the parts for the Gordon mic preamp, see Wilson Audio Specialties, "Excellence in All Things—Wilson's Crossover Capacitors," YouTube video, June 8, 2020, https://www.youtube.com/watch?v=Dk2WwV-znl4.

8. Reichenbach Engineering was connected with the company Altec (known most widely through the Altec Lansing brand of home stereo equipment) and later with Jensen Audio, and ultimately was purchased by Cine-Mag (https://cinemag.biz/cork_board/corkboard_history-1.php). On the relation between Electrodyne, Quad Eight, Sphere, and Reichenbach, see "Electrodyne History," *Tape Op*, accessed December 2020, https://tapeop.com/interviews/49/electrodyne-history/.

9. For example, one of the main output transformers used in Neve designs is the LO1166. Rupert Neve designed and hand-wound the original batches in 1965–1966 before they were manufactured for Neve by St Ives, Marinair, and at least one other unspecified company in England. Today, Carnhill exclusively manufactures these for AMS Neve. See "Getting Wound Up about Neve Transformers," AMS Neve, accessed December 2020, https://web.archive.org/web/20131009105233/http://ams-neve.com/news-and-events/getting-wound-about-neve-transformers.

10. Interview, Peter Montessi, 2017.

11. "Coil Audio," Gearspace, accessed August 2021, https://gearspace.com/board/high-end/1260519-coil-audio-18.html.

12. "Does the Vintech X73i Really Have St Ives Transformers?," Gearspace, accessed April 2023, https://gearspace.com/board/so-much-gear-so-little-time/93468-does-vintech-x73i-really-have-st-ives-transformers.html.

13. We don't discuss the history of vacuum tube opamps or the many other applications of opamps for analog computing. See Lojek (2007) for further information.

14. The equivalent of $413 in 2020. See Harrison (2005: 3).

15. Lojek (2007: 315).

16. Lojek (2007: 92).

17. Nakamura (2014: 928).

18. Henderson (1989: 51).

19. Cowie (1999).

20. Jung (2005: 819).

21. For an engineer-written survey of the opamp "problem," see "Op Amps: Myth and Facts," *NwAvGuy* (blog), August 10, 2011, http://nwavguy.blogspot.com/2011/08/op-amps-myths-facts.html. For an exhaustive comparison of opamps available in 2004 and their performance within a headphone amplifier, see "Notes on Audio Op-Amps," TangentSoft, accessed December 2020, https://tangentsoft.com/audio/opamps.html.

22. "MELCOR and API," Google Groups, May 17, 2000, https://groups.google.com/g/rec.audio.pro/c/WTu6r6m2zmo/m/qnVL-TTHHBAJ.

23. For a brief history of API and its designs, see "A Brief History of API," Dan Alexander Audio, accessed December 2020, https://web.archive.org/web/20150711195907/https://danalexanderaudio.com/apihistory.html.

24. Interview, Ben Sneesby, 2019.

25. For the Soyuz walkthrough, see Production Expert, "A Tour of the Soyuz Microphone Factory - Part 1 - Heavy Lifting," YouTube video, October 14, 2018, https://youtu.be/8-8LQnS3lTQ?t=528. For the Sennheiser/Neumann walkthrough, see Learn Audio Engineering, "Sennheiser/Neumann Factory Tour in Germany," YouTube video, September 14, 2019, https://youtu.be/-H3-s7mPIxc?t=230.

26. For information on Josephson's microphones patent, see "David Josephson," *Tape Op*, accessed December 2020, https://tapeop.com/interviews/btg/48/david-josephson/.

27. See "NOS Telefunken VF14 Microphone Tube," TubeDepot, accessed April 2023, https://www.tubedepot.com/products/vf14.

28. Interview, Peter Montessi, 2019.

29. "Meet the Maker: EveAnna Manley," KMR, January 18, 2019, https://www.kmraudio.com/news/designer-talk-eveanna-manley/.

30. On quantities of replacement tubes, see "How Many U67 Backup Tubes Do You Own," Gearspace, accessed May 2021, https://gearspace.com/board/so-much-gear-so-little-time/1318557-how-many-u67-backup-tubes-do-you-own.html. On the glow potential of tubes, see "Do All Tubes Glow?," Gearspace, accessed September 2022, https://gearspace.com/board/so-much-gear-so-little-time/1385854-do-all-tubes-glow.html. On "touching a tube" and "tube in a box," see "Adding a Tube Touch on the Way In," Gearspace, accessed December 2023, https://gearspace.com/board/so-much-gear-so-little-time/1398099-adding-tube-touch-way.html?highlight=tubes#post16438306.

31. Fortune (2016).

32. Interview, Robin Porter, 2017.

33. Interview, Ted Fletcher, 2020.

34. Interview, Marshall Terry, 2019.

35. Interview, Hutch Hutchison, 2017.
36. Interview, Jakob Erland, 2016.
37. Interview, Robin Porter, 2017.
38. "The Neve 1064/66/73/84 Design Points and Failings," Aurora Audio, accessed December 2020, https://web.archive.org/web/20161014010044/http://auroraaudio.net/forums/topic/the-neve-1064667384-design-points-and-failings.
39. "Getting Wound Up about Neve Transformers," AMS Neve, accessed December 2020, https://web.archive.org/web/20131009105233/http://ams-neve.com/news-and-events/getting-wound-about-neve-transformers.
40. Interview, Peter Montessi, 2017.
41. Goffman (1956).

CHAPTER 5

1. Interview, EveAnna Manley, 2020.
2. "F.A.Q.," Manley, accessed December 2022, https://www.manley.com/faq.
3. Bates (2016).
4. Masco (2006); Simondon (2012).
5. Akrich (1992).
6. Winner (1993); Feenberg (1999); Steen (2014).
7. On circuitbending, see Pinch (2016).
8. Sheller (2014: 27).
9. Lockwood (2019).
10. Simondon (2012: 2).
11. Bijker (1997).
12. Gregory changed the marketing of and text on the PCBs from "made in America" to "designed in Kiqötsmovi," which interested us but does not seem to have been discussed yet within the Drip or GroupDIY user communities. Hence, we avoid overinterpreting the significance of this shift.
13. Interview, Ted Fletcher, 2021.
14. On February 3, 1967, Meek shot and killed his landlady, Violet Shenton, before committing suicide.
15. "Clariphonic Manual," Thomann, 2010, https://images.thomann.de/pics/prod/clariphonic_manual.pdf.
16. Malsky (2003).
17. Interview, Ted Fletcher, 2021.

18. Interview, Ben Sneesby, 2021.
19. Parisi (2011: 190–191).
20. Parisi (2011: 206, 209).
21. Rutenberg (1939).
22. Payne and Altman (1962).
23. Jain (1999).
24. Plotnick (2018: 209).
25. Plotnick (2017: 1646).
26. Paterson (2007: 80).
27. Paterson (2007: 86).
28. For the case of Turkey, see Bates (2016). For Canada, see Sokil (2022).
29. Interview, Marshall Terry, 2019.
30. Jahrmann (2011).
31. Moseley (2015: 152).
32. Bates and Bennett (2022).

CHAPTER 6

1. Blesgraaf (2014: 75).
2. Palov (2011); Subotnik (1970).
3. On concretization, see Simondon (2017). On durability, see Law (2009). On inscription, see Akrich (1992). On trade show social bonds, see Andreae et al. (2013: 200).
4. We learned that Cable Gal had a brother, Cable Guy, but there had not been room in the back of the station wagon for him, so he was left back in Connecticut.
5. Cornfeld (2018).
6. Held in Frankfurt since 1980, Musikmesse is one of the world's largest music instrument trade shows and was expanded to include audio production equipment. In 2019, 56,000 trade visitors and 1,606 exhibitors attended (85,000 total visitors), although instrument makers account for more of that number than pro audio exhibitors. "Facts & Figures," Musikmesse, accessed December 2020, https://web.archive.org/web/20190130035507/https://musik.messefrankfurt.com/frankfurt/en/facts-figures.html.
7. Both Samantha and Eliot are longtime members of AES, have attended and presented at numerous conferences and conventions, and have published works in the AES e-Library.
8. AES's main history is documented at http://www.aes.org/aeshc/docs/aeshist/plunkett_aes-founding.pdf, with supplemental information at http://www.aes.org/aeshc/docs/aeshist/history-of-the-aes.html

9. Lojek (2007).

10. Mix with the Masters is technically an independent entity, but the AES has provided them with the largest and most prominent stage at the trade show since MWTM's inception, and events are produced "in association" with *Sound on Sound* magazine, who is the most prominent press entity at all AES shows.

11. Eshun (1998: 13).

12. Katz (2010: 126–127).

13. Marvin (1988).

14. Howard (2015).

15. Pinch (2001).

16. "About," TEC Awards, accessed December 2023, https://www.tecawards.org/about. For general info on NAMM's past trade shows, see "NAMM Show Location & Date History 1901–2025," NAMM, accessed August 2024, https://www.namm.org/library/blog/namm-show-location-date-history-1901–2018. For their oral history interviews, see "Oral History Interviews," NAMM, accessed August 2024, https://www.namm.org/library/oral-history.

17. Furlett (2015: 543). See also "NAMM Members Advocate for Music Education in Washington, D.C.," NAMM, accessed August 2024, https://ww1.namm.org/news/press-releases/namm-members-advocate-music-education-washington; and "Restart Act," Michael Bennett, accessed December 2020, https://www.bennet.senate.gov/public/_cache/files/c/8/c841064b-5a96-4d0e-84f6-e2cc06da00c8/336A2BFB17A35A84B84AD3BD40B641B2.restart-act-summary-v2.pdf.

18. NAMM (2020: 294).

19. Lawrence (2012).

20. Bennett (2016: 406).

21. Sokil (2022).

22. This includes but goes beyond what Cornfeld refers to as the "promosexuality" of the trade show (2018).

23. Dan Daley. "Biograph: Fletcher," *Mix Online*, January 1, 1999, https://www.mixonline.com/recording/biographfletcher-376408.

CHAPTER 7

1. These forums have, however, been inactive for years.

2. Ray Brigleb, "Larry Crane," June 11, 2015, in *The Job PDX*, produced by Needmore Designs, podcast, 51:28, http://thejobpdx.com/podcast/53-larry-crane/.

3. Annetts (2015: 231).

4. Théberge (1997: 104).

5. Annetts (2015: 56).
6. Bennett (2012b).
7. de Carvalho (2012).
8. All reviews quoted are from *Tape Op* issues 100, 101, 102, 103, and 105.
9. Bates and Bennett (2022).
10. Moors (1998: 218).
11. Moors (2013: 81).
12. Jhally (1987: 6).
13. Annetts (2015).
14. Annetts (2015: 91).
15. Interview, Larry Crane, 2017.
16. Annetts (2015: 97).
17. Latvia is the home of multiple gear manufacturers, perhaps the best known being Blue Microphones. Riga was formerly a Soviet engineering cluster, which remained in the country following the fall of the USSR.
18. David Mellor, quoted in Bennett (2012b).
19. Anderson (1991).

CHAPTER 8

1. "Neumann U67 2018 Reissue Internal View," Gearspace, accessed December 2020, https://www.gearslutz.com/board/high-end/1204485-neumann-u67-2018-reissue-internal-view.html.
2. On the relational ontology of objects, see Gell (1998). On studio talk, see Gay (1998). On taste and aesthetics, see Hennion (2007).
3. Murawski (2018).
4. Samantha has never participated in or contributed to any online gear forum at any time during her professional or academic career.
5. Bates and Bennett (2022).
6. https://groups.google.com/forum/#!forum/rec.audio.pro archived the text, but this doesn't capture its look in the 1990s.
7. Baym (2000: 6).
8. Hauben and Hauben (1997).
9. Jan Schauman, "Usenet Is Still a Strange Place," accessed December 2023, https://www.netmeister.org/news/usenet/.

10. For reminisces on pre-2006 Harmony Central forums, see "How Many of You Guys Used to Hang at Harmony Central?," The Gear Page, accessed December 2023, https://www.thegearpage.net/board/index.php?threads/how-many-of-you-guys-used-to-hang-at-harmony-central.1511506/page-4. The site, a student project, was originally a subdomain of MIT; Scott announced it on the rec.music.makers newgroup: https://groups.google.com/forum/#!searchin/rec.music.makers.synth/scott$20lehman|sort:date/rec.music.makers.synth/HGww2-aoI2k/ZI52lOWH_r8J. To get a sense of the site's typical look, see "Effects Resources," Harmony Central, accessed December 2023, https://web.archive.org/web/19961105185352/http://harmony-central.com:80/Effects/.

11. API, AMS Neve, and Abbey Road (studios) launched their websites in 1996. See API's dealer map here: "Sales," API Audio, accessed December 2023, https://web.archive.org/web/19970126021456/http://www.apiaudio.com/sales.htm.

12. On user groups, see Théberge (1997). On ham radio "tribes," see Douglas (1999: 330).

13. The "experimental" nature of the forum organization is evinced by the many times that subforums were created, renamed, assigned new moderators, deleted, or converted to "archive-only" status.

14. "The ProAudio Community Portal," ProSoundWeb, accessed December 2023, https://web.archive.org/web/20001109134900/http://www.prosoundweb.com/.

15. "Rec Pit," ProSoundWeb, accessed December 2023, https://web.archive.org/web/20031208131710/http://recpit.prosoundweb.com/viewforum.php?f=9&sid=798a04e500a13953ac247bd25b81d6f3.

16. Mixerman (2004).

17. For some reminisces on The Womb, see "The Womb: Just . . . Gone?," Gearspace, accessed July 2018, https://www.gearslutz.com/board/gear-free-zone-shoot-the-breeze/1172363-womb-just-gone.html.

18. For one blogger's take on The Womb Forums, see "Live Recording," Songs from So Deep, accessed December 2023, https://songsfromsodeep.wordpress.com/tag/the-womb-forums/. The last homepage capture we could find was July 17, 2017, which shows "Threads 19,568 Posts 384,986 Members 10,723." See "The Womb," The Womb Forums, accessed December 2023, https://web.archive.org/web/20170717053218/http://thewombforums.com/.

19. "Forum: Slipperman's Distorted Gtrs and Satanic Drums from Hell," The Womb Forums, accessed December 2023, https://web.archive.org/web/20160122164100/http://thewombforums.com/forumdisplay.php?10-Slipperman-s-Distorted-Gtrs-And-Satanic-Drums-From-Hell.

20. Leyshon (2009).

21. It is worth remembering that video streaming services such as Vimeo (2004–) or YouTube (2005–) had limited adoption during this period, and only subsequently became places to share audio-related video tutorials.

22. Švelch (2015).

23. Since Gearslutz prohibits multi-accounting and uses automated means and user reporting to enforce this, each registered user is actually a discrete person or business.

24. https://gearslutz.com prior to February 2021; https://gearspace.com subsequently.

25. "Are Readers Really Abandoning Pitchfork and Gearslutz?," *Trust Me I'm a Scientist*, accessed December 2023, https://web.archive.org/web/20171116070115/http://www.trustmeimascientist.com/2011/10/03/are-readers-abandoning-pitchfork-and-gearslutz/.

26. This sentiment was shared with Eliot at two trade show–situated Gearslutz meetups by (at that time) Gearslutz-curious people who had yet to join; it often surfaces in the first few posts by new users to the general gear subforums, too. That said, of the many professional audio engineers working today who we personally know, none actively participate in Gearslutz—or are willing to admit that they do!

27. Bourdieu (1970).

28. Womxn Audio, "Meg Lee Chin Interview," YouTube video, February 25, 2021, https://www.youtube.com/watch?v=QUT1AXEr67Y.

29. Farmelo (2014).

30. Farvid et al. (2017).

31. Rentschler (2014).

32. Gong and Hoffman (2012); Papp et al. (2015).

33. Gear Club Podcast, "Gear Club Podcast #69: Gearspace Joins the Club with Jules Standen," YouTube video, April 14, 2020, https://www.youtube.com/watch?v=PTm5MARPOto.

34. "Namm is Over as We've Known It," Gearspace, accessed December 2023, https://web.archive.org/web/20230419024010/https://gearspace.com/board/the-namm-show-2023-a/1403621-namm-over-we-ve-known.html.

35. Gay (1998); Schmidt-Horning (2004).

36. Bishop (2014).

37. On consumption battles, see Veblen (1899). On war metaphors, see Lakoff and Johnson (1980).

38. The same pseudonymous user subsequently asked the same question of noted electronic music blogger Peter Kirn (Create Digital Music) on Twitter. Kirn's supportive retweet led to dozens of CDM followers trying to scour the web to find and promote such businesses—indicating divergent values within two adjacent gear cultures.

39. To clarify, if a user refers to a musician in the third person as LGBTQIA+, that doesn't trigger a moderation incident, although it might trigger ridicule by other users. But beginning a post with some variant of "As a transgender mixing engineer, I" can lead to an incident, and discussing being actively discriminated against due to gender is not permitted—which played out in several threads during the peak of the 2017 #MeToo movement.

40. Gardy Stein, "RIP Sam Clayton Jr," *Reggaeville*, accessed May 2020, https://www.reggaeville.com/artist-details/brain-damage/news/view/rip-sam-clayton-jr/.

41. DeCook (2019).

42. Interview, Larry Crane, 2017.

43. GroupDIY was formerly named Prodigy-pro.

44. Interview, Jakob Erland, 2016.

45. The quantitative data behind this comes from analysis of Alexa and SimpleWeb site ranking stats, and Google Trends analysis of terms such as "gearslutz" and "gearslutz.com."

46. Rauch (2018).

CHAPTER 9

1. Gell (1998).
2. Pickering (1995).
3. Wyatt (2003).
4. Meintjes (2003: 94)
5. Pickering (1995).
6. Bortolotti (2020).
7. Erofeeva (2019: 600).
8. In the 1980s and 1990s, some digital signal processors did feature parameter save/recall functionality through small disk systems.
9. Akrich (1992).
10. Gibson (1979).
11. Akrich (1992).
12. Pentzold and Bischof (2019: 3).
13. Benjamin (1968: 220).
14. Veblen (1899).
15. Bates (2020a).
16. Bates (2020a).
17. However, there have recently been women-personified guitars and guitar pedals: St. Vincent's guitar was codesigned with Ernie Ball, and Robert Keeley released a Bad Snacks Artist Series version of one of their delay pedals.
18. "Least Developed Countries (LDCs)," United Nations, accessed December 2023, https://www.un.org/development/desa/dpad/least-developed-country-category.html.

19. Paul Abbott, "An Argument for Specialization," *Tape Op*, accessed December 2022, https://tapeop.com/columns/end-rant/77/ glosses this from the *Tape Op* point of view.

20. Théberge (1997).

CHAPTER 10

1. In case you are wondering, Samantha did not win the $100.

2. Thompson (1995).

3. Barad (2007).

4. Hui et al. (2020: 5).

5. Barad (2007: 67).

6. A related test was subsequently published on Crane Song's website: "Jitter," Crane Song, accessed July 2021, http://www.cranesong.com/jitter_1.html.

7. Interview, Robin Porter, 2017.

8. Interview, Hutch Hutchison, 2017.

9. Interview, Rick Perotta, 2020.

10. Interview, Jakob Erland, 2016.

11. John LaGrou of Millennia Media, interviewed in Swanke (2002: 26–27).

12. For additional biographical details about Winston, see Larry Crane, "Terri Winston," *Tape Op*, accessed December 2002, https://tapeop.com/interviews/78/terri-winston/.

13. Interview, Terri Winston, 2017.

14. Interview, Tim de Paravicini, 2019.

15. Not all web forum users concur with prevalent testing methodologies. For a critique of the expectation bias underpinning to these, see Gearspace, accessed December 2023, https://gearspace.com/board/showpost.php?p=2254993.

16. On the allegedly different "sound" of Windows XP and Mac OSX, see "Why Can't Daws Sum?," Gearspace, accessed December 2022, https://gearspace.com/board/music-computers/159119-why-cant-daws-sum-4.html#post1680111. On the sound of hard drives and the effects of interface connectors on them, see "Digital Controversy!," Gearspace, accessed December 2022, https://gearspace.com/board/high-end/240302-digital-controversy.html. For a derivative argument that audio stored on different parts of the hard drive sounds different, see "Certain Areas of Your H/Drive Sound Better!!," Gearspace, accessed December 2022, https://gearspace.com/board/music-computers/639206-certain-areas-your-h-drive-sound-better.html. Lyn Fuston's tests did demonstrate differences between ProTools's 48-bit floating point algorithm and the 32-bit floating point algorithms found in most other DAWs, but they did not demonstrate that the difference was significant. The forum set up to discuss Fuston's tests is "3D Comparison CD Discussions," 3dB, accessed December 2022,

https://web.archive.org/web/20160822053039/http://www.3daudioinc.com/3db/forum display.php?24-3D-Comparison-CD-Discussions. See also contemporaneous discussion of those tests on Gearspace: "DAW Summing Comparison - A New Testing Angle," Gearspace, accessed December 2023, https://gearspace.com/board/music-computers/178659-daw-summing-comparison-new-testing-angle-3.html.

17. Gearshoot, accessed February 2024, https://www.gearshoot.com/about/.
18. Ortoleva (2012: 16).
19. Ortoleva (2012: 16).
20. Interview, Larry Crane, 2017.

CHAPTER 11

1. audio_hertz, Instagram, December 9, 2022, https://www.instagram.com/reel/Cl8v_EHPbDW/?igshid=YmMyMTA2M2Y=.
2. Davison (2020: 122).
3. Davison (2020: 127).
4. Massey (2015).
5. Bates (2012b).
6. Bennett (2012).
7. Marvin (1988: 173).
8. Kealy (1979); Bates (2016).
9. Williams (2010); Bennett (2018).
10. Bates (2020b).
11. Bates (2012b).
12. Veblen (1899: 56).
13. Hurdley (2007).
14. Pearce (1999: 16).
15. Bisset (2013).
16. Clark (2011).
17. Gyüre and Jánosi (2009).
18. Bates and Bennett (2022: 12).
19. Marvin (1988: 165, 160–164).
20. Kheshti (2019: 69).
21. Kheshti (2019: 71).

22. Veblen (1899: 89).

23. Potvin (2010: 5).

CHAPTER 12

1. See "AES New York 2019 Exhibition Floor Recording Stage Will Inspire and Educate," Audio Engineering Society, accessed December 2019, http://www.aes.org/blog/2019/10/aes-new-york-2019-exhibition-floor.

2. See "AES New York 2019 Project Studio Expo Recording Stage Session Details," AES New York, accessed December 2019, http://www.aes.org/events/147/projectstudioexpo/?displayall.

3. See *Cambridge Dictionary*, s.v. "On the Run," accessed December 2020, https://dictionary.cambridge.org/dictionary/english/on-the-run.

4. Secrets and revelatory discourse are not confined to gear cultures; such discourses are found in the broader music industry as well as in society more generally, particularly in terms of business and economics.

5. *Oxford English Dictionary*, s.v. "Secret," March 2023, https://www.oed.com/view/Entry/174537.

6. Kelly (2002: 10).

7. Vrij et al. (2003).

8. Afifi and Caughlin (2006).

9. Kuhn (1995).

10. Kelly (2002: 10–15).

11. Williams (2010).

12. Marvin (1988: 12).

13. Marvin (1988: 14).

14. Bennett (2018).

15. Devine (2019).

16. Morton (2006: 42).

17. Bennett and Bates (2018: 3).

18. Schmidt-Horning (2013: 49, 120).

19. Schmidt-Horning (2013: 166).

20. Schmidt-Horning (2013: 199).

21. Zak (2001: 165).

22. Bennett (2018: 83).

23. Williams (2010: 166).

24. Théberge (1997).

25. For example, RMIT University in Melbourne offers a course entitled "How to Make a Hit Record and Other Production Secrets." See RMIT University, accessed December 2022, https://web.archive.org/web/20201029002216/https://www.rmit.edu.au/study-with-us/short-courses/courses/s375001-how-to-make-a-hit-record#overview.

26. Mix with the Masters, "Program," accessed December 2020, https://web.archive.org/web/20200202164141/https://mixwiththemasters.com/seminars/weeklong#program.

27. Leyshon (2009); Arditi (2020).

28. LaPolt (2022).

29. Bregitzer (2009).

30. Senior (2011).

31. Senior (2014).

32. Senior (2014).

33. Théberge (1997).

34. LCraneTapeOp, Twitter, "Why do people use the word 'arsenal,'" January 22, 2020, https://twitter.com/LCraneTapeOp/status/1220086510545825793.

35. Lakoff and Johnson (1980: 5–23).

36. Bomb Factory plugins was started by Dave Amels in 1997 and named after the eponymous studio in LA.

37. Leyshon (2009).

CHAPTER 13

1. Leonard and Knifton (2017).

2. Harvey (2016).

3. UNESCO (1972).

4. Howard (2003: 1).

5. Smith (2015).

6. Howard (2003: 4).

7. "Our Story—With the Vision of Peter Rodriguez," Heritage Audio, accessed December 2022, https://heritageaudio.com/about/.

8. Message and Whitcomb (2015: xxxvi).

9. Fforde (2007).

10. Keane (2003: 414).

11. Fairchild (2021: 91).

12. Leonard and Knifton (2017: 12).
13. Interview, Annie Jamieson, 2020.
14. Leonard and Knifton (2017: 12).
15. Versluys (2022: 38).
16. Holzhauser (2021).
17. On workplaces, see Bennett (2018: 11–14). On landmark recordings, see von Appen and Doehring (2006) and Jones (2008).
18. Microtech Gefell, "75th Anniversary UM75 Microphone from Gefell," Harmony Central, October 28, 2004, https://www.harmonycentral.com/news/microphones/75th-anniversary-um75-microphone-from-gefell-r7697/.
19. "2024 Winners," NAMMTEC, accessed December 2022, https://www.tecawards.org/.
20. "2021 Technology Hall of Fame Inductees," NAMMTEC, accessed December 2022, https://www.tecawards.org/tecnology-hall-of-fame.
21. Gell (1998: 25).
22. "Leading the Way in Sound," Solid State Logic, accessed December 2022, https://www.solidstatelogic.com/.
23. Lanier and Rader (2021: 599).
24. Fairchild (2021: 136).
25. Auner (2003).
26. Interview, Annie Jamieson, 2020.
27. Interview, Annie Jamieson, 2020.
28. Straughan (2015: 364).
29. Poliquin (2008: 123–5).
30. Youdelman (2017: 39).
31. Bennett (2016).
32. Straughan (2015: 374).

CHAPTER 14

1. Interview, Dave Harries, 2020.
2. On revival, see Rauch (2018). On nostalgia, see Schrey (2014). On technostalgia, see Taylor (2001).
3. Boym (2001: xiii).
4. Boym (2001: xv).
5. Boym (2001: 346).

6. Schrey (2014: 28, 34).
7. O'Grady (2019: 456).
8. Taylor (2001: 96).
9. Derrida (2011).
10. Fisher (2012: 19).
11. Sexton (2012: 562).
12. Acland (2007: xviii–xiv).
13. Acland (2007: xx).
14. Rauch (2018: 571).
15. Rauch (2018: 572).
16. Bennett (2012: 130).
17. Interview, Kelli Coyne, 2017.
18. Jones (2006: 19–20).
19. Interview, Tim de Paravacini, 2019.
20. Lewis (1986: 202).
21. Wichterman (2018).
22. Marx (1990: 995).
23. Bennett (2012).
24. Bell et al. (2015).
25. O'Grady (2019: 127).
26. O'Grady (2019: 131).
27. Interview, Robin Porter, 2017.
28. "The Dark Side of the Moon Recording Console Realised $1.8 Million at Bonhams Rock Memorabilia Sale," Bonhams, accessed February 2024, https://www.bonhams.com/press_release/23582/.
29. "About Us," Analogr, accessed February 2024, https://analogr.com/about.
30. Massey (2015).
31. "UA Founder Bill Putnam Jr.," Universal Audio, accessed May 2023, https://www.uaudio.com/blog/bill-putnam-ua-interview/.
32. Moore (2012).
33. Moore and Wakefield (2017: 4).
34. "The Monster Extreme Dynamic Processor, Slate Digital, accessed December 2023, https://slatedigital.com/the-monster-extreme-dynamic-processor/.

35. "Abbey Road Reverb Plates," Waves Creative Access, accessed February 2024, https://www.waves.com/plugins/abbey-road-reverb-plates.

CHAPTER 15

1. Named after the Cliff Richard song.
2. Herbst and Menze (2021).
3. Bates (2021).
4. Graeber (2015: 110).
5. Lanier (2010).
6. Graeber (2015: 120).
7. Graeber (2015: 114).
8. Brereton et al. (2020); Brooks et al. (2021).
9. Lazar et al. (2023).
10. Sokil (2022).
11. Lazar et al. (2022); Smith et al. (2021).
12. Bates and Troitski (2024).

References

UNPUBLISHED INTERVIEWS

Coyne, Kelli. October 19, 2017. AES Convention New York.

Crane, Larry. October 19, 2017. AES Convention New York.

Crane, Stephen. March 23, 2019. Videoconference.

de Paravicini, Tim. March 16, 2019. Email.

Erland, Jakob. June 5, 2016. AES Paris.

Fletcher, Ted. March 18, 2020. Videoconference.

Harries, Dave. April 14, 2020. Videoconference.

Hutchison, Hutch. October 20, 2017. AES Convention New York.

Jamieson, Annie. April 17, 2020. Videoconference.

Manley, EveAnna. January 21, 2020. Manley Labs. Chino, CA.

Montessi, Peter. October 20, 2017. AES Convention New York.

Perotta, Rick. January 21, 2020. Royer Mics. Burbank, CA.

Porter, Robin. October 19, 2017. AES Convention New York.

Sneesby, Ben. November 6, 2018. Videoconference.

Terry, Marshall. October 18, 2019. AES Convention New York.

Winston, Terri. October 19, 2017. AES Convention New York.

PUBLISHED SOURCES

Acland, Charles R. 2007. *Residual Media*. Minneapolis: University of Minnesota Press.

Afifi, Walid A., and John P. Caughlin. 2006. "A Close Look at Revealing Secrets and Some Consequences That Follow." *Communication Research* 33 (6): 467–488.

Akrich, Madeleine. 1992. "The De-Scription of Technical Objects." In *Shaping Technology/ Building Society: Studies in Sociotechnical Change*, edited by Wiebe E. Bijker and John Law, 205–24. Cambridge: MIT Press.

Anderson, Benedict. 1991. *Imagined Communities: Reflections on the Origin and Spread of Nationalism*. London: Verso.

Andreae, Michael, Jinn-yuh Hsu, and Glen Norcliffe. 2013. "Performing the Trade Show: The Case of the Taipei International Cycle Show." *Geoforum* 49 (October): 193–201.

Annetts, Alex. 2015. "Masculinity and Gear Fetishism in Audio Technology Community Discourse." PhD diss., Anglia Ruskin.

Arditi, David. 2020. *Getting Signed: Record Contracts, Musicians, and Power in Society*. Cham: Palgrave.

Argabrite, Zak, Jim Murhy, Sally Jane Norman, and Dale Carnegie. 2022. "Technology Is Land: Strategies towards Decolonisation of Technology in Artmaking." *NIME 2022*. https://doi.org/10.21428/92fbeb44.68f7c268.

Ascarza, William. 2020. "Mine Tales: Rosemont Open Pit Would Cap Long History of Copper Mining in District." *Arizona Daily Star*, August 10, 2020.

Asmarhansyah, A., Rodrigo B. Badayos, Pearl B. Sanchez, Pompe C. Sta Cruz, and Leonardo M. Florece. 2017. "Land Suitability Evaluation of Abandoned Tin-Mining Areas for Agricultural Development in Bangka Island, Indonesia." *Journal of Degraded and Mining Lands Management* 4 (4): 907–918. https://doi.org/10.15243/jdmlm.2017.044.907.

Auner, Joseph. 2003. "'Sing it For Me': Posthuman Ventrilioquism in Recent Popular Music." *Journal of the Royal Music Association* 128 (1): 98–122.

Auty, Richard M. 2007. "Natural Resources, Capital Accumulation and the Resource Curse." *Ecological Economics* 61 (4): 627–634.

Barad, Karen. 2007. *Meeting the Universe Halfway: Quantum Physics and the Entanglement of Matter and Meaning*. Durham: Duke University Press.

Bates, Eliot. 2012a. "The Social Life of Musical Instruments." *Ethnomusicology* 56 (3): 363–395.

Bates, Eliot. 2012b. "What Studios Do." *Journal on the Art of Record Production* 7.

Bates, Eliot. 2016. *Digital Tradition: Arrangement and Labor in Istanbul's Recording Studio Culture*. New York: Oxford University Press.

Bates, Eliot. 2020a. "Agency in Contemporary Recording Production." In *The Oxford Handbook of the Creative Process in Music*, edited by Nicolas Donin. New York: Oxford University Press.

Bates, Eliot. 2020b. "Recording Studios Since 1970." In *The Bloomsbury Handbook of Music Production*, edited by Simon Zagorski-Thomas and Andrew Bourbon, 125–39. New York: Bloomsbury Academic.

Bates, Eliot. 2020c. "Resource Ecologies, Political Economies and the Ethics of Audio Technologies in the Anthropocene." *Popular Music* 39 (1): 66–87.

Bates, Eliot. 2021. "The Interface and Instrumentality of Eurorack Modular Synthesis." In *Rethinking Music through Science and Technology Studies*, edited by Christophe Levaux and Antoine Hennion, 170–88. London: Routledge.

Bates, Eliot, and Samantha Bennett. 2022. "Look at All Those Big Knobs! Online Audio Technology Discourse and Sexy Gear Fetishes." *Convergence* 28 (5): 1241–1259.

Bates, Eliot, and Arseni Troitski. 2024. "Grid Culture." In *Modular Synthesis: Patching Machines and People*, edited by Andreas Kitzman and Ezra Teboul. London: Routledge.

Baudrillard, Jean. 1998. *The Consumer Society: Myths and Structures*. London: SAGE.

Baym, Nancy. 2000. *Tune in, Log on: Soaps, Fandom, and Online Community*. Thousand Oaks: Sage.

Bell, Adam, Ethan Hein, and Jarrod Ratcliffe. 2015. "Beyond Skeuomorphism: The Evolution of Music Production Software User Interface Metaphors." *Journal on the Art of Record Production* 9.

Benjamin, Walter. 1968. *Illuminations: Essays and Reflections*. Translated by Harry Zohn. New York: Schoken Books.

Bennett, Samantha. 2012a. "Endless Analogue: Technological Precursors in Contemporary Recording Practice." *Journal on the Art of Record Production* 7.

Bennett, Samantha. 2012b. "Revisiting the 'Double Production Industry': Advertising, Consumption and 'Technoporn' Surrounding the Music Technology Press." In *Music, Business and Law: Essays on Contemporary Trends in the Music Industry*, edited by Antti-Ville Kärjä, Lee Marshall, and Johannes Brusila, 117–45. Helsinki: International Institute for Popular Culture.

Bennett, Samantha. 2015. "Never Mind the Bollocks: A Tech-Processual Analysis." *Popular Music and Society* 38 (4): 466–486.

Bennett, Samantha. 2016. "Behind the Magical Mystery Door: History, Mythology and the Aura of Abbey Road Studios." *Popular Music* 35 (3), 396–417.

Bennett, Samantha. 2017. "Songs About Fucking: John Loder's Southern Studios and the Construction of a Subversive Sonic Signature." *Journal of Popular Music Studies* 29 (2): e12209.

Bennett, Samantha. 2018. *Modern Records, Maverick Methods: Technology and Process in Popular Music Record Production 1978–2000*. New York: Bloomsbury Academic.

Bennett, Samantha, and Eliot Bates. 2018. *Critical Approaches to the Production of Music and Sound*. New York: Bloomsbury Academic.

Bijker, Wiebe E. 1997. *Of Bicycles, Bakelites, and Bulbs: Toward a Theory of Sociotechnical Change*. Cambridge: The MIT Press.

Bishop, Jonathan. 2014. "Trolling for the Lulz? Using Media Theory to Understand Transgressive Humour and Other Internet Trolling in Online Communities." In *Transforming Politics and Policy in the Digital Age*, edited by Jonathan Bishop, 155–72. Hershey: IGI Global.

Bisset, Colin. 2013. "How the Lava Lamp Became a Symbol of Sex and the Space Age." *ABC*, May 29, 2013, sec. By Design.

Blesgraaf, Peter. 2014. "Redesigning the Theme Park Queue." *Journal of Motivation, Emotion, and Personality* 2 (2): 74–77.

Bonham's. 2017. "The Dark Side of the Moon Recording Console Realised $1.8 Million at Bonham's Rock Memorabilia Sale." Bonhams, March 27, 2017. https://www.bonhams.com/press_release/23582/

Bortolotti, Lisa. 2020. *The Epistemic Innocence of Irrational Beliefs*. Oxford: Oxford University Press.

Bourdieu, Pierre. 1970. "The Berber House or the World Reversed." *Social Science Information* 9 (2): 151–170.

Boym, Svetlana. 2001. *The Future of Nostalgia*. New York: Basic Books.

Bregitzer, Lorne. 2009. *Secrets of Recording: Professional Tips, Tools and Techniques*. Amsterdam: Focal Press.

Brereton, Jude, Helena Daffern, Kat Young, and Michael Lovedee-Turner. 2020. "Addressing Gender Equality in Music Production." In *Gender in Music Production*, edited by Russ Hepworth-Sawyer, Jay Hodgson, Liesl King, and Mark Marrington, 219–50. New York: Focal Press.

Bronfen, Elizabeth. 2002. "Noir Hysteria: Figures of Masculinity in Crisis in Contemporary Hollywood Cinema." *Figurationen* 3 (1): 65–80.

Brooks, Grace, Amandine Pras, Athena Elafros, and Monica Lockett. 2021. "Do We Really Want to Keep the Gate Threshold That High?" *Journal of the Audio Engineering Society* 69 (4): 238–260.

Brounley, David Matthew. 2022. "That Sound in Your Head: Guitar Tone Values in the Entrepreneurial Age." PhD diss., Stony Brook University.

Bruns, Axel, and Jan-Hinrik Schmidt. 2011. "Produsage: A Closer Look at Continuing Developments." *New Review of Hypermedia and Multimedia* 17 (1): 3–7.

Cheng, William. 2019. *Loving Music Till It Hurts*. New York: Oxford University Press.

Clare, Anthony. 2001. *On Men: Masculinity in Crisis*. London: Arrow Books.

Clark, Thomas Alan. 2011. "Lava Lamp Optics." *Applied Optics* 50 (28): F16–F20.

Clark-Flory, Tracy. 2020. "The Manosphere's Existential Crisis Is Building the Future of the Far-Right." *Jezebel*, October. https://jezebel.com/the-manospheres-existential-crisis-is-building-the-futu-1844673454.

Connell, R. W., and James W. Messerschmidt. 2005. "Hegemonic Masculinity: Rethinking the Concept." *Gender & Society* 19 (6): 829–859.

Cornfeld, Li. 2018. "Babes in Tech Land: Expo Labor as Capitalist Technology's Erotic Body." *Feminist Media Studies* 18 (2): 205–220.

Cowie, Jefferson. 1999. *Capital Moves: RCA's Seventy-Year Quest for Cheap Labor*. Ithaca: Cornell University Press.

Daley, Dan. 1999. "Biograph: Fletcher." *Mix Online* (blog), January 1, 1999.

Davis, Tony. 2016a. "Arizona Jaguar 'Biologically Insignificant,' Wildlife Manager Says." *Arizona Daily Star*, June 11, 2016.

Davis, Tony. 2016b. "Jaguar Video Sparked Controversy That Rages Six Months Later." *Arizona Daily Star*, August 15, 2016.

Davis, Tony. 2018a. "Tribes Sue Feds over Rosemont Mine, Citing 'Irreversible' Damage to Sacred Sites." *Arizona Daily Star*, April 14, 2018.

Davis, Tony. 2018b. "Jaguar That Roamed Huachuca Mountains Shown Dead in Photo, Officials Say." *Arizona Daily Star*, June 22, 2018.

Davis, Tony. 2021. "Hudbay Says It Found Copper That Could Result in Open Pit Mines on Santa Ritas' West Side." *Arizona Daily Star*, April 1, 2021.

Davison, Patrick. 2020. "The Language of Internet Memes." In *The Social Media Reader*, edited by Michael Mandiberg, 120–34. New York: New York University Press.

de Carvalho, Alice Tomaz. 2012. "The Discourse of Home Recording: Authority of 'Pros' and the Sovereignty of the Big Studios." *Journal on the Art of Record Production* 7.

DeCook, Julia. 2019. "How Deep Does the Rabbit Hole Go? The 'Wonderland' of r/TheRedPill and Its Ties to White Supremacy." *B2o*, November.

Departments of the Army and the Air Force. 1952. *Basic Theory and Application of Electron Tubes.* TM11–662. USA.

Derrida, Jacques. 2011. *Specters of Marx: The State of the Debt, the Work of Mourning and the New International.* Translated by Peggy Kamuf. London: Routledge.

Devine, Kyle. 2019. *Decomposed: The Political Ecology of Music*. Cambridge: The MIT Press.

Dion, Michelle L., Jane Lawrence Sumner, and Sara McLaughlin Mitchell. 2018. "Gendered Citation Patterns across Political Science and Social Science Methodology Fields." *Political Analysis* 26 (3): 312–327.

Douglas, Susan J. 1999. *Listening in: Radio and the American Imagination: From Amos "n" Andy and Edward R. Murrow to Wolfman Jack and Howard Stern*. New York: Times Books.

Downing, Lisa. 2008. *The Cambridge Introduction to Michel Foucault*. Cambridge: Cambridge University Press.

Earhart, Amy E, Roopika Risam, and Matthew Bruno. 2020. "Citational Politics: Quantifying the Influence of Gender on Citation in Digital Scholarship in the Humanities." *Digital Scholarship in the Humanities* 36 (3): 581–594.

Ebitz, David. 1988. "Connoisseurship as Practice." *Artibus et Historiae* 9 (18): 207–212.

Elliott, Charlene. 2006. "Considering the Connoisseur: Probing the Language of Taste." *Canadian Review of American Studies* 36 (2): 229–236.

Erofeeva, Maria. 2019. "On Multiple Agencies: When Do Things Matter?" *Information, Communication & Society* 22 (5): 590–604.

Eshun, Kodwo. 1998. *More Brilliant Than the Sun: Adventures in Sonic Fiction*. London: Quartet Books.

Fairchild, Charles. 2021. *Musician in the Museum: Display and Power in Neoliberal Popular Culture*. New York: Bloomsbury.

Farmelo, Allen. 2014. "Show Us Your Racks! The Sexualization of Recording Equipment." *Pink Noise* (blog), October 26.

Farvid, Panteá, Virginia Braun, and Casey Rowney. 2017. "'No Girl Wants to Be Called a Slut!': Women, Heterosexual Casual Sex and the Sexual Double Standard." *Journal of Gender Studies* 26 (5): 544–560.

Feenberg, Andrew. 1999. *Questioning Technology*. London: Routledge.

Fforde, Cressida. 2007. "Collection, Repatriation, and Identity." In *Museums and Their Communities*, edited by Sheila Watson, 244–53. London: Routledge.

Fisher, Mark. 2012. "What Is Hauntology?" *Film Quarterly* 66 (1): 16–24.

Forelle, MC. 2022. "The Material Consequences of 'Chipification': The Case of Software-Embedded Cars." *Big Data & Society* 9 (1): 1–12.

Fortune, Evan. 2016. "Remembering Audio Pioneer Saul Walker." *Mix Online* (blog), October 21, 2016.

Fox, Barry. 1980. "Nice Timing: Secrets of Successful Invention." *Audio Engineering Society Convention* 65, paper 1601. https://aes2.org/publications/elibrary-page/?id=3770.

Furlett, Joseph. 2015. "The Insufficiency of the Musical Instrument Passport Program Under CITES and the Lacey Act: The Need for a Centralized Wood Title Certification System for Manufactured Wood Products and Wooden Musical Instruments, 48 J. Marshall L. Rev. 495." *UIC Law Review* 48 (2): 495–537.

Gabrys, Jennifer. 2013. *Digital Rubbish: A Natural History of Electronics*. Ann Arbor: University of Michigan Press.

Gaston-Bird, Leslie. 2019. *Women in Audio*. New York: Routledge.

Gay, Leslie C. 1998. "Acting up, Talking Tech: New York Rock Musicians and Their Metaphors of Technology." *Ethnomusicology* 42 (1): 81–98.

Gell, Alfred. 1998. *Art and Agency*. London: Clarendon.

Gibson, J. J. 1979. *The Ecological Approach to Visual Perception*. Boston: Houghton Mifflin.

Gilberthorpe, Emma, and Dinah Rajak. 2017. "The Anthropology of Extraction: Critical Perspectives on the Resource Curse." *The Journal of Development Studies* 53 (2): 186–204.

Godoy, Ricardo. 1985. "Mining: Anthropological Perspectives." *Annual Review of Anthropology* 14 (1): 199–217.

Goffman, Erving. 1956. *The Presentation of Self in Everyday Life*. Edinburgh: University of Edinburgh Social Sciences Research Center.

Gong, Lijia, and Alina Hoffman. 2012. "Sexting and Slut-Shaming: Why Prosecution of Teen Self-Sexters Harms Women." *Georgetown Journal of Gender and the Law* 13: 577–589.

Graeber, David. 2005. "Value: Anthropological Theories of Value." In *A Handbook of Economic Anthropology*, edited by James Carrier, 439–54. Cheltenham: Edward Elgar.

Graeber, David. 2015. *The Utopia of Rules: On Technology, Stupidity, and the Secret Joys of Bureaucracy*. Brooklyn, NY: Melville House.

Grajeda, Tony. 2015. "The 'Sweet Spot': The Technology of Stereo and the Field of Auditorship." In *Living Stereo: Histories and Cultures of Multichannel Sound*, edited by Paul Théberge, Kyle Devine, and Tom Everrett, 37–64. New York: Bloomsbury Academic.

Gyüre, Balázs, and Imre M. Jánosi. 2009. "Basics of Lava-Lamp Convection." *Physical Review E* 80 (4): 046307.

Haring, Kristen. 2007. *Ham Radio's Technical Culture*. Cambridge: The MIT Press.

Harkins, Paul. 2019. *Digital Sampling: The Design and Use of Music Technologies*. Boca Raton: CRC Press.

Harrison, Linden T. 2005. *Current Sources & Voltage References*. Oxford: Newness.

Harvey, David C. 2016. "The History of Heritage." In *The Routledge Research Companion to Heritage and Identity*, edited by Brian Graham and Peter Howard, 19–36. London: Routledge.

Hauben, Michael, and Ronda Hauben. 1997. *Netizens: On the History and Impact of Usenet and the Internet*. Los Alamitos, CA: IEEE Computer Society Press.

Henderson, Jeffrey. 1989. *The Globalisation of High Technology Production: Society, Space and Semiconductors in the Restructuring of the Modern World*. London: Routledge.

Hennion, Antoine. 2007. "Those Things That Hold Us Together: Taste and Sociology." *Cultural Sociology* 1 (1): 97–114.

Herbst, Jan-Peter, and Jonas Menze. 2021. *Gear Acquisition Syndrome: Consumption of Instruments and Technology in Popular Music*. Huddersfield: University of Huddersfield Press.

Herring, Scott. 2014. *The Hoarders: Material Deviance in Modern American Culture*. Chicago: University of Chicago Press.

Holland, Janet, Caroline Ramazanoglu, Sue Sharpe, and Rachel Thomson. 1998. *The Male in the Head: Young People, Heterosexuality and Power*. London: The Tufnell Press.

Holzhauser, Nicole. 2021. "Quantifying the Exclusionary Process of Canonisation, or How to Become a Classic of the Social Sciences." *International Review of Sociology* 31 (1): 97–122.

Hornborg, Alf. 2014. "Technology as Fetish: Marx, Latour, and the Cultural Foundations of Capitalism." *Theory, Culture & Society* 31 (4): 119–140.

Hornborg, Alf. 2018. "Relationism as Revelation or Prescription? Some Thoughts on How Ingold's Implicit Critique of Modernity Could Be Harnessed to Political Ecology." *Interdisciplinary Science Reviews* 43 (3–4): 1–11.

Howard, P., 2003. "Heritage." *Management, Interpretation, Identity*. Accessed December 2022. https://jamestownhistoricalsociety.org/wp-content/uploads/2015/04/JHS-2002-1-spring-1.pdf.

Hui, Alexandra, Mara Mills, and Viktoria Tkaczyk. 2020. "Testing Hearing: An Introduction." In *Testing Hearing: The Making of Modern Aurality*, edited by Viktoria Tkaczyk, Mara Mills, and Alexandra Hui, 1–19. New York: Oxford University Press.

Hurdley, Rachel. 2007. "Focal Points: Framing Material Culture and Visual Data." *Qualitative Research* 7 (3): 355–374.

Hutton, Jane. 2020. *Reciprocal Landscapes: Stories of Material Movements*. Abingdon: Routledge.

Jacka, Jerry K. 2018. "The Anthropology of Mining: The Social and Environmental Impacts of Resource Extraction in the Mineral Age." *Annual Review of Anthropology* 47 (1): 61–77.

Jahrmann, Margarete. 2011. "Ludics for a Ludic Society. The Art and Politics of Play." PhD diss., University of Plymouth.

Jain, Sarah S. 1999. "The Prosthetic Imagination: Enabling and Disabling the Prosthesis Trope." *Science, Technology, & Human Values* 24 (1): 31–54.

Jhally, Sut. 1987. *The Codes of Advertising: Fetishism and the Political Economy of Meaning in the Consumer Society*. London: Routledge.

Jones, Carys Wyn. 2008. *The Rock Canon: Canonical Values in the Reception of Rock*. Aldershot: Ashgate.

Jones, Steven E. 2006. *Against Technology: From the Luddites to Neo-Luddism*. London: Routledge.

Jung, Sun. 2010. *Korean masculinities and transcultural consumption: Yonsama, Rain, Oldboy, K-pop idols*. Hong Kong: Hong Kong University Press.

Jung, Walt. 2005. *Op Amp Applications Handbook*. Amsterdam: Newnes.

Jung, Walter G. 1986. *IC Op-Amp Cookbook*. 3rd ed. Indianapolis, IN: H.W. Sams.

Katz, Mark. 2010. *Capturing Sound: How Technology Has Changed Music*. Rev. ed. Berkeley: University of California Press.

Katz, Mark. 2015. "The Persistence of Analogue." In *Musical Cultures of the 21st Century*, edited by Gianmario Borio, 275–87. Farnham: Ashgate.

Kealy, Edward R. 1979. "From Craft to Art: The Case of Sound Mixers and Popular Music." *Sociology of Work and Occupations* 6 (1): 3–29.

Keane, Webb. 2003. "Semiotics and the Social Analysis of Material Things." *Words and Beyond: Linguistic and Semiotic Studies of Sociocultural Order* 23 (3): 409–425.

Kehew, Brian, and Kevin Ryan. 2006. *Recording the Beatles*. USA: Curvebender Publishing.

Kelly, Anita E. 2002. *The Psychology of Secrets*. New York: Kluwer Academic/Plenum Publishers.

Kheshti, Roshanak. 2019. *Wendy Carlos's Switched-On Bach (33 1/3)*. New York: Bloomsbury.

Knierzinger, Johannes. 2014. "The Socio-Political Implications of Bauxite Mining in Guinea: A Commodity Chain Perspective." *The Extractive Industries and Society* 1 (1): 20–27.

Kuhn, Annette. 1995. *Family Secrets: Acts of Memory and Imagination*. London: Verso.

Lakoff, George, and Mark Johnson. 1980. *Metaphors We Live By*. Chicago: Chicago University Press.

Lanier, Clinton D., and C. Scott Rader. 2021. "Synthesizers: An Exploration into the Iconicity of Marketplace Icons." *Consumption Markets & Culture* 24 (6): 596–610.

Lanier, Jaron. 2010. *You Are Not a Gadget: A Manifesto*. New York: Alfred A. Knopf.

Law, John. 2004. *After Method: Mess in Social Science Research*. London: Routledge.

Law, John. 2009. "Actor Network Theory and Material Semiotics." In *The New Blackwell Companion to Social Theory*, edited by Bryan S Turner, 141–58. Chichester: Blackwell Publishing.

Lawrence, Alistair. 2012. *Abbey Road: The Best Studio in the World*. New York: Bloomsbury.

Lazar, Emily, Carolyn Malachi, Jordan Hamlin, Gabriela Rodriguez Bonilla, Jasmine Kok, Meghan Smyth, Beverly Keel, and Beth Appleton. 2023. "Lost in The Mix: An Analysis of Credited Technical Professionals in the Music Industry Highlighting Women and Non-Binary Producers and Engineers Across DSP Playlists, Genres, Awards, Record Certifications & Distributors." Fix The Mix Annual Report.

Leonard, Marion, and Robert Knifton. 2017. "A Critical Survey of Museum Collections of Popular Music in the United Kingdom." *Popular Music History* 10 (2): 171–191.

Levick, Lainie, Julia Fonseca, Darius J. Semmens, Juliet Stromberg, Melanie Tluczek, Robert A. Leidy, Melissa Scianni, D. Phillip Guertin, and William G. Kepner. 2008. *The Ecological and Hydrological Significance of Ephemeral and Intermittent Streams in the Arid and Semi-Arid American Southwest*. EPA/600/R-08/134 ARS/233046. Washington, DC: US Environmental Protection Agency.

Lewis, David K. 1986. *On the Plurality of Worlds*. Oxford: Blackwell.

Leyshon, Andrew. 2009. "The Software Slump?: Digital Music, the Democratisation of Technology, and the Decline of the Recording Studio Sector within the Musical Economy." *Environment and Planning A: Economy and Space* 41 (6): 1309–1331.

Lockwood, Dave. 2019. "Phil Dudderidge of Focusrite: The Man Behind the Brand." *Sound on Sound*, August 2019.

Loe, Meika. 2002. (De) constructing the Viagra phenomenon: Claims, Markets, and the Science of Sex. PhD diss., University of California, Santa Barbara.

Loibl, Antonia, and Luis A. Tercero Espinoza. 2021. "Current Challenges in Copper Recycling: Aligning Insights from Material Flow Analysis with Technological Research Developments and Industry Issues in Europe and North America." *Resources, Conservation and Recycling* 169 (June).

Lojek, Bo. 2007. *History of Semiconductor Engineering*. Berlin: Springer.

Malhi, Yadvinder. 2017. "The Concept of the Anthropocene." *Annual Review of Environment & Resources* 42 (October):77–104.

Malm, Andreas. 2018. *The Progress of This Storm: Nature and Society in a Warming World*. London: Verso.

Malsky, Matthew. 2003. "Stretched from Manhattan's Back Alley to MOMA: A Social History of Magnetic Tape and Recording." In *Music and Technoculture*, edited by René Lysloff and Leslie Gay Jr., 233–63. Middletown: Wesleyan University Press.

Marsh, Amy. 2010. "Love among the Objectum Sexuals." *Electronic Journal of Human Sexuality* 13 (1).

Marvin, Carolyn. 1988. *When Old Technologies Were New: Thinking About Electrical Communication in the Late Nineteenth Century*. New York: Oxford University Press.

Marx, Karl. 1996. *Capital: A Critique of Political Economy. Vol. 1*. Translated by Samuel Moore and Edward Aveling. London: Lawrence & Wishart.

Masco, Joseph. 2006. *The Nuclear Borderlands: The Manhattan Project in Post-Cold War New Mexico*. Princeton: Princeton University Press.

Massey, Howard. 2015. *The Great British Recording Studios*. Milwaukee, WA: Hal Leonard Books.

Massy, Sylvia, and Chris Johnson. 2016. *Recording Unhinged: Creative and Unconventional Music Recording Techniques*. Milwaukee: Hal Leonard.

Matory, James Lorand. 2018. *The Fetish Revisited: Marx, Freud, and the Gods Black People Make*. Durham: Duke University Press.

McRobbie, Angela. 1980. "Settling Accounts with Subcultures: A Feminist Critique." *Screen Education* 34: 37–49.

Meintjes, Louise. 2003. *Sound of Africa!: Making Music Zulu in a South African Studio*. Durham: Duke University Press.

Message, Kylie, and Andrea Whitcomb. 2015. *Museum Theory*. Hoboken: Wiley-Blackwell.

Miller, Toby. 2005. "A Metrosexual Eye on Queer Guy." *GLQ: A Journal of Lesbian and Gay Studies* 11 (1): 112–117.

Mitchell, Robert. 2012. "Simondon, Bioart and the Milieu of Biotechnology." *Inflexions* 5 (March): 68–110.

Mixerman [Eric Sarafin]. 2004. *The Daily Adventures of Mixerman*. Los Angeles: Mixerman Multimedia.

Moehn, Frederick. 2012. *Contemporary Carioca: Technologies of Mixing in a Brazilian Music Scene*. Durham: Duke University Press.

Mogan, Christopher, Michael Kyrios, Isaac Schweitzer, Keong Yap, and Richard Moulding. 2012. "Phenomenology of Hoarding—What Is Hoarded by Individuals with Hoarding Disorder?" *Journal of Obsessive-Compulsive and Related Disorders* 1: 306–311.

Moore, Austin. 2012. "All Buttons in: An Investigation into the Use of the 1176 FET Compressor in Popular Music Production." *Journal on the Art of Record Production* 6.

Moore, Austin, and Jonathan Wakefield. 2017. "An Investigation into the Relationship Between the Subjective Descriptor Aggressive and the Universal Audio 1176 FET Compressor." *142nd Convention of the Audio Engineering Society*, paper 9749. https://aes2.org/publications/elibrary-page/?id=18625.

Moors, Annelies. 1998. "Wearing Gold." In *Border Fetishisms: Material Objects in Unstable Spaces*, edited by Patricia Spyer, 208–23. London: Routledge.

Moors, Annelies. 2013. "Wearing Gold, Owning Gold: The Multiple Meanings of Gold Jewelry." *Etnofoor* 25 (1): 79–89.

Moreton-Robinson, Aileen. 1998. "When the Object Speaks, A Postcolonial Encounter: Anthropological Representations and Aboriginal Women's Self-Presentations." *Discourse: Studies in the Cultural Politics of Education* 19 (3): 275–289.

Morton, David. 2006. *Sound Recording: The Life Story of a Technology*. Baltimore: Johns Hopkins.

Moseley, Roger. 2015. "Digital Analogies: The Keyboard as Field of Musical Play." *Journal of the American Musicological Society* 68 (1): 151–228.

Motschenbacher, Heiko. 2014. "Focusing on Normativity in Language and Sexuality Studies." *Critical Discourse Studies* 11 (1): 49–70.

Mumby, Dennis K. 1998. "Organizing Men: Power, Discourse, and the Social Construction of Masculinity(s) in the Workplace." *Communication Theory* 8 (2): 164–183.

Murawski, Michał. 2018. "Marxist Morphologies: A Materialist Critique of Brute Materialities, Flat Infrastructures, Fuzzy Property, and Complexified Cities." *Focaal* 2018 (82): 16–34.

Nader, Laura. 1972. "Up the Anthropologist: Perspectives Gained from Studying Up." In *Reinventing Anthropology*, edited by Dell H. Hymes, 284–11. New York: Pantheon Books.

Nakamura, Lisa. 2014. "Indigenous Circuits: Navajo Women and the Racialization of Early Electronic Manufacture." *American Quarterly* 66 (4): 919–941.

Nash, June. 1979. *We Eat the Mines and the Mines Eat Us: Dependency and Exploitation in Bolivian Tin Mines*. New York: Columbia University Press.

NAMM. 2020. *The 2020 NAMM Global Report*. National Association of Music Merchants: Carlsbad, CA.

Newell, Sasha. 2014. "The Matter of the Unfetish: Hoarding and the Spirit of Possessions." *HAU: Journal of Ethnographic Theory* 4 (3): 185–213.

Nordsletten, Ashley E, and David Mataix-Cols. 2012. "Hoarding versus Collecting: Where Does Pathology Diverge from Play?" *Clinical Psychology Review* 32 (3): 165–176.

O'Grady, Pat. 2019. "The Analogue Divide: Interpreting Attitudes towards Recording Media in Pop Music Practice." *Continuum* 33 (4): 446–459.

Ortoleva, Peppino. 2012. “Homo ludicus. The Ubiquity of Play and Its Roles in Present Society.” *G| A| M| E Games as Art, Media, Entertainment* 1 (1): 5–17.

Osiceanu, Maria-Elena. 2015. “Psychological Implications of Modern Technologies: ‘Technofobia’ versus ‘Technophilia.’” *Procedia—Social and Behavioral Sciences* 180 (5 May): 1137–1144.

Palov, Rebekkah. 2011. “Harald Bode—A Short Biography.” *EContact!* 13 (4).

Papp, Leanna J., Charlotte Hagerman, Michelle A. Gnoleba, Mindy J. Erchull, Miriam Liss, Haley Miles-McLean, and Caitlin M. Robertson. et al. 2015. “Exploring Perceptions of Slut-Shaming on Facebook: Evidence for a Reverse Sexual Double Standard.” *Gender Issues* 32 (1): 57–76.

Parisi, David. 2011. “Tactile Modernity: On the Rationalization of Touch in the Nineteenth Century.” In *Media, Technology, and Literature in the Nineteenth Century*, edited by Collette Colligan and Margaret Linley, 189–214. Farnham: Ashgate.

Pascoe, Bruce. 2014. *Dark Emu Black Seeds: Agriculture or Accident?* Australia: Magabala Books.

Paterson, Mark. 2007. *The Senses of Touch Haptics, Affects and Technologies*. Oxford: Berg.

Payne, Dan, and James W. Altman. 1962. *An Index of Electronic Equipment Operability: Report of Development.* AIR-C-43-1/62-FR. Signal Corps Technical Requirements EWD. Pittsburgh: American Institute for Research.

Pearce, Susan M. 1999. *On Collecting: An Investigation into Collecting in the European Tradition.* London: Routledge.

Pels, Peter J. 2010. “Magical Things: On Fetishes, Commodities, and Computers.” In *The Oxford Handbook of Material Culture Studies*, edited by Dan Hicks and Mary C. Beaudry, 613–33. Oxford: Oxford University Press.

Pentzold, Christian, and Andreas Bischof. 2019. “Making Affordances Real: Socio-Material Prefiguration, Performed Agency, and Coordinated Activities in Human–Robot Communication.” *Social Media + Society* 5 (3): 1–11. Pickering, Andrew. 1995. *The Mangle of Practice: Time, Agency, and Science*. Chicago: University of Chicago Press.

Pietz, William. 1985. “The Problem of the Fetish, I.” *RES: Anthropology and Aesthetics* 9: 5–17.

Pietz, William. 1987. “The Problem of the Fetish, II.” *RES: Anthropology and Aesthetics* 13: 23–45.

Pinch, Trevor. 2001. “Why You Go to a Piano Store to Buy a Synthesizer: Path Dependence and the Social Construction of Technology.” In *Path Dependence and Creation*, edited by R Garud and P Karnoe, 381–401. New Jersey: LEA Press.

Pinch, Trevor. 2016. “‘Bring on Sector Two!’: The Sounds of Bent and Broken Circuits.” *Sound Studies* 2 (1): 36–51.

Pinch, Trevor J., and David Reinecke. 2009. “Technostalgia: How Old Gear Lives on in New Music.” In *Sound Souvenirs: Audio Technologies, Memory and Cultural Practices*, edited by Karin Bijsterveld and José van Dijck, 152–68. Amsterdam: Amsterdam University Press.

Pinch, Trevor, and Frank Trocco. 2002. *Analog Days: The Invention and Impact of the Moog Synthesizer*. Cambridge: Harvard University Press.

Pink, Sarah. 2015. *Doing Sensory Ethnography*. 2nd ed. Los Angeles: Sage.

Pink, Sarah, Elisenda Ardèvol, and Débora Lanzeni, eds. 2016. *Digital Ethnography: Principles and Practice*. London: Sage.

Plotnick, Rachel. 2017. "Force, Flatness and Touch without Feeling: Thinking Historically about Haptics and Buttons." *New Media & Society* 19 (10): 1632–1652.

Plotnick, Rachel. 2018. *Power Button: A History of Pleasure, Panic, and the Politics of Pushing*. Cambridge: The MIT Press.

Poliquin, Rachel. 2008. "The Matter and Meaning of Museum Taxidermy." *Museum and Society* 6 (2): 123.

Potvin, John. 2010. "The Velvet Masquerade: Fashion, Interior Design and the Furnished Body." In *Fashion, Interior Design and the Contours of Modern Identity*, edited by Alla Myzelev and John Potvin, 1–18. London: Routledge.

Rauch, Jennifer. 2018. *Slow Media: Why Slow Is Satisfying, Sustainable, and Smart*. New York: Oxford University Press.

Reese, Lyn. 1988. "Nimble Fingers: From 19th Century New England Mills to 20th Century Global Assembly Lines." *OAH Magazine of History* 3 (3–4): 45–49.

Rentschler, Carrie A. 2014. "Rape Culture and the Feminist Politics of Social Media." *Girlhood Studies* 7 (1): 65–82.

Richards, Barry. 1993. "Technophobia and Technophilia." *British Journal of Psychotherapy* 10 (2): 188–195.

Richardson, Diane. 2010. "Youth Masculinities: Compelling Male Heterosexuality." *The British Journal of Sociology* 61 (4): 737–756.

Roy, Elodie A. 2020. "'Total Trash': Recorded Music and the Logic of Waste." *Popular Music* 39 (1): 88–107.

Ruby, Jay. 2000. *Picturing Culture: Explorations of Film & Anthropology*. Chicago: University of Chicago Press.

Rutenberg, D. 1939. "The Early History of the Potentiometer System of Electrical Measurement." *Annals of Science* 4 (2): 212–243.

Schmidt, Leigh Eric. 2000. *Hearing Things: Religion, Illusion, and the American Enlightenment*. Cambridge: Harvard University Press.

Schmidt-Horning, Susan. 2004. "Engineering the Performance: Recording Engineers, Tacit Knowledge and the Art of Controlling Sound." *Social Studies of Science* 34 (5): 703–731.

Schmidt-Horning, Susan. 2013. *Chasing Sound: Technology, Culture, and the Art of Studio Recording from Edison to the LP*. Baltimore: The Johns Hopkins University Press.

Schoneveld, George, Maisory Chacha, María Njau, Jesper Jønsson, Paolo Omar Cerutti, and Xiaoxue Weng. 2018. *The New Face of Informality in the Tanzanian Mineral Economy: Transforming Artisanal Mining through Foreign Investment?* London: IIED Research Report.

Schrey, Dominik. 2014. "Analogue Nostalgia and the Aesthetics of Digital Remediation." In *Media and Nostalgia: Yearning for the Past, Present and Future*, edited by Katharina Niemeyer, 27–38. London: Palgrave Macmillan.

Schrock, Douglas, and Michael Schwalbe. 2009. "Men, Masculinity, and Manhood Acts." *Annual Review of Sociology* 35: 277–295.

Seidman, Steven. 2001. "From Identity to Queer Politics: Shifts in Normative Heterosexuality and the Meaning of Citizenship." *Citizenship Studies* 5 (3): 321–328.

Self, Douglas. 2010. *Small Signal Audio Design*. Oxford, UK: Focal Press.

Senior, Mike. 2011. *Mixing Secrets*. New York: Routledge.

Senior, Mike. 2014. *Recording Secrets for the Small Studio*. New York: Routledge.

Sexton, Jamie. 2012. "Weird Britain in Exile: Ghost Box, Hauntology, and Alternative Heritage." *Popular Music and Society* 35 (4): 561–584.

Sheller, Mimi. 2014. *Aluminum Dreams: The Making of Light Modernity*. Cambridge: The MIT Press.

Simondon, Gilbert. 2012. "On Techno-Aesthetics." Translated by Arne De Boever. *Parrhesia* 14: 1–8.

Simondon, Gilbert. 2014. *Sur la technique (1953–1983)*. Paris: Presses Universitaires de France.

Simondon, Gilbert. 2017. *On the Mode of Existence of Technical Objects*. Translated by Cécile Malaspina and John Rogove. Minneapolis: Univocal Publishing.

Simpson, Mark. 1994. "Here Come the Mirror Men." *Independent*, November 15, 1994.

Smith, James H. 2021. *The Eyes of the World: Mining the Digital Age in the Eastern DR Congo*. Chicago: University of Chicago Press.

Smith, James H., and Jeffrey W. Mantz. 2006. "Do Cellular Phones Dream of Civil War? The Mystification of Production and the Consequences of Technology Fetishism in the Congo." In *Inclusion and Exclusion in the Global Arena*, edited by Max Kirsch, 71–93. New York: Routledge.

Smith, Laurajane. 2015. "Intangible Heritage: A Challenge to the Authorised Heritage Discourse?" *Revista d'etnologia de Catalunya* 40: 133–142.

Smith, Stacy, Katherine Pieper, Marc Choueti, Karla Hernandez, and Kevin Yao. 2021. *Inclusion in the Recording Studio? Gender and Race/Ethnicity of Artists, Songwriters & Producers across 900 Popular Songs from 2012–2020*. Los Angeles: USC Annenberg.

Sokil, Allison. 2022. "The Hums: Feminist Listening and Gendered Affects in Music Production and Audio Engineering in Canada." PhD dissertation., University of Toronto.

Solnit, Rebecca. 2008. "Men Explain Things to Me (Facts Didn't Get in Their Way)." TomDispatch.com. Accessed December 2022. http://www.tomdispatch.com/blog/174918/tomgram%3A_rebecca_solnit%2C_the_archipelago_of_arrogance.

Stavrou, Michael Paul. 2003. *Mixing with Your Mind: Closely Guarded Secrets in Sound Engineering*. UK: Flux Research.

Steen, Marc. 2014. "Upon Opening the Black Box and Finding It Full: Exploring the Ethics in Design Practices." *Science, Technology, & Human Values* 40 (3): 389–420.

Sterne, Jonathan. 2003. *The Audible Past: Cultural Origins of Sound Reproduction*. Durham: Duke University Press.

Sterne, Jonathan. 2012. *MP3: The Meaning of a Format*. Durham: Duke University Press.

Straughan, Elizabeth R. 2015. "Entangled Corporeality: Taxidermy Practice and the Vibrancy of Dead Matter." *GeoHumanities* 1 (2): 363–377.

Straw, Will. 1991. "Systems of Articulation, Logics of Change: Communities and Scenes in Popular Music." *Cultural Studies* 5 (3): 368–388.

Subotnik, Morton. 1970. "The Synthesizer: Is It the Ultimate Musical Weapon?" *Recording Engineer/ Producer* 1 (1): 17–21.

Švelch, Jaroslav. 2015. "Excuse My Poor English: Language Management in English-Language Online Discussion Forums." *International Journal of the Sociology of Language* 232: 143–175.

Swanke, Ian. 2022. "Behind the Gear: John LaGrou." *Tape Op*, December 2022.

Sweeney, Brian N. 2014. "Masculine Status, Sexual Performance, and the Sexual Stigmatization of Women." *Symbolic Interaction* 37 (3): 369–390.

Taylor, Timothy D. 2001. *Strange Sounds: Music, Technology & Culture*. New York: Routledge.

Théberge, Paul. 1997. *Any Sound You Can Imagine: Making Music/ Consuming Technology*. Hanover: Wesleyan University Press.

Thompson, Emily. 1995. "Machines, Music, and the Quest for Fidelity: Marketing the Edison Phonograph in America, 1817–1925." *The Musical Quarterly* 79 (1): 131.

Tucker, Anthony, William L. Martens, Glenn Dickens, and Michael P. Hollier. 2013. "Perception of Reconstructed Sound-Fields: The Dirty Little Secret." *Audio Engineering Society 52nd International Conference*, paper 4-3. https://aes2.org/publications/elibrary-page/?id=16906.

UNESCO. 1972. "Convention Concerning the Protection of the World Cultural and Natural Heritage." https://www.refworld.org/docid/4042287a4.html.

Vágnerová, Lucie. 2017. "'Nimble Fingers' in Electronic Music: Rethinking Sound through Neo-Colonial Labour." *Organised Sound* 22 (2): 250–258.

Veblen, Thorstein. 1899. *The Theory of the Leisure Class: An Economic Study in the Evolution of Institutions*. New York: MacMillan.

Versluys, Miguel John. 2022. "Mémoire Volontaire? Canonisation as Cultural Innovation in Antiquity." In *Canonisation as Innovation: Anchoring Cultural Formation in the First Millennium BCE*, edited by Damien Agut-Labordère and Miguel John Versluys, 34–79. Leiden: Brill.

von Appen, Ralf, and André Doehring. 2006. "Nevermind The Beatles, Here's Exile 61 and Nico: 'The Top 100 Records of All Time' – a Canon of Pop and Rock Albums from a Sociological and an Aesthetic Perspective." *Popular Music* 25 (1): 21–39.

Vrij, Aldert, Beth Paterson, Karl Nunkoosing, Stavroula Soukara, and Annerieke Oosterwegal. 2003. "Perceived Advantages and Disadvantages of Secrets Disclosure." *Personality and Individual Difference* 35 (3): 593–602.

Wallach, Jeremy. 2008. *Modern Noise, Fluid Genres Popular Music in Indonesia, 1997–2001*. Madison: University of Wisconsin Press.

Ware, Chris. 1998a. *Acme Novelty Library #10*. Seattle: Fantagraphic Books.

Ware, Chris. 1998b. *Acme Novelty Library #11*. Seattle: Fantagraphic Books.

Wheaton, Michael G. 2016. "Understanding and Treating Hoarding Disorder: A Review of Cognitive-Behavioral Models and Treatment." *Journal of Obsessive-Compulsive and Related Disorders* 9: 43–50.

Whiteman, Paul. 1948. *Records for the Millions*. New Jersey: Hermitage Press.

Wichterman, Jenna. 2018. "Two Accounts of a Change in Properties: Perdurantism and Endurantism." *Sapere Aude* (blog), October 16. https://sapereaude.voices.wooster.edu/2018/10/16/two-accounts-of-a-change-in-properties-perdurantism-and-endurantism/.

Williams, Alan. 2010. "'Pay Some Attention to the Man Behind the Curtain'—Unsung Heroes and the Canonization of Process in the Classic Albums Documentary Series." *Journal of Popular Music Studies* 22 (2): 166–179.

Winner, Langdon. 1993. "Upon Opening the Black Box and Finding It Empty: Social Constructivism and the Philosophy of Technology." *Science, Technology, & Human Values* 18 (3): 362–378.

Wolfe, Paula. 2019. *Women in the Studio: Creativity, Control and Gender in Popular Music Sound Production*. Farnham: Ashgate.

Wyatt, Sally. 2003. "Non-Users Also Matter: The Construction of Users and Non-Users of the Internet." In *How Users Matter: The Co-construction of Users and Technologies*, edited by N. Oudshoorn and T. J. Pinch, 67–79. Cambridge: MIT Press.

Youdelman, Rachel. 2017. "Iconic Eccentricity: The Meaning of Victorian Novelty Taxidermy." *PsyArt* 21: 38–68.

Zak, Albin J. III. 2001. *The Poetics of Rock: Cutting Tracks, Making Records*. Berkeley: University of California Press.

Index

Abbey Road Studios, 1, 18, 43, 145, 154–159, 165, 212, 275, 291, 303, 305, 309. *See also* EMI
ABX tests. *See* Testing, ABX
Acoustics, studio, 43, 65, 146, 163, 194–195, 207, 238, 243, 287
AD/DA conversion (analog-digital conversion), 85, 94, 222–223, 263
A-Designs, 92, 107, 224. *See also* Montessi, Peter
Advertising
  gear, 39, 44–45, 163–166, 169, 171–175, 182, 194, 235, 266
  minerals, 117
  use of sexual imagery, 37, 40
AES (Audio Engineering Society), 3, 40, 79–80, 141–148, 150–154, 158–159, 257, 267, 269–271, 281, 329
  Diversity and Inclusion Committee, 39, 328
  e-Library, 259
  trade show floor, 4, 15, 50, 130–131, 219, 300, 324
Aesthetics. *See also* Technoaesthetics
  audible, 151, 176, 199, 210, 213, 295, 329
  interfacial, 59, 69, 77, 110, 112–113, 117–123, 129, 184, 248, 306, 316
  military and war, 30, 124, 134
  vintage and heritage, 6, 125, 250
Agency, 7, 18, 24–25, 34, 123, 175, 196, 203, 207–218, 209, 245, 322–323. *See also* Control; Intra-action
  attribution of, 10, 35, 181, 294, 299, 316, 332
  configurations, 205, 210, 211
  dance of agency, 209
  displaced, 43, 174, 206, 208, 213–215, 321, 325
  misattribution, 229
  and secrets, 254, 258
AKG Acoustics, 97, 101, 138, 246, 275–276, 280, 283, 314
Akrich, Madeline, 211, 346n3
Albini, Steve, 175, 188, 190, 207–208, 275
Alctron, 97, 283, 328. *See also* Clones
Aluminum, 65–66, 69–77, 75t, 88, 105, 111, 113, 117–118, 128, 168, 255, 325
Amalgamate, 66, 73. *See also* Mining
Ampex
  Mailing List, 189
  tape machines, 51, 260, 275–276, 302, 314
Amplifiers, guitar, 52, 84, 99, 101, 119, 252
AMS Neve, 104–105, 122, 170, 173–174, 213–214, 223, 281–282, 300, 313, 343n9

Analog tape, 1, 2, 51, 124, 138, 155, 157, 189, 276, 278, 291–292, 302, 310, 314, 317
plugin emulations, 26
Annetts, Alex, 164–165, 176
Anthropocene, 23, 63–67, 71, 322, 329
Anthropomorphism, 11, 32, 42–43
API (Automatic Processes Inc.), 2–3, 6, 50–51, 84–85, 96, 101, 130, 233, 235, 254, 275–276, 344n23, 349n11
Architecture, studio, 146, 238
ASM mining (artisanal and small-scale mining), 66–67. *See also* Mining
Associated milieu, 16, 222–223
Audiophilia, 11, 16, 53, 199, 223
AudioSEX (message forum), 190, 201
Australasian Sound Recording Association, 1
Auto-Tune, 99, 151, 259
Avalon VT-737sp, 2, 69, 314
Awards, gear, 7, 246, 250, 270, 275–278, 289. *See also* TEC Awards, TECnology Hall of Fame

Bakelite, 73, 113, 117–118, 128, 306
Bangka (Indonesia), 64–67
Barad, Karen, 10, 222
Barnes, Matt, 79, 123
Becker, Walter, 52, 341n49. *See also* GAS
BeesNeez Microphones, 97–98, 125, 324, 329. *See also* Sneesby, Ben and Veronica
Bijker, Wiebe, 118
Bode, Harald, 141
Boxes
black, 114, 117
microphone, 186, 246
presence, 120–121, 155, 275
Boym, Svetlana, 294–295
Breadboards, 14, 106, 224. *See also* PCBs
Bunkers, 6, 25, 238–240, 247–248. *See also* Man cave, womb
Burning in (manufacturing stage), 232
Buttons, tactile, 46, 113, 115, 120, 122–124, 127–130, 134, 138, 210, 297, 314
"all buttons in" mode, 275, 307–308

Canonization, 270–282, 286–289, 327. *See also* Golden age
of recordings, 45, 54, 140, 151, 158, 174, 208, 267, 304
of studios, 43
of technologies, 7, 18, 24, 101, 149, 300
Capacitance (electrical property), 90
Capacitors, 23, 48, 53, 83, 88–90, 95, 102, 106, 174, 225, 291, 342n33, 343n7
electrolytic, 69–75, 72f, 105, 198, 342n34
film, 79, 167
tantalum, 67
CAPI (Classic Audio Products Inc.), 84–85, 115, 201
Capital
convertible, 9, 57, 300
cultural and social, 19, 57, 184, 185t, 187, 198, 201, 212, 307
fossil, 342n21
Capitalism, 8, 33, 34–35, 322, 326, 331
Carnhill Group, 91–92, 343n9. *See also* Transformers
Cats, studio, 25, 243f, 245
Cheng, William, 31, 42
Chin, Meg Lee, 194, 196
CITES, 153
Clayton, Sam, Jr., 200
Clearmountain, Bob, 241–242, 245
Clipping, 122, 152
Clones (gear), 85, 90, 92, 113, 154, 174, 194–195, 214, 270–272, 282–284, 306–307, 321
Coauthorship, 20–22, 39, 316–317
Cobain, Kurt (singer), 207–208, 214–215
Colonialism, 32–35, 41, 61, 244, 286, 322, 327
artifact theft and repatriation, 272–273
Portuguese, 32–33

Coltan, 67–68, 74. *See also* Tantalum
Components, electronic, 8, 23, 55, 59–60, 114–116, 127, 134, 174, 177–178, 201, 210, 224–225, 322. *See also* Capacitors; Diodes; Modding; Opamps; Resistors; Tubes, vacuum
  in electronics assembly, 81–91, 103–108, 167, 169, 195, 198, 216, 282
  manufacture of, 5, 34, 39, 44, 64–66, 70–73, 77, 153, 316
  NOS (new old stock), 53, 83–84, 89, 99, 198, 270
  surface mount (SMD), 81, 146
  through-hole, 60, **80**, 81, 108
Compressors, dynamic range, 18–19, 70, 114, 194, 225, 255, 298, 303, 308. *See also* Limiters, FET; UREI 1176
  vari-mu, 18, 314
Concretization, 141, 219, 346n3
Connell, Raewyn and James Messerschmidt, 36. *See also* Masculinity, hegemonic
Connoisseurship, 31, 38, 159, 171–172, 198, 228, 232, 237, 241, 270, 321–322, 326
Consoles, mixing, 1, 18, 50–51, 85, 89–90, 104, 138, 148–149, 157–158, 270, 276, 280, 285–287, 293, 303–305. *See also* Meter bridges
Consumption, conspicuous, 11, 35, 38, 51–52, 55, 213, 215, 240, 245, 350n37
Control, 24–25, 112–117, 119, 126–133, 166, 203, 205, 209–214, 237, 239. *See also* Agency
  of women, 36, 197, 323
Copper, 61–68, 77, 93, 119
COVID-19 (coronavirus pandemic), 9, 15, 92, 145, 153, 200, 240
Cowie, Jefferson, 39
Crane, Larry, 163, 175–176, 200–201, 232, 263–264
Crane, Stephen, 1–2, 291, 310
Cultures, gear, **8**, 9–17. *See also* Gearslutz; GroupDIY; ProSoundWeb
  durability, 6, 8, 11, 136, 141, 203, 315, 337n5

Daking, Geoff, 69, 125, 147, 224
Davis, Tony, 62
DAWs (digital audio workstations), 15, 67, 137–140, 216, 228, 261, 352n16. *See also* ProTools
dbx, Inc., 2, 69, 233, 235, 276, 314
Democratic Republic of Congo, 32, 63–64, 67–68, 74–75, 83
Derr, Dave, 121–122, 179, 224
Devine, Kyle, 64
Diodes, 88, 94
Discourse analysis, 14, 165
Dissipation factor, 88
Distortion, 225, 227. *See also* Jitter
  guitar, 52
  harmonic, 96, 104–105, 130, 225
  intermodulation, 199, 223
DIY (do it yourself), 13, 79, 86, 151, 163, 282, 284, 319. *See also* GroupDIY; Modding
DUC (Digidesign User Conference), 189
Durability (network theory). *See* Cultures, gear

EMI, 120, 124, 155–158, 174, 212, 275–277, 293, 309. *See also* Abbey Road Studios
  TG12345 console, 1, 138, 157, 293, 303
  Emojis, 183–187
  Empirical Labs, 2, 121, 130, 179, 194, 224, 233, 302
  Encoder, continuous, 112, 129
  Episteme, 22, 326
  Epistemic innocence, 209, 213, 217
  Epistemology, 10, 133
  Equalizers (EQs), 10, 15, 84–85, 94, 96, 99, 104–105, 123–124, 175, 237, 256, 302, 313–314. *See also* Neve 1073

EMI (*continued*)
Manley Massive Passive, 110–111, 310
Pultec EQP-1A, 2, 131, 235
Erland, Jakob, 102–103, 154, 201, 224–225
ESR (equivalent series resistance), 88–89
Eventide, Inc., 170, 276, 278, 284
E-waste, 67
Exhibitionism, technological, 31, 50, **51**, 52–56, 303–305
in museums, 155, 270, 273–275, 286–288
Extractivism, 5, 23, 39, 59, 61–77, 83, 117, 165, 255, 316, 322–323, 329

Faders, 1, 121, 127–129, 134, 138, 210, 245, 280, 305. *See also* Potentiometers
Fairchild Recording Equipment Corporation, 2, 18, 43, 119, 236, 275, 289, 302–303, 321
Fairchild Semiconductor, 94–95
Feilo, Shanghai, 283, 328. *See also* Clones
Fetishization, 5–7, 17, 19, 22–26, 27–57, 59, 84, 112, 206, 236–237, 247, 275, 295–296, 307–308, 318–319, 322, 330
of fascia colors, 124–125
of material components, 69, 84, 99, 121, 216–217
in print media, 164–165, 172–174, 176–182
sexual, 6, 221, 289, 293, 300
Fingers, nimble, 23, 84, 95, 324
Fisher, Mark, 296
Fletcher (Mercenary Audio), 159, 179, 188
Fletcher, Ted, 101, 122–123, 125, 159, 224
Focusrite Audio Engineering Ltd., 317
Red series, 3, 13, 69, 113, 117–118, 125, 248
Forums, message, 11, 14, 46, 135–136, 142, 183–203, 228, 237, 318–319, 322, 325, 331. *See also* Gearslutz; GroupDIY; ProSoundWeb; Womb Forums
magazine-run, 163–164, 182
moderation policies, 168, 199, 200, 216, 350n39
Fuston, Lynn, 220, 228, 352n16
Future Producers (message forum), 190, 201

GAS (gear acquisition syndrome), 31, 50–57, 134–135, 164–166, 170–171, 212, 289, 306, 318, 331. *See also* Hoarding; Objectophilia; Technophilia
Gaston-Bird, Leslie, 150
Gear
advertising (*see* Advertising, gear)
definition, 4–8
forums (*see* Forums, message)
outboard (*see* Compressors, dynamic range; Equalizers; Preamplifiers, microphone; Reverberation; Signal processors)
parts (*see* Components, electronic)
pimps, 159, 178, 187, 196, 315
sexualization of (*see* Fetishization, sexual)
as toys, 49, 133, 230
by type (*see* Analog tape; Consoles, mixing; Microphones)
vintage (*see* Aesthetics, vintage and heritage; Golden age; Heritagization)
as weapons (*see* Metaphors, military; Weapons, secret)
Gear cultures. *See* Cultures, gear
Gearshoot, 229–231
Gearslutz/Gearspace (web forum), 11, 14, 100, 111, 183–188, 190, 193–202, 227, 240, 322, 325, 331, 341n52
and sexy gear discourse, 41, 46
Geeks, 147, 161, 197, 202
Gell, Alfred, 280, 348n2
Gender, 35–38, 323–325. *See also* Masculinities, hegemonic; Masculinity; Women in electronics manufacturing
booth babes, 142, 198, 322, 328
cis-gender, 37, 111,145, 150, 200, 248, 322
diversity, 21, 26, 39, 150, 152, 158, 226, 315, 328–329
LGBTQIA+, 36–37, 41, 111, 196–197, 310, 323, 325–326

magazine representation, 86, 175–176, 180–181, 323
role-play, 37, 220–222
trade shows, 11, 198, 323–324, 328
Gerbers, 82. *See also* PCBs
Global accounting, 67–69, 71, 77
Gold, 28, 49, 63, 66–67, 98, 125, 172, 246, 277
Golden age, 7, 24, 51, 101, 151, 206, 212–213, 217–218, 242, 249, 272, 282, 289, 295, 309–310, 320–321, 327. *See also* Heritagization
Golden ears, 224, 228, 237. *See also* Audiophiles
Gold standard, 2
Graeber, David, 318, 327, 339n2
Grandmaster Flash, 150
Grohl, Dave, 18
GroupDIY (web forum), 14, 190, 201–202, 345n12, 351n43
GUI (graphical user interface), 115, 139, 231
Guinea, Republic of, 70, 74, 75t, 76
Guitars, electric, 42, 52, 119, 193, 196, 232, 266, 274, 277, 284–287, 318, 351n17

Ham radio, 11, 81, 190
Haptics, 5, 23, 60, 112–113, 126–129, 134, 282, 308
haptic ideation, 13, 46, 158, 285, 293
Haring, Kristen, 11
Harmony Central (web site), 189–190, 349n10
Harries, Dave, 291
Heritagization, 10, 13, 26, 35, 84, 131, 138, 181, 208, 242, 250, 269–289, 315, 318–323, 327–328. *See also* Golden age
in advertising and print, 6, 158, 171, 174, 179
and obdurance, 294, 296, 299–300, 304–306, 311
Hi-fidelity, 11, 119. *See also* Audiophilia
Hill, Dave, 210, 222–223, 327
Hoarding, 2, 6, 22, 31, 44, 47–51, 57, 81, 130, 212, 239, 248, 300, 314, 320, 328, 330–331
Hornborg, Alf, 34–35, 47, 339n2
HudBay Minerals Inc., 62–63, 341n9
Hutchison, Hutch, 102, 105–106, 108, 147, 224, 327
Hutton, Jane, 68, 71

Iconicity, technological, 7, 11, 270, 279–282, 289, 293–294, 303–305, 336. *See also* Shure 55SH
IMF structural adjustment programs, 59, 68, 70, 76. *See also* Colonialism
Indigeneity, 21, 41, 61–63, 77, 254, 272, 310
Inequality, income, 35, 255, 327
Interfaces. *See also* Aesthetics, interfacial; DAW; Skeuomorphism
audible, 129–130
controllability, 210
digital audio, 6, 15, 85, 90, 138–139, 307
haptic, 126–129
material, 116–119
playful, 130–134
theory of, 112–116
visual, 119–126
Interviews, 13, 18, 133, 147, 164, 166, 175–182, 219, 226, 301, 303, 324
Intra-action, 10, 134, 221–222, 325, 339n14. *See also* Agency
iTRI scheme, 68

Jaguars (Panthera onca), 23, 61–62
Jamieson, Annie, 273–274, 285
Javits Center (New York), 4, 79, 130, 146, 148
Jitter (digital audio), 199, 222–223, 228
Josephson, David, 79–80, 99, 175, 177, 189, 276, 283, 327, 344n26
Jung, Walter, 106

Kehew, Brian and Kevin Ryan, 154–155, 159
Kheshti, Roshanak, 244
Knobs, 69, 84–85, 112–123, 126–134, 211, 280, 285, 291–293, 310, 325. *See also* Bakelite; Potentiometers

Lacey Act, 153
LaGrou, John, 178, 226
Lakoff, George and Mark Johnson, 231, 263, 339n2
Lazar, Emily, 39
Lexicon, 1, 138, 314
Lights
- fairy, 25, 206, 239, 243–244, 247–248, 322
- lava lamps, 25, 206, 242, 247, 314, 322
- LEDs (light-emitting diodes), 56, 113, 121, 122, 157, 243–244, 284, 322
- VU meters, 30, 51, 113, 119–121, 239, 287, 306–307

Limiters, FET, 2, 18, 43, 174, 212, 236, 265, 266t, 308, 321. *See also* Compressors, dynamic range
Lojek, Bo, 94, 147, 343n13
Lolcows, 199, 315
Lomayesva, Gregory, 119
Lord-Alge, Chris (CLA), 123, 148–149, 214, 307
Loss tangent, 88
Loudspeakers, 3–4, 10, 16, 48, 51, 65, 94, 148, 237, 240–242, 305, 314, 330
Ludicity, 19, 22, 60, 132–134, 210, 232, 285, 288, 299. *See also* Play
- in online interactions, 206
- and weapons, 230–231

Lunchboxes (equipment case), 122, 233, 237, 278
Lundahl Transformers, 91. *See also* Transformers

Man caves, 6, 25, 238–240, 247–248. *See also* Bunkers; Wombs
Manley, EveAnna, 39, 100, 109–112, 177, 224, 327, 329
Manley Labs, 69, 86f, 91, 99, 105, 109–112, 120–122, 170–172, 231, 276, 283, 300, 310, 314
Manosphere, 12, 37–38, 200, 298. *See also* Masculinities, hegemonic
Marinair Radar Company, 92, 197, 343n9. *See also* AMS Neve; Transformers
Marvin, Carolyn, 39, 243, 254, 324
Marx, Karl, 33–35, 49, 65, 300
Masculinities, hegemonic, 31, 36–42, 136, 165, 197, 202, 322–324, 326
Masculinity. *See also* Geeks; Man caves; Manosphere
- heterosexuality, 12, 200
- toxic, 11, 46

Massy, Sylvia, 39, 148, 251–252, 257, 267, 269–271, 281
Materiality, 12, 21–23, 38, 40, 48, 54, 57, 67, 93, 113, 117–118, 133, 243, 256, 277, 293, 304, 309, 331
- irreducible, 5, 33

McGlynn, Matt, 79
Media, slow, 203, 297, 301
Meek, Joe, 101, 122–123, 125, 224, 345n14
Mellor, David, 180, 182
Metaphors, 10, 42, 64, 168, 171, 244, 249–250, 263–264, 319–320
- military, 30–31, 49, 56, 169, 199, 231, 350n37
- NSFW (not safe for work), 197
- sexual, 248, 253

Meter bridges, 51, 206, 240–243, 247, 273, 314–315
Microphone capsules, 23, 79, 82, 84, 97–99, 167, 246, 277
Microphone diaphragm, 98, 172, 198, 324
Microphone Parts (company), 79, 89, 146, 167–168, 284

Microphones. *See also* AKG; Microtech Gefell; Neumann; Schoeps Mikrofone; Shure
condenser, 79, 84, 97, 104, 167, 280, 314
dynamic, 207
ribbon, 11, 98, 157, 166–167, 246
tube, 99, 100, 207, 225, 246, 277
Microtech Gefell, 97, 246, 269, 276, 277–278, 283, 356n18
Mine tailings, 66
Mining, 23, 35, 61–71, 74, 77, 83, 136, 153, 255, 316, 342n17
Mixerman (Eric Sarafin), 188, 190–192. *See also* Womb Forums
Mixers. *See* Consoles, mixing
Mix with the Masters, 147–148, 159, 257–258, 262, 347n10
Modding, 60, 95–96, 114, 119, 169, 227, 284, 305
Monitors, studio. *See* Loudspeakers
Montessi, Peter, 92, 99, 107, 224
Mullard Limited, 99–100. *See also* Tubes, vacuum
Museums, 1–2, 6, 7, 13, 26, 31, 47–48, 139, 155–157, 212, 246, 250, 272–275, 278, 281, 289
Bob Paquette Microphone Museum, 269–270
Queen Studio Experience, 284–288
Musikmesse Frankfurt, 141, 145, 318, 346n6

NAB (National Association of Broadcasters), 146–147
Nakamura, Lisa, 95
NAMM (North American Music Merchants), 6, 15, 137–145, 152–154, 198, 219, 246, 288, 328–329. *See also* TEC Awards
NAMM Foundation, 83
Netizens, 189. *See also* Newsgroups
Neumann, Georg GmbH, 97–98, 174, 246, 276–283, 344n25
U47 microphone, 1, 99
U67 microphone, 183–187, 305–306
U87 microphone, 3
Neve, Rupert, 1–2, 18, 108, 122, 173, 177, 282, 343n9
Neve Electronics (1961–1985), 91–92, 101–102, 105, 121, 138, 270, 272, 276, 280, 325, 338n33, 343n5. *See also* AMS Neve; Rupert Neve Designs
Neve 1073 Channel Amp, 25, 82, 85, 104–105, 138, 174, 227, 233, 314, 343n5
Newsgroups (Usenet), 14–15, 96, 102, 188–189, 192, 203, 318
Nichicon Corporation, 70–3, 342n34
Noronha, Annelise, 158
NOS (new old stock). *See* Components, NOS
Nostalgia, 35, 84, 87, 118, 140, 149, 155, 244, 284, 294–295, 302, 327. *See also* Golden age; Technostalgia

Obdurance, 7, 10, 26, 249–250, 291–311
Objectophilia, 23, 31, 44–47, 50, 233–234. *See also* Technophilia
Objects, 348n2
hoarded, 48–49
problematic, 186–187
properties of, 38, 42, 210–213, 299
Obsolescence, 7, 15, 26, 67, 81–82, 85, 105, 108, 288, 319
O'Grady, Pat, 295, 301–302
Opamps (operational amplifiers), **83**, 93–97, 106, 114–115, 343n13, 344n21
Ornamentation, 235, 238, 240–248, 287, 322. *See also* Cats, studio; Lights; Rugs, Persian
Overburden, 66

Panasonic, 70–73, 342n34
Paradoxes, 8, 70, 103, 186, 205–206, 230–231, 269, 286, 308

Parisi, David, 126
Patchbays, 15
PCBs (printed circuit boards), 79–81, 96, 201, 345n12
  materiality of, 63, 65, 107, 119
Pedals, guitar, 52, 189–190, 195, 310, 318, 351n17
Pels, Peter, 35, 47
Perotta, Rich, 225. *See also* Royer Labs
Pietz, William, 32, 47
Pinch, Trevor, 152, 345n7
Play, 126, 128, 130–135, 159, 210, 293, 320. *See also* Ludicity
  gender role-play, 37, 220–222
  and testing, 228, 231
  word, 197
Plotnick, Rachel, 127, 130, 338n25
Plugins, software, 15, 45, 115–116, 142, 151–152, 194, 264–266, 299, 330
  effects, 16
  hardware emulation, 7, 26, 100, 138, 214, 230, 282, 302, 306–309, 321
  signature and personified, 149, 215
Podcasting, 9, 151, 180–181, 198
Pornography, 36, 196–197, 202, 221
Porter, Robin, 101, 104, 147, 223, 303–304. *See also* AMS Neve
Potentiometers, 71, 111–112, 126–129, 210. *See also* Faders; Knobs
  standardization, 13
Preamplifiers, microphone, 2, 10, 33, 69, 88–89, 92, 110–111, 114–116, 125, 171, 178, 210, 232, 237, 276. *See also* Neve 1073
ProSoundWeb, 190–194
ProTools (software), 148, 189, 191, 281, 300, 302, 352n16. *See also* DAWs
Pseudonymity, 53–54, 136, 183, 186–188, 191–193, 240, 325, 350n38. *See also* Forums
Pseudoscience, 223, 227, 230, 326

Quad Eight, 92, 97, 276, 343n8
Quality factor (Q factor), 88

Real World Studios (England), 148–149
Recapping, 83, 89–90, 96. *See also* Capacitors
rec.audio.pro (newsgroup), 14–15, 96, 102, 188–189, 348n6
Reciprocal landscapes, 59, 67–76, 329
rec.music.makers (newsgroup), 189, 349n10
Recorder, analog tape. *See* Tape, analog
Recording (magazine), 165, 190
Rec Pit. *See* ProSoundWeb
Reichenbach Engineering Company, 91–92, 343n8. *See also* Cinemag Group
Repatriation, artifact, 272–273
Resistance (electrical property), 88. *See also* ESR
Resistors, 71, 79, 88, 94, 102, 106, 127, 167, 342n33
  color codes, 324
  force-sensing, 113
Reverberation (signal processing). *See also* Signal processors
  digital, 1, 15, 213–214, 266t, 314
  plate, 163, 308–309, 321
Rosemont (copper mine), 61–64, 68
Royer Labs, 9, 98f, 174, 224–225, 246, 338n24
Rugs, Persian, 25, 239, 244, 247
Rupert Neve Designs, 105, 122, 278. *See also* Neve, Rupert; Neve Electronics

Schmidt-Horning, Susan, 255–256
Schmitt, Al, 305–306, 309
Schoeps Mikrofone, 79, 104, 113, 246, 276
Second Congo War, 63, 67–68, 74, 83
Secrets, 10, 25–26, 155, 179–182, 193, 195, 241, 249–267, 307, 318. *See also* Weapons, secret
  in manufacturing, 84, 91, 96, 169
  military, 29–30, 230

Self, Douglas, 106
Semiotics, 10–11, 30, 112–113, 117
797 Audio, 97, 283, 328. *See also* Clones
Shadow Hills Industries, 18, 27–33, 39, 131, 182, 224, 288, 320
Sheller, Mimi, 69–70
Shootouts, 25, 53f, 54, 183, 186–187, 190, 195, 205–206, 213, 221, 229–232, 307, 320. *See also* Testing, ABX
Shure Incorporated, 3, 264, 269, 271
  55SH microphone, 280–281 (*see also* Iconicity, technological)
Signal processors, 2, 4, 123, 138, 169–171, 189, 230, 265, 276, 278, 284, 314, 327, 351n8. *See also* Compressors; Equalizers; Reverberation
Simondon, Gilbert, 9, 337n7, 346n3
Skeuomorphism, 16, 115, 214, 282, 296, 299, 301–302, 306–309
Slew rate, 95–96, 199
Slurry, 67
SMD (surface mount devices). *See* Components, electronic
Smith, James, 67, 71
Sneesby, Ben and Veronica, 97–98, 125–126, 324
Sokil, Allison, 346n28
Sound City (studio), 18
Sound on Sound (magazine), 24, 163–185, 190, 260, 265, 266t, 277
Sowter Transformers, 91. *See also* Transformers
Spectral repair, 152
SSL (Solid State Logic), 95, 149, 151, 170, 242, 275–276, 281, 305
  console emulation, 214
  as prop, 148
Standen, Julian, 194–198
Stands, microphone, 146, 186
St Ives, 92, 343n9. *See also* Carnhill; Transformers
Studer, 2, 124, 138, 275–276, 284, 302
Studios, recording. *See also* Architecture, studio; *Names of individual studios*
  home, 11, 152, 165, 173, 235–239, 261t, 313
Supply chains, 5, 13, 23, 35, 42, 59, 64–65, 71, 73–77, 81–82, 107, 153, 316, 329–330
  knowledge of, 34
Sweetwater Sound, 161–164, 171–173, 181–182, 235, 265, 300
Switches, rotary, 112–113, 127–129, 131, 291. *See also* Buttons; Potentiometers
Synesthesia, 112, 132, 142
Synthesizers (instruments), 87, 95, 151, 169, 194, 265, 278, 281, 296, 310, 330
  modular, 141, 318
  Moog, 138, 244
  retail, 152, 169

Tanner, Geoff, 105, 343n5
Tantalum, 63–64, 67–68, 73–74, 75t, 83, 88. *See also* Coltan
Tape, analog, 1–2, 51, 124, 138, 155, 157, 189, 260, 266t, 276, 278, 291–292, 302, 310, 314, 317. *See also* Ampex; Studer
  emulation of, 26, 85
Tape Op (magazine), 24, 163, 165–182, 190, 196, 265t, 284, 332
Taxidermy, technological, 26, 250, 284–289, 299
Taylor, Timothy, 296
TEC Awards, 51, 152–153, 276–278, 307, 347n16
Technoaesthetics, 8, 118, 247, 308, 323, 329, 337n7. *See also* Aesthetics, interfacial
Technobabble, 24, 54–55, 89, 95, 100, 186, 198–203, 216, 227, 231, 308, 321, 326

Technologies. *See also* Exhibitionism, gear; Iconicity; Taxidermy; Voyeurism
alleged democratization, 165, 208, 216, 256–257, 323
home stereo, 16
instruments as, 7, 11, 141, 319
personified, 25, 32, 121, 208, 213–215, 322
Technophilia, 31, 44–46, 50–51, 57. *See also* Objectophilia
Technoporn, 6, 36, 196, 202. *See also* Pornography
Technostalgia, 7, 11, 18, 24, 135, 157, 270, 294, **296**, 328, 337n6
TECnology Hall of Fame, 278
Telefunken, 97, 99–100, 114–116, 126, 167, 210, 264, 276, 344n27
Teletronix, 2, 235, 276. *See also* UAD
Terry, Marshall, 91, 103, 122, 131–134, 189, 224, 329
Testing, 39, 205–206, 219–232, 352n15. *See also* Shootouts
ABX, 25, 178, 221, 228, 231, 289
beta-testing, 39
in design process, 89, 93, 177–179
dramaturgical, 219–221
Théberge, Paul, 42, 162–165, 169, 216, 262, 317, 337n11
Theories, conspiracy, 200
3D Audio, Inc., 190, 220, 228, 352n16. *See also* Fuston, Lynn
301 Studios, 1, 2, 241, 291, 338n24
Through-hole. *See* Components, through-hole
Tin, 64, 66–67, 73–74, 75t
Tohono O'Odham, 62, 341n9
Trade shows, 15–16, 24–25, 40, 79, 135–136, 141–154, 159, 194, 219, 222, 315–318. *See also* AES; Musikmesse; NAMM
and gender, 11, 198, 323–324, 328 (*see also* Booth babes)
Transformers, 63, 82–84, 90–93, 104–105, 107, 114–116, 131, 174, 197, 278, 343n9
gendered winding labor, 11, 91, 93, 108
Transient response, 95, 199
Transient sculpting, 152
Transistors, 71, 93–94, 105–106
Tubes, vacuum, 48, 71, 81, 84, 92, 99–100, 106, 110, 114, 198, 225, 236, 246, 306. *See also* Microphones, tube

UAD (Universal Audio), 6–7, 137–140, 159, 214, 231, 246, 306–307
UREI (United Recording Electronics Industries), 69, 235
UREI 1176 (compressor), 2, 225, 233, 275–276, 306–307, 314
Usenet, 188–189
US Geological Survey, 73

Valorization (of objects), 7, 293, 296–298, 300–306
Vari-mu. *See* Compressors
Veblen, Thorstein, 35, 56, 240, 245, 287, 339n2, 350n37
Voyeurism, technological, 2, **6**, 13, 31, 44–46, 50–52, 55f, 56, 184, 231, 246, 270, 304
VU meters. *See* Lights, VU meters

Walker, Saul, 96, 101
Weapons, secret, 6, 53, 111, 129, 181–182, 250, 252, 262–267, 275
Winston, Terri, 226, 352n12
Womack, Paul "Willie Green," 150
Womb Forums, The, 190, 192, 349n17, 349n18, 349n19
Wombs (studio type), 6, 25, 238–240, 247–248. *See also* Bunkers; Man caves
Women, in electronics manufacturing, 11, 23, 39, 80, 86–87, 91, 98, 100, 108, 316, 323
Women's Audio Mission, 226, 297, 323

Young Guru, 148